Series:

Advances in AgriGenomics

Series editor: Chittaranjan Kole, Kolkata, West Bengal, India

1. **Potato Improvement in the Post-Genomics Era**
   Jagesh Kumar Tiwari, 2023
2. **Genetics, Genomics and Breeding of Bamboos**
   Malay Das, Liuyin Ma, Amita Pal, and Chittaranjan Kole (eds.), Forthcoming
3. **The Avocado Genome: Genomics, Genetics and Breeding**
   J.I. Hormaza, Mary Lu Arpaia, Neena Mitter, and Alice Hayward (eds.), Forthcoming
4. **The Asparagus Genome: Genetics, Genomics and Breeding**
   Francesco Mercati, Jim Leebens-Mack, and Francesco Sunseri (eds.), Forthcoming

Advances in AgriGenomics
Series editor: Chittaranjan Kole

Potato Improvement in the Post-Genomics Era

Jagesh Kumar Tiwari
Senior Scientist, Division of Crop Improvement
ICAR-Central Potato Research Institute, Shimla
Himachal Pradesh, India

CRC Press
Taylor & Francis Group
Boca Raton London New York

CRC Press is an imprint of the
Taylor & Francis Group, an **informa** business
A SCIENCE PUBLISHERS BOOK

First edition published 2023
by CRC Press
6000 Broken Sound Parkway NW, Suite 300, Boca Raton, FL 33487-2742

and by CRC Press
4 Park Square, Milton Park, Abingdon, Oxon, OX14 4RN

CRC Press is an imprint of Taylor & Francis Group, LLC

Library of Congress Cataloging-in-Publication Data (applied for)

ISBN: 978-1-032-19656-5 (hbk)
ISBN: 978-1-032-19657-2 (pbk)
ISBN: 978-1-003-26023-3 (ebk)

DOI: 10.1201/9781003260233

Typeset in Times New Roman
by Radiant Productions

Preface to the Series

Agricultural genomics, popularly called as AgriGenomics, has emerged as one of the most frontier fields in life-sciences since the beginning of the 21st century. Whole genome sequencing of the leading crop plant rice in 2002 marked the begining of genome sequencing in an array of crop plant species. Elucidation of structural genomics of the nuclear genomes of the crop plants and their wild allied species included enumeration of the sequences and their assembly, repetitive sequences, gene annotation and genome duplication, synteny with other sequences and comparison of gene families. Sequencing of organellar genomes including chloroplasts and mitochondria enriched the structural genomics database. At the same time, researches on functional genomics such as transcriptomics, proteomics and metabolomics presented useful information. All these advances in genome sciences facilitated complementation of the erstwhile strategies of genome elucidation and improvement employing molecular breeding and genetic transformation with genomics-assisted breeding and most recent technique of genome editing. All these concepts, strategies, techniques and tools will be deliberated in the 'stand-alone' volumes of this book series.

Chittaranjan Kole
Kolkata, India

Preface

Potato (*Solanum tuberosum* L.; $2n = 4x = 48$) is the third most non-cereal food crop of the world, after rice and wheat, in terms of human consumption. In 2020, global annual potato production was 359.07 million tons from an area of 16.49 million hectares with productivity of 21.76 tons per hectare (FAOSTAT, 2022). Owing to the importance of potato in food and nutritional security, the year 2008 was declared as the 'International Year of the Potato' by the United Nations. Potato is a clonally propagated crop, tetraploid, highly heterozygosity and suffers from acute inbreeding depression. Besides, biotic and abiotic stresses, tuber quality and processing parameters are the major challenges of potato crop under the climate change scenario where underground plant part (tuber) is economically important. To address these various issues at genomics level, the potato genome was deciphered in 2011 (*Nature*) by the Potato Genome Sequencing Consortium (PGSC) to dissect genes underlying various agronomically important traits. Since then, considerable advancements have been made in the post-genomics era in potato research and development.

The book, *Potato Improvement in the Post-Genomics Era* describes the recent updates on post-genome research in potato. Chapter 1 discusses 'Potato: A Global Scenario and Omics Research' where status of potato is described in terms of production, consumption, challenges and ways to overcome these challenges through omics research. Chapter 2 outlines the 'Potato Genome Sequencing and Resequencing of Wild/Cultivated Species', mainly genome sequencing technologies, genome sequences of potato and various wild/cultivated species. Chapter 3 describes 'Genomics in Potato Germplasm Management and Utilization', the various wild and cultivated potato species, conservation, major world genebanks, and characterization using genomics tools. Chapter 4 on 'Molecular Markers, Mapping and Genome-Wide Characterization' illustrates important molecular markers technologies in plant science, high-throughput marker system, genome mapping and gene discovery by Single Nucleotide Polymorphism (SNP) and genome wide association studies for various traits. Chapter 5 highlights 'Conventional to Genomics-assisted Breeding' in potato. This chapter covers conventional breeding to marker-assisted selection to rapid the breeding process by reducing field exposure. Later, with the advancement of genomics, genomic prediction and genomic selection have been illustrated in potato. Gene discovery is an important aspect of any crop breeding and biotechnology. Chapter 6 on 'Omics Approaches in Potato' describes the use of multi-omics approaches in potato, such as transcriptomics (total RNA), microRNAs (small RNA), proteomics (proteins), metabolomics (metabolites) and ionomics (ions or nutrients) for biotic, abiotic and quality traits in potato along with multi-omics approaches for interaction studies. Chapter 7 on 'Phenomics in Potato' outlines briefly phenomics, an automated high-throughput phenotyping technology for precision phenotyping and speed breeding for potato traits improvement. Chapter 8 on 'Genome Editing (CRISPR/Cas) Technology in Potato' describes the use of CRISPR/Cas technology in potato next-generation breeding technology and rapid advancement. Chapter 9 on 'Conventional True Potato Seed (TPS) to Diploid Hybrid Potato Technologies' illustrates the true potato seed technology and prospects on recent diploid hybrid technology in potato. Chapter 10 on 'Pre-breeding Genomics in Somatic Hybridization' describes the pre-breeding technology, like somatic hybridization and use of genomics in crop improvement. Finally Chapter 11 on 'Potato Transgenics' outlines the transgenics development in potato in the post-genomics era.

I hope this reference book will prove a highly useful resource to obtain updated information for academicians, researchers, students, stakeholders and the farming community involved in potato research and development. I am grateful to the Director, ICAR-Central Potato Research Institute, Shimla, and all scientists/technicians/ research fellows and administrative staffs/other colleagues of the institute for their support under the institute research programs. I express my gratitude to Dr. Swarup K. Chakrabarti who has always been an instrumental and source of inspiration for me. I express my special thanks to my scientist colleagues Drs. Vinod Kumar, R.K. Singh, S.K. Luthra, Tanuja Buckseth, and Dalamu; and hard working lab staff Sh. Sheeshram Thakur, Ms. Poonam, Rasna, Sapna, Nilofer, Sharmistha, Nisha, etc. I am thankful to the funding agencies (ICAR/DBT) for their support under CABin project, ICAR-IASRI, New Delhi; and ICAR-LBS Young Scientist Award Project, and DBT (Govt. of India). I am thankful to series editor, Prof. Chittaranjan Kole and Dr. Vijay, Science Publishers, CRC Press for providing the opportunity to write this book. I extend my special thanks and gratitude to the whole team of the CRC Press for editing of the book.

Jagesh Kumar Tiwari
Shimla, India

I dedicate this book to my family Alpana, Shivangi, Omkunj & parents and ICAR-CPRI family for their continuous support during the professional and personal journey.

Foreword

Potato (*Solanum tuberosum* L.; $2n = 4x = 48$) is one of the staple food crops of the world, after rice and wheat in terms of human consumption. As per the FAOSTAT (2022) data, global potato production in 2020 was 359.07 million tons from an area of 16.49 million hectares with productivity of 21.76 t/ha. Application of plant breeding techniques contributed largely in transformation of Andean potatoes to the modern-day potato varieties being commercially cultivated throughout the globe. It is presumed that rational application of modern-day genomic tools would take the crop to its next level of commercialization. It is inevitable now to harness the potential of genomic tools to address imminent challenges to this wonder crop posed by climate change and global warming, input-intensive production system, biotic and abiotic stresses, yield plateauing and risk-prone post-harvest management. It is possible now to address these apparently formidable problems due to rapid advancement in genomics-aided technologies after the completion of potato genome sequencing in 2011 by the The Potato Genome Sequencing Consortium (PGSC). Availability of potato genome sequence data in the public database has paved the way for re-sequencing of several wild and cultivated species, thereby facilitating post-genomics research in potato.

The book entitled *Potato Improvement in the Post-Genomics Era* covers advancements in potato genomic research while providing the latest developments in potato in terms of sequencing technologies, genome sequencing/resequencing of potatoes, new reference assembly, genome-wide germplasm characterization, mapping, gene/marker discovery and pre-breeding genomics using SNP markers and prospects of next-generation breeding by applying genomic selection and genome editing technologies. Further, the book also describes application of multi-omics approaches, such as transcriptomics, proteomics, metabolomics, phenomics and ionomics for integrated gene management and system biology approach. The use of genomics tools, like high-throughput SNP genotyping, genome-wide association mapping, genotyping by sequencing, genomic selection and genome editing have immense opportunities to advance potato breeding for complex inherited traits. The technique of genome editing has recently been demonstrated to overcome the self-incompatibility problem in diploid potato, thereby paving the way for designing diploid potato F_1 hybrid.

I congratulate the author for his hard work and efforts in putting up the potato post-genomics research in the right perspective. I am sure the book will be an important reference resource on potato post-genomics research for researchers, students, academicians, farmers and policy makers.

Dr. Swarup Kumar Chakrabarti
Vice-Chancellor
Uttar Banga Krishi Viswavidyalaya,
Cooch Behar, West Bengal, India
and
Former Director, ICAR-Central Potato Research Institute,
Shimla, Himachal Pradesh, India

Contents

Chapter 1

Potato

A Global Scenario and Omics Research

1. Introduction

Potato (*Solanum tuberosum* L.) is the third most important food crop in the world, after rice and wheat in terms of human consumption. Potato is an economically important staple food in both developed and developing countries because of its high-yield potential, high nutritive value, easily digestible and wholesome food (Chakrabarti et al., 2017). It contains high protein-calorie ratio (17 g protein:1000 kcal) and yields more energy, protein and dry matter per unit area and time as compared to cereal crops. Also potato protein has a very high biological value (98) due to the presence of balanced amino acids (Singh et al., 2020). Besides its table purpose, potato is drawing greater attention of the processing industry for chips, French fries and other products. The United Nations declared the year 2008 as the 'International Year of the Potato'. Henceforth, potato is known as 'hidden treasure of the Andes' and has been identified as a 'food for the future' by Food and Agriculture Organization. Potato holds a great potential in providing both food and nutritional security at the global level (Global Potato Conference, 2008).

The cultivated potato originated in the Andes of Peru and Bolivia in South America, more specifically in the basin of Lake Titicaca (Bradshaw and Mackay, 1994). Evidence shows that hybridization of wild species, *S. stenotomum* with *S. sparsipilium* and subsequent chromosome doubling, produced tetraploid *S. tuberosum* Group Andigena in the central Andes and from them, the cultivated potato originated (Hawkes, 1990). Potato is grown widely from temperate to sub-tropical climates in the world. In temperate climates, potato growth entails planting in summer season at low temperature (15–25°C), low temperature during harvesting time (15–25°C), longer crop duration (140–180 days), longer photoperiod (14–16 h light), optimum day (15–30°C)/night (14–22°C) temperature during crop growth for producing high yields of tubers, high dry matter and low reducing sugars. On the other hand, in sub-tropical climates, potato is grown during the winter season by being planted during high temperature (25–30°C), low temperature (10–20°C) during harvesting time, shorter crop duration (90–100 days), shorter photoperiod

(9–11 h light), optimum day (15–32°C)/low night (2–15°C) temperature during crop growth and frost occurrence. Tubers are produced with low yields, less dry matter and more reducing sugars than those in temperate climate (Singh et al., 2018). On the other hand, potato crop faces serious biotic/abiotic stresses and climate change issues. The potato genome sequence was made available in 2011 (The Potato Genome Sequencing Consortium, 2011), and so potato improvement is now feasible for multiple traits through rapid-breeding applying genomics.

2. Global Scenario of Potato Production

In 2020, the global annual potato production was 359.07 million tons from 16.49 million hectares area with 21.76 tons per hectare productivity (FAOSTAT, 2022) (Table 1). In 2020, out of the total world potato production (359.07 mt), maximum production was recorded in Asia (178.60 mt) followed by Europe (107.69 mt), Americas (44.92 mt), Africa (26.23 mt) and Oceania (1.63 mt). India (51.30 mt) is the second largest potato producer, after China (78.23 mt) from an area of 2.15 and 4.21 m ha, with productivity of 23.77 and 18.54 t/ha in India and China, respectively. India and China were not only the major contributors to the Asian growth in potato production but produced one-third of global potato, contributing significantly to world potato production. The USA is on the top with the highest productivity (50.79 t/ha) followed by New Zealand (50.73 t/ha). In the past 20 years (2000–2020), the trend in potato production increased in the world (11.58%) with maximum increase in India (105.20%), followed by Africa (103.22%), Asia (47.01%) and China (17.97%), while it decreased considerably in Europe (–24.96%), Poland (–67.61%) and Russian Federation (–33.45%) (Fig. 1). In the past six decades (1961–2020), an increasing trend of potato production was observed in the world, including Asia, Americas, Africa and Oceania, while decreasing trend was found in Europe (Table 2, Fig. 2).

Table 1. Major potato producing countries in the world (2020).

Country	Production (mt)	Area (mha)	Yield (t/ha)
China	78.23	4.21	18.54
India	51.30	2.15	23.77
Ukraine	20.83	1.32	15.72
Russian Federation	19.60	1.17	16.64
USA	18.78	0.36	50.79
Germany	11.71	0.27	42.83
Bangladesh	9.60	0.46	20.82
France	8.69	0.21	40.52
Poland	7.84	0.22	34.76
Netherlands	7.02	0.16	42.67
United Kingdom	5.52	0.14	38.87
Peru	5.46	0.33	16.47
Canada	5.29	0.14	36.87
Belarus	5.23	0.25	20.64
World	359.07	16.49	21.76

(*Source*: FAOSTAT, 2022), mt: million tonnes, mha: million hectare, t/ha: ton per hectare

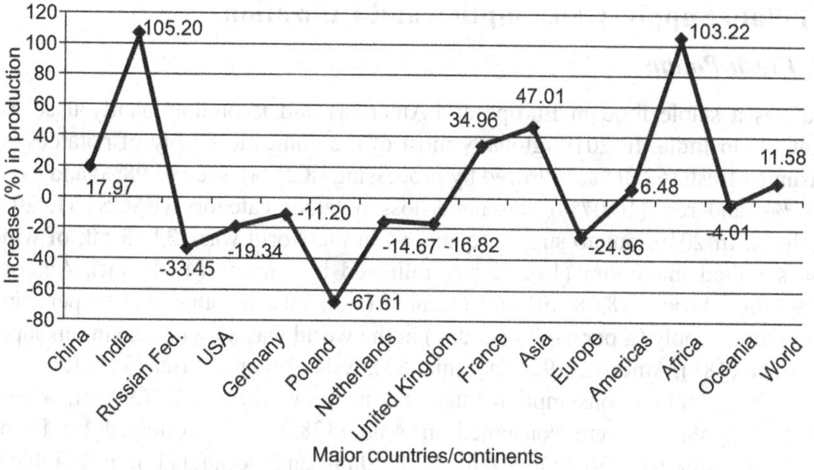

Fig. 1. Increase (%) in potato production over 2000 mt in the major potato producing nations and continents during the last two decades (2000–2020) (*Source*: FAOSTAT, 2022).

Table 2. World potato production (mt) in the world during 1961–2020.

Region	**Year (mt)**						
	1961	1970	1980	1990	2000	2010	2020
Africa	2.10	3.03	5.16	8.22	12.91	24.66	26.23
Americas	22.57	26.86	26.75	32.71	42.19	40.57	44.92
Asia	23.36	34.71	47.42	64.18	121.49	154.00	178.60
Europe	221.83	232.39	160.09	160.27	143.51	107.31	107.69
Oceania	0.69	1.06	1.08	1.44	1.70	1.81	1.63
World	270.55	298.05	240.50	266.83	321.80	328.35	359.07

(*Source*: FAOSTAT, 2022)

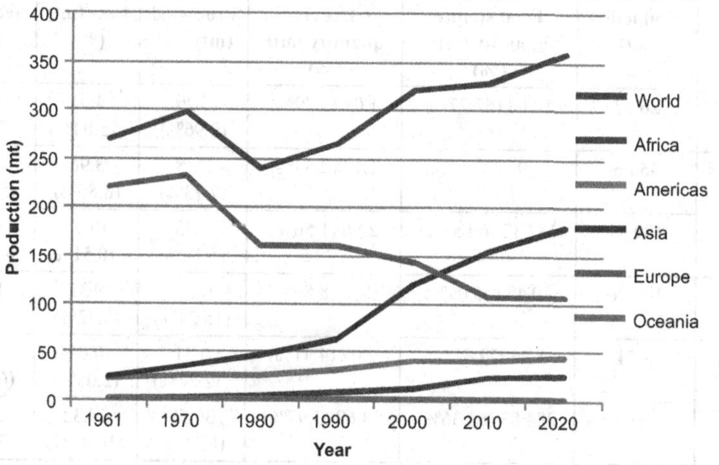

Fig. 2. Trend in potato production during 1961–2020.

3. Potato Supply, Consumption and Utilization

3.1 Fresh Potato

Potato is a staple food in Europe and Americas and is predominantly used as a vegetable in India. In 2019, globally most of the domestic supply of potatoes was consumed fresh (63.36%) followed by processing (8.25%), seed (7.98%) and export (3.97%), and rest (10.07%) was under-loss potatoes category (FAOSTAT, 2022) (Table 3). In 2019, potato supply (quantity) in the world was 234.83 mt, of which Asia supplied maximum (132.12 mt) followed by Europe (49.43 mt), Americas (33.93 mt), Africa (18.08 mt) and Oceania (1.26 mt). In other words, per capita potato food supply (g per capita per day) in the world was 84 with maximum supply in Europe (181), Americas (92), Oceania (83), Asia (79) and Africa (39) (Table 4).

In 2019, potato consumption (quantity) in the world was 247.99 mt, wherein maximum potatoes were consumed in Asia (138.49 mt), followed by Europe (53.18 mt), Americas (36.37 mt), Africa (18.50 mt) and Oceania (1.43 mt) (Table 4). The per capita potato consumption (kg per capita per year) in the world was 32.45 with the highest in Europe (71.23), followed by Americas (36.01), Oceania (34.63), Asia (30.31) and Africa (14.52). The variation in food supply distribution is affected due to the rising population in Asia followed by Africa, Americas, Europe and Oceania. Thus, during 2010–2019, overall per capita potato consumption (kg per capita per year) fell in the world (2.84%), Africa (20.09%), Oceania (12.73%) and Europe (10.47%), while it increased in Asia (7.56%) and marginally in Americas (0.81%) (Table 5). A few top per capita potato consumer (kg per capita per year) countries are Belarus (175.98), Latvia (112.20) and Kazakhstan (110.91), whereas consumers in terms of total quantity are China (91.74 mt) on top, followed by India (51.62 mt) and Russian Federation (23.34 mt) (Table 6). This could be due to the world total population (over 7.63 billion) with maximum pressure in Asia followed by Africa, Americas, Europe and Oceania. Potato consumption in India and China is accelerating due to increasing

Table 3. Different forms of potato utilization (2019).

Region	Production (mt)	Food supply quantity (mt) (%)	Export quantity (mt) (%)	Processed (mt) (%)	Loss (mt) (%)	Seed (mt) (%)
Africa	26.67	18.08 (67.77%)	1.01 (3.79%)	0.79 (2.96%)	4.01 (15.04%)	2.12 (7.95%)
Americas	45.06	33.93 (75.30%)	1.14 (2.53%)	3.48 (7.73%)	3.96 (8.80%)	2.57 (5.72%)
Asia	189.91	132.12 (69.57%)	2.96 (1.56%)	9.75 (5.14%)	19.96 (10.51%)	9.35 (4.93%)
Europe	107.26	49.43 (46.09%)	9.50 (8.86%)	16.34 (15.24%)	9.35 (8.72%)	15.40 (14.36%)
Oceania	1.74	1.26 (72.32%)	0.07 (4.41%)	0.21 (12.25%)	0.03 (2.05%)	0.12 (6.94%)
World	370.66	234.83 (63.36%)	14.69 (3.97%)	30.58 (8.25%)	37.32 (10.07%)	29.58 (7.98%)

(*Source*: FAOSTAT, 2022)

Table 4. Potato food supply and consumption in the world (2019).

Region	Food supply				Consumption		
	Production (mt)	kcal/ capita/day	Quantity (g/ capita/day)	Quantity (mt) (%)	Population (m)	kg/capita/ yr	Quantity (mt)
Africa	26.67	28	39	18.08 (67.77%)	1275.92	14.52	18.50
Americas	45.06	61	92	33.93 (75.30%)	1006.51	36.01	36.37
Asia	189.91	55	79	132.12 (69.57%)	4560.66	30.31	138.49
Europe	107.26	122	181	49.43 (46.09%)	746.41	71.23	53.18
Oceania	1.74	49	83	1.26 (72.32%)	41.57	34.63	1.43
World	370.66	58	84	234.83 (63.36%)	7631.09	32.45	247.99

(*Source*: FAOSTAT, 2022)

Table 5. Trend of potato consumption in the world (2019).

Region	Consumption (kg/capita/yr)		Increase (%)
	2010	2019	
Africa	18.17	14.52	–20.09
Americas	35.72	36.01	0.81
Asia	28.18	30.31	7.56
Europe	79.56	71.23	–10.47
Oceania	39.68	34.63	–12.73
World	33.4	32.45	–2.84

(*Source*: FAOSTAT, 2022)

Table 6. Top potato consumers in the world (2019).

Country	kg/capita/yr	Country	Quantity (mt)
Belarus	175.98	China	91.74
Latvia	112.2	India	51.62
Kazakhstan	110.91	Russian Federation	23.34
Poland	99.42	Ukraine	20.69
Ukraine	97.3	USA	18.68
Romania	96.75	Bangladesh	10.28
Rwanda	96.61	Germany	8.52
Kyrgyzstan	93.65	France	6.14
Peru	91.94	Poland	6.03
Russian Federation	89.37	United Kingdom	5.73
World	32.45	World	247.99

(*Source*: FAOSTAT, 2022)

industrialization and participation of women in the job market, thus creating a demand for processed, ready-to-eat convenient food, particularly in urban areas.

3.2 Processed, Seed and Loss of Potatoes

Agri-processing sector experiences very fast growth rate when an economy transforms from developing to developed economy. Potato is always the front-runner when processing sectors are taken into consideration. The processing sector is an emerging sector for potato entrepreneurs to add value addition through potato products, such as chips, French fries, flakes, etc. Potato starch industry is still becoming popular in some parts of the world, like China and Europe. In terms of total quantity processed, 30.58 mt (8.25%) potatoes are processed and the major processing countries are China (8.27 mt), the Netherlands (4.63 mt) and Germany (3.69 mt) (Table 7). In other words, top countries with maximum percentage of their produce used in processing are Belgium (94.13%), the Netherlands (65.95%), Canada (32.19%) and Germany (30.32%) (FAOSTAT, 2022). Seed is the most important input of potato production, and in 2019 worldwide, 29.58 mt (7.98%) potatoes were used as seed potatoes (FAOSTAT, 2022) (Table 7), with maximum amount of seed being used in Ukraine (5.53 mt), Russian Federation (4.37 mt) and India (3.68 mt). With the availability of

Table 7. Top countries in export, processed, seed, and loss potatoes (2019).

Export potatoes			**Processed potatoes**		
Country	Production (mt)	Export Quantity (mt) (%)	Country	Production (mt)	Processed (mt) (%)
France	8.69	2.32 (26.70%)	China	78.18	8.27 (10.58%)
Netherlands	7.02	2.28 (32.48%)	Netherlands	7.02	4.63 (65.95%)
Germany	11.71	1.87 (15.97%)	Belgium	3.92	3.69 (94.13%)
Belgium	3.92	0.99 (25.26)	Germany	11.71	3.55 (30.32%)
Egypt	5.21	0.68 (13.05%)	Canada	5.29	1.70 (32.19%)
India	51.30	0.41 (0.80%)	India	51.30	0.10 (0.19%)
World	370.66	14.69 (3.96%)	World	370.66	30.58 (8.25%)
Loss potatoes			**Seed potatoes**		
Country	Production (mt)	Loss (mt) (%)	Country	Production (mt)	Seed (mt) (%)
India	51.30	11.39 (22.20%)	Ukraine	20.83	5.53 (26.55%)
China	78.18	4.59 (5.87%)	Russian Federation	19.60	4.37 (22.30%)
Ukraine	20.83	3.27 (15.70%)	India	51.30	3.68 (7.17%)
Russian Federation	19.60	1.55 (7.91%)	China	78.18	2.91 (3.72%)
Algeria	4.65	1.44 (30.97%)	Belarus	5.23	1.33 (25.43%)
France	8.69	1.24 (14.27%)	USA	18.78	0.91 (4.85%)
World	370.66	37.32 (10.07%)	World	370.66	29.58 (7.98%)

(*Source*: FAOSTAT, 2022)

quality planting materials, seed replacement rate may increase in future. In 2019, the maximum potato loss was found in India (11.39 mt), i.e., 22.20% of total produce (51.30 mt), followed by China (4.59 mt, 5.87%) and Ukraine (3.27 mt, 15.70%) (Table 7). This resulted in annual loss of 37.32 mt (10.07%) of potatoes with total world production of 370.66 mt (FAOSTAT, 2022). This is an unfortunate situation that such a huge amount of potatoes was wasted in a developing country like India, where the population is increasing exponentially. It is seen that due to the high summer temperatures during harvest, lack of state-of-the-art cold storage facilities and inadequate transportation facilities are the causes of high wastage of potato in India.

3.3 Potato Export

In 2019, the world's potato export was 14.69 mt (3.96%) out of a total production (370.66 mt), where the maximum exporting countries were France (2.32 mt) followed by the Netherlands (2.28 mt) and Germany (1.87 mt). As potato is a semi-perishable and bulky agri-commodity, its export should be guided by a long-term policy. Also potato is a politically sensitive crop; therefore targeted steps are taken to keep its retail prices at affordable level for the ordinary consumer. Export is generally a crisis management tool during the years of oversupply. However, in international markets, exports can only be increased through building credibility and entering into long-term contracts. It is anticipated that the future food policy will concentrate more and more on ensuring national food security. Since China and India are the massive producers of potatoes, a healthy growth in processed potato products is anticipated. However, unknown and complex future developments at the international level are difficult to be assessed.

4. Biotic Stress in Potato

4.1 Late Blight

Late blight, caused by the oomycete *Phytophthora infestans* (Mont.) de Bary, is the most devastating disease of potato worldwide (Fig. 3). In 1845, late blight caused a complete loss of crops in the European countries, particularly in Ireland and was popularly known as 'Irish Famine'. The favorable conditions for disease development are mild temperature (18 ± 2°C) and high relative humidity (> 90%) (Wang et al., 2008). Late blight is controlled mainly through fungicide application and the use of resistant varieties. Besides, cultural practices, like sanitation, crop rotation, fertilizers and crop geometry can be successfully applied. Several resistance (R) genes from wild or cultivated potatoes have been exploited through either conventional breeding approaches or biotechnological tools (Tiwari et al., 2013). Some of the late-blight resistant wild species are *S. demissum*, *S. bulbocastanum*, *S. microdontum*, *S. pinnatisectum*, *S. cardiophyllum* and *S. verrucosum*. Biopesticides and biocontrol agents, like *Trichoderma viride* and *Pseudomonas aeruginosa*, have also been attempted as safer options to manage this disease (Bradshaw et al., 2006; Tiwari et al., 2015, 2018).

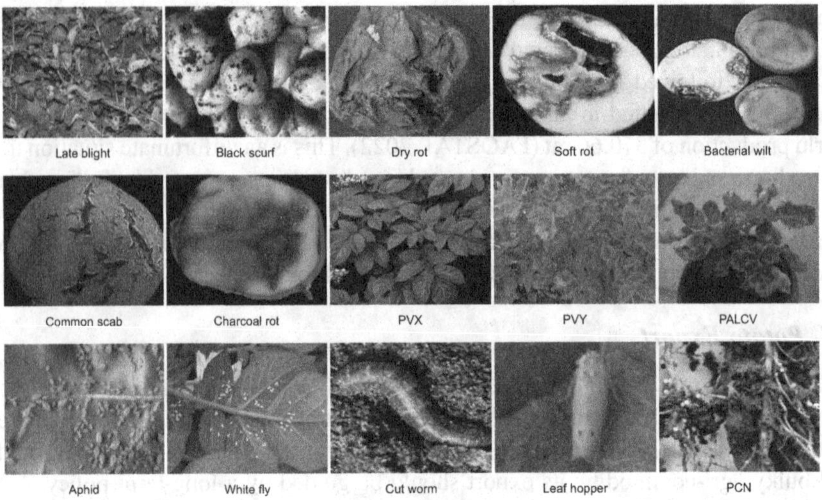

Fig. 3. Major diseases and insect-pests in potato.

4.2 *Viruses*

Potato is infected by more than 30 viruses, which lead to yield reduction, depending upon disease severity. The major potato viruses are *potato virus X* (PVX), *potato virus Y* (PVY), *potato virus S* (PVS), *potato virus M* (PVM), *potato leaf roll virus* (PLRV), *tomato leaf curl New Delhi virus-potato* (ToLCNDV) in India, and *potato spindle tuber viroid* (PSTVd) (Kumar et al., 2020) (Fig. 3). In general, crop losses are higher in the case of mixed infections. PVY and PLRV are the most devastating viruses, causing up to 80% yield losses, whereas PVX, PVS and PVM are mild, causing up to 30% yield loss. In general, virus management methods include prevention of viral transmission, eradication of infected sources, control and avoidance of vectors and use of virus-free healthy seeds, resistant varieties and biotechnological tools. Conventional and molecular breeding approaches have been applied to breed resistant varieties. The Ry_{adg} gene from *S. tuberosum* Gp. Andigena has been found effective against several strains of PVY. Besides, there are many resistance genes, like Ry_{sto} from *S. stoloniferum* conferring PVY resistance and Rlr_{etb} from *S. etuberosum* providing resistance to PLRV (Singh et al., 2020).

4.3 *Soil and Tuber-borne Diseases*

Potato is affected by several soil and tuber-borne diseases, causing heavy losses, particularly during storage. Dry rot, charcoal rot and bacterial soft rot cause losses during storage, whereas black scurf and common scab impact tuber appearance and reduce marketable value (Sagar and Sanjeev, 2020) (Fig. 3). Dry rot, caused by fungus *Fusarium oxysporum*, is an important post-harvest disease responsible for losses during transport and storage when the favorable temperature range is between 15–28°C. Charcoal rot disease, caused by fungus *Macrophomina phaseolina*, is prevalent at high soil moisture combined with high temperature (28–30°C). Bacterial soft rot, caused by *Pectobacterium atrosepticum* and *P. carotovorum*, is another devastating disease of potato during harvest, transport and storage. Black scurf, caused by fungus *Rhizoctonia solani*, affects tuber quality and causes moderate yield losses. Common scab, caused by fungus *Streptomyces scabies*, causes lesions on tuber skin and congenial environments are soil pH (5.2 to > 8.0), temperature (20–30°C) and low soil moisture. Bacterial wilt, caused by *Ralstonia solanacearum*, or brown rot is one of the most damaging diseases to develop between 15–35°C. Potato wart, caused by *Synchytrium endobioticum*, is a problem faced in hilly regions. These soil and tuber-borne diseases are managed by sanitation, use of disinfected and healthy quality seeds and boric acid (3%) treatment (Singh et al., 2020). Hence, genomics-aided research is necessary to address the above diseases.

4.4 Insect-pests

Potato is infested by many insect-pests, such as aphids, whiteflies, thrips, white grubs, cutworms, leaf hopper, potato tuber moth and mites (Shah et al., 2020) (Fig. 3). Aphid, mainly *Myzus persicae*, is the vector of potato viruses (PVY, PVA, PLRV, PVS and PVM). Whitefly (*Bemisia tabaci*) is a severe emerging problem in potato under climate change scenario and transmits ToLCNDV-potato virus and causes mosaic, chlorosis and curling of apical leaves symptoms. Potato leaf hopper (*Amrasca biguttula biguttula*) causes hopper burn symptoms. Thrips (*Thrips palmi*) is a vector of groundnut bud-necrosis virus, causing stem necrosis. White grub (*Brahmina coriacea*) is destructive in hilly regions, cutworm (*Agrotis segetum*) is again a destructive pest of potato in both hills and plains. Potato tuber moth (*Phthorimaea operculella*) is a serious pest causing damage in storage and fields. Mite (*Polyphagotarsonemus latus*) damages early planted crop when the temperature is high. Potato cyst nematodes (PCN) (*Globodera rostochiensis* and *G. pallida*) are major problems in the world, particularly in temperate hilly regions. Hence, advance research based on genomics is required to address these problems (Singh et al., 2020).

5. Abiotic Stresses

5.1 Heat Stress

Potato is considered primarily a crop of cool and temperate climates. High temperature inhibits crop yield by overall reduction in plant development due to heat stress or by reduced partitioning of assimilates in tubers (Aksoy et al., 2015). Minimum night temperature plays a crucial role during tuberization in potato and largely determines whether plants will tuberize or not. Tuberization is reduced at night temperatures above 20°C, with complete inhibition of tuberization above 25°C (Khan et al., 2015) (Fig. 4). The most important effect of high night temperatures is on the partitioning of assimilated carbon between leaves and tubers. Exposure of potato plants to high temperature alters the hormonal balance in the plants. As a result, most of assimilated carbon is partitioned to the shoots and not to the tubers. Hence, heat stress breeding programs should consider the ability to tuberize at higher night temperature (above 22°C) (Dahal et al., 2019). Genomics-assisted improvement can address the issues involved in potato tuberization.

Fig. 4. Heat and nitrogen stress in potato.

5.2 Drought Stress

Potato is mostly an irrigated crop in the plains and rain-fed in the hilly regions. Owing to being a shallow root crop, potato is more prone to drought stress. Drought is an emerging problem in potato production due to unavailability of irrigation water. Drought may occur due to erratic rainfall, inadequate irrigation and lack of water supply. Drought may affect potato growth and production by reducing foliage biomass, decreasing photosynthesis rate per unit leaf area, early crop maturity and shorter vegetative growth period. Potato plant is highly sensitive to water stress and the decline in photosynthesis is fast and substantial even at relatively low water potentials of –3 to –5 bars (Monneveux et al., 2013). Tuber traits, such as shape, cracking, dry matter content and reducing sugars are highly influenced by the availability of soil moisture during the vegetative period. Hence, breeding of drought stress tolerance is now a priority in potato applying genomics approaches.

5.3 Nutrient Stress and Salinity

Major and minor nutrients are essential for good vegetative growth, yield and quality of potato. Among all, nitrogen is the most important nutrient for plant growth and development, including for potato. Potato is a shallow-rooted crop and mostly irrigated cultivations are followed on the sandy-loam soils with excessive application of N fertilizers. This practice increases the chance of nitrate leaching, contamination in the groundwater and high cost of cultivation. In addition, improving P and K use efficiency of plants would also save additional costs. Hence, improving N use efficiency of plant is one of the key options to minimize N losses, save the cost of production and improve the environmental quality to achieve sustainable crop yield (Fig. 4). Besides, salinity is another issue in highly irrigated conditions. Salinity could be either due to soil or irrigation water. Salinity causes nutritional imbalance, restricts plant growth and development, early senescence and severely reduces tuber yield in semi-arid and arid regions (Ahmed et al., 2020). Variability for salinity stress tolerance exists in potato germplasm, which can be harnessed for development of varieties.

5.4 Frost Stress

Frost is the major problem of temperate growing regions of potato. Temperatures below –2°C can result in partial or complete loss of the crop. In temperate zones, frosts can occur during spring season when the crop is in the initial stage of vegetative growth or during autumn when it is nearing maturity. Higher crop losses occur in tropical highlands and subtropical plains where frosts can occur any time during the crop growth period. Acclimation or hardening may increase the resistance to frosts in many plants. Exposure of the plants to prolonged low temperatures is effective in increasing resistance to frost injury in wild potato species (Ahmed et al., 2020). Since there is very limited research, there is a need of genomics-based work on frost tolerance in potato.

6. Challenges in Potato

6.1 Climate Change and Rising Temperature

Climate change scenario is supposed to adversely affect potato production and productivity in the most parts of the potato growing countries (Dahal et al., 2019). For example, modeling research at the Indian Council of Agricultural Research-Central Potato Research Institute, Shimla suggests that by the year 2020, potato yield is estimated to fall by 19.65% in the state of Karnataka followed by Gujarat (18.23% fall) and Maharashtra (13.02% fall) with an overall fall of 9.56% at the national level if the requisite steps are not taken to mitigate the effects of climate change. The situation is expected to further worsen by the year 2050, when the national-level potato production is expected to fall by 16% in the absence of needed steps. However, the potato production fall may be much severer in states like Karnataka (45% fall), Gujarat (32% fall), Maharashtra (24.5% fall) and Madhya Pradesh (16.5% fall) if preventive steps are not taken (Dua et al., 2020).

In India, ICAR-CPRI, Shimla is internationally acclaimed for developing suitable varieties and technologies for sub-tropical regions. This has virtually transformed potato crop from temperate to sub-tropical crop and enabled the spread from the cooler hilly regions to the vast Indo-Gangetic Plains as a *rabi* crop, where crop duration varies between 75–90 days and covers nearly 90% of total potato production in the country. This has triggered a revolution in potato production, causing very fast growth in area, production and productivity in the country. However, the impact of global warming started manifesting during 1990s and it became imperative that further adaptation of potato from sub-tropical to more warmer growing conditions would be necessary in the near future to sustain its cultivation in the plains. In fact, the Intergovernmental Panel on Climate Change (IPCC) in its *Fourth Assessment Report* predicted that the potato growing season in 2055 is likely to be warmer by 2.41–3.16°C. Potato production may decline by 15.32% in the year 2050. To overcome this situation and to satisfy the projected demand, it is necessary to strengthen research work on developing varieties and production technologies for cultivating potato under warmer conditions. Research emphasis should be on developing short-duration, early-maturing varieties with heat tolerance for both fresh consumption and processing. Concomitantly, safe and sustainable technologies should be developed for heat- and water-stress management, nutrient management, management of invasive and range-expanding pests and diseases, and cold chain management.

6.2 Broadening of Genetic Base and Productivity Enhancement

The food basket of most countries is changing due to economic growth, lifestyle and dietary preference. There is a perceptible shift in food preference from cereals to vegetables and fruits, in both developed and developing countries. As per the projection made by ICAR-CPRI, Shimla, India would require about 125 million tons of potato annually by 2050. This enormous jump in production has to come from broadening the genetic base, breaking the yield barriers and productivity enhancement. The availability of additional cultivable land for potato cultivation would be virtually nil due to unfavorable changes in land utilization pattern. On the contrary, yield stagnation has emerged as a roadblock in achieving productivity enhancement in a sustainable manner. Hence, innovative technologies, such as genomics tools are immediately required for breaking the yield barriers. Research emphasis should be targeted on harnessing maximum yield potential of potato by broadening genetic base of varieties, improving photosynthetic energy-conversion efficiency, conferring atmospheric nitrogen-fixation ability, root system architecture for input use efficiency and improving sink strength. Besides, emphasis should be given on diploid breeding to exploit hybrid vigor in potato, which is yet to be exploited. Precision breeding tools, such as genomic selection, genome editing and genomics-assisted breeding should be fully integrated with conventional breeding to break the yield barrier in potato.

6.3 Sustainable Production System

The sustainable crop production system is much required for developing countries, where resources are limited than in developed countries. For example, in India, the Green Revolution during the 1960s, took the country out of the 'ship-to-mouth' existence to a status of 'food-surplus nation'. On the other hand, all natural resources including soil, water and energy are under severe constraints now. Inputs for agricultural production will also become scarcer and dearer with time. It is also imperative that future food-production technologies should be carbon neutral and sustainable. Moreover, the income of farmers in real terms has not appreciated adequately during recent years, causing widespread agrarian distress. Therefore, a paradigm shift is necessary now from the policy of mere food production to income security of farmers, particularly who are small and marginal farmers. All our future technologies should aim at addressing the farming system in an integrated manner. Research emphasis should be given on integrated farming system approach for technology development, water use efficiency, nutrient responsive technologies, conservation agriculture and bio-intensive crop management. Use of information technology-enabled advisories and Artificial Intelligence (AI) should be encouraged for technology dissemination.

6.4 Post-harvest Management

Post-harvest loss of potato is highest in India (22.20%) (FAOSTAT, 2022), where it is harvested at the onset of summer season. About 50% of these losses can be prevented, using appropriate post-harvest measures. Establishing on-farm primary processing facilities would capacitate small farmers in a big way. The family farmers can be trained to undertake post-harvest processing and packaging of farm produce, preferably on-farm or close to the production site. Such technologies would promote entrepreneurship in rural areas by strengthening forward linkage in agriculture. This would generate additional working days to farm for family members, add value to harvest and generate additional income. The following areas should be given a thrust for lowering post-harvest losses: development of processing varieties and technologies, on-farm storage and primary processing units, energy-efficient storage structure, technologies for cold chipping, managing bruising injuries and technologies for export facilitation.

6.5 Integrated Disease and Insect-pest Management

Potato is affected by late blight, viruses, bacterial wilt, aphids and other common soil and tuber-borne pests and diseases. Movement of plant pests, pathogens and invasive weeds does not recognize physical and political boundaries. Globalization of commodity and food trade has increased the bio-security risk, which threatens food security of the nation. Pests and pathogens have their strength in number and are capable of adapting to changing climate much faster than our efforts to breed new varieties. Keeping in view the future challenges, emphasis on the following areas should be given for effective plant health management: robust diagnostic tools for effective interception of alien invasive pests and pathogens, application of pathogenomics for understanding epidemiology and management, breeding of multiple stress tolerance, use of chemicals for integrated pest management, emphasis on biological control of pests and diseases, decision support system and disease forecasting for environmental safety.

7. Conclusion

The stagnating growth rates of cereals' productivity, large-scale diversion of food-grains to feed and bio-fuel and expected steep rise in per capita consumption of pulses, edible oil, fruits, vegetables, milk, sugar and non-vegetarian food in the regime of steadily rising population is bound to put pressure on the existing cultivable land. Since, cultivable land is expected to remain more or less constant in the coming decades, the role of crops, like potato, having higher production potential per unit land and time will become imperative. In this context, the potato crop has very high probability of making crucial contribution to ensuring global food security. In developing countries, like India, currently potato is being consumed mainly as a vegetable by the entire population. A higher proportion of working women and nuclear families would totally transform the demographic structure of India. Incidentally this change will be conducive for higher demand for potato (as vegetable and fast food ingredient) and processed potato products in future. Fast increment in per capita as well as total household consumption of potato in the recent past is expected to sustain in the foreseeable future; hence, there is great potential of enhancing potato production in the developing countries. The gradual shift of potato cultivation from developed nations to the developing ones provides great scope for potato export from the developing countries.

A large number of challenges in potato make potato more complex and difficult to address. So our preparedness has to be much more rigorous, precise and should extend over longer periods of time. With every new addition to our knowledge on impending issues, like climate change and their implications on potato, we need to envisage more robust and superior strategies to deal with them. Therefore, in the post-genomics era, there is a need to make omics-based research and development to address food security and industrial application of potato in the future. Strong potato research and development base is expected to provide requisite support to the potato industry in the world. The applications of modern technologies, such as biotechnology, omics approaches (genomics, transcriptomics, phenomics, etc.), diagnostics, precision agriculture, information technology, remote sensing and Artificial Intelligence have already been successfully applied in various developed countries and are now being applied in developing countries.

Acknowledgement

I am thankful to the Director, ICAR-Central Potato Research Institute, Shimla, and scientists/technical/research fellows and other colleagues of the institute for their support under the institute research projects on biotechnology, germplasm, breeding and seed research on aeroponics. I am also grateful to the funding agencies for support under the externally funded projects (CABin, ICAR-IASRI, New Delhi; ICAR-LBS Young Scientist Award Project, and DBT, Government of India).

References

Ahmed, H.A.A., Şahin, N.K., Akdoğan, G., Yaman, C., Köm, D. and Uranbey, S. (2020). Variability in salinity stress tolerance of potato (*Solanum tuberosum* L.) varieties using *in vitro* screening. *Ciência e Agrotecnologia*, 44: e004220.

Aksoy, E., Demirel, U., Öztürk, Z.N., Çalişkan, S. and Çalişkan, M.E. (2015). Recent advances in potato genomics, transcriptomics, and transgenics under drought and heat stresses: A review. *Turkish J. Bot.*, 39: 920–940.

Bradshaw, J.E. and Mackay, G.R. (1994). *Potato Genetics*. CAB International, UK, p 552.

Bradshaw, J.E., Bryan, G.J. and Ramsay, G. (2006). Genetic resources (including wild and cultivated Solanum species) and progress in their utilisation in potato breeding. *Potato Res.*, 49: 49–65.

Chakrabarti, S.K., Conghua, X. and Tiwari, J.K. (2017). *The Potato Genome*. Springer Nature, Switzerland. p 326.

Dahal, K., Li, X.-Q., Tai, H., Creelman, A. and Bizimungu, B. (2019). Improving potato stress tolerance and tuber yield under a climate change scenario—A current overview. *Front. Plant Sci.*, 10: 563.

Dua, V.K., Sharma, J., Govindakrishnan, P.M., Arora, R.K. and Kumar, S. (2020). Impact of climate change on potato cultivation. pp. 19–42. *In*: Singh et al. (eds.). *Potato Science and Technology for Sub-Tropics*. New India Publishing Agency, New Delhi, India.

FAOSTAT. (2022). https://www.fao.org/faostat/en/#data; accessed on 2nd February, 2022.

Global Potato Conference. (2008). *Global Potato Conference, 2008*. Opportunities and Challenges in New Millennium w.e.f. 9–12 December, 2008, New Delhi, India.

Hawkes, J.G. (1990). *The Potato, Evolution, Biodiversity and Genetic Resources*. Belhaven Press, London, p 259.

Khan, M.A., Munive, S. and Bonierbale, M. (2015). Early generation *in vitro* assay to identify potato populations and clones tolerant to heat. *Plant Cell Tiss. Organ Cult.*, 121: 45–52.

Kumar, R., Jeevalatha, A., Baswaraj, R. and Tiwari, R.K. (2020). Viral and viroid diseases of potato and their management. pp. 267–294. *In*: Singh et al. (eds.). *Potato Science and Technology for Sub-Tropics*. New India Publishing Agency, New Delhi, India.

Monneveux, P., Ramírez, D.A. and Pino, M.T. (2013). Drought tolerance in potato (*S. tuberosum* L.): can we learn from drought tolerance research in cereals? *Plant Sci.*, 205-206: 76–86.

Sagar, V. and Sanjeev, S. (2020). Soil and tuber borne diseases of potato and their management. pp. 247–266. *In*: Singh et al. (eds.). *Potato Science and Technology for Sub-Tropics*. New India Publishing Agency, New Delhi, India.

Shah, M.A., Bairwa, A., Naga, K.C., Subhash, S., Raghavendra, K.V. et al. (2020). Important potato pests and their management. pp. 295–326. *In*: Singh et al. (eds.). *Potato Science and Technology for Sub-Tropics*. New India Publishing Agency, New Delhi, India.

Singh, A.K., Janakiram, T., Chakrabarti, S.K., Bhardwaj, V. and Tiwari, J.K. (2018). *Indian Potato Varieties*. ICAR-Central Potato Research Institute, Shimla, Himachal Pradesh, India. p 179.

Singh, A.K., Chakrabarti, S.K., Singh, B., Sharma, J. and Dua, V.K. (2020). *Potato Science and Technology for Sub-tropics*. New India Publishing Agency, New Delhi. p. 382.

The Potato Genome Sequencing Consortium. (2011). Genome sequence and analysis of the tuber crop potato. *Nature*, 475: 189–195.

Tiwari, J.K., Sundaresha, S., Singh, B.P., Kaushik, S.K., Chakrabarti, S.K. et al. (2013). Molecular markers for late blight resistance breeding of potato: An update. *Plant Breed.*, 132: 237–245.

Tiwari, J.K., Devi, S., Sharma, S., Chandel, P., Rawat, S. and Singh, B.P. (2015). Allele mining in *Solanum* germplasm: Cloning and characterization of *RB*-homologous gene fragments from late blight resistant wild potato species. *Plant Mol. Biol. Rep.*, 33: 1584–1598.

Tiwari, J.K., Devi, S., Ali, N., Luthra, S.K., Kumar, V. et al. (2018). Progress in somatic hybridization research in potato during the past 40 years. *Plant Cell Tiss. Organ Cult.*, 132: 225–238.

Wang, M., Allefs, S., van den Berg, R.G., Vleeshouwers, V.G.A.A., van der Vossen, E.A.G. and Vosman, B. (2008). Allele mining in *Solanum*: conserved homologues of *Rpi-blb1* are identified in *Solanum stoloniferum*. *Theor. Appl. Genet.*, 116: 933–943.

Chapter 2

Potato Genome Sequencing and Resequencing of Wild/Cultivated Species

1. The Potato

Potato (*Solanum tuberosum* L.) is the third most important food in the world, after rice and wheat. In 2020, the world potato production was 359.07 million tonnes over a 16.49 million hectare area with an average productivity of 21.76 t/ha (FAOSTAT, 2022). *Solanum* species are one of the richest sources of genetic diversity in plants. Cultivated and wild potato species are widely distributed in American continents, ranging from south-western region of the USA to southern Chile and Argentina, from sea level to highlands of the Andes mountain ranges (Spooner and Hijmans, 2001). The cultivated potato (*S. tuberosum* Group Tuberosum) has narrow genetic diversity, is auto-tetraploid ($2n = 4x = 48$), highly heterozygous, suffers from inbreeding depression and is affected with many biotic/abiotic stresses and climate change issues. Although a few wild *Solanum* species can be crossed with the cultivated potato, but most of them are non-crossable due to differences in Endosperm Balance Number (EBN) and ploidy number. Notably, wild potato species have several desirable traits, such as resistance to diseases and insect-pests. Hence, introgression of desirable genes from wild species is important for potato improvement. Thus, knowledge of the potato genome sequence would facilitate advance breeding with target traits, though potato is a vegetatively propagated crop through tubers but the genetics of many traits is poorly understood. Unlike diploid crops, cultivated tetraploid potato contains four copies of each chromosome, which makes it difficult to follow inheritance patterns. As a result, several agronomic traits are less clear and hinder its improvement. The use of modern genomics tools and genome sequencing has immense potential in uncovering genes' underlying traits for fast breeding. There is a need to develop high quality and annotated genome of potato to discover new genes/alleles for desirable traits underlying the quantitative/qualitative traits.

2. Genome Sequencing Technologies

2.1 Structural Genomics

Structural genomics uncovers the nucleotide sequences of genome structure. It provides information about genome sequence, architecture and evolutionary aspects of the genome of any crop. Plant genomes generally vary in their genome size and are characterized by a large number of genome duplication and repetitive elements (Bennett and Leitch, 2011). Currently, 249 assembled and annotated plant genomes are available in public on the Phytozome (v13) (https://phytozome-next.jgi.doe.gov/). With advancements in sequencing technologies and reduced per sample costs, the number of plant genomes sequences has increased at a rapid rate in the past one decade (Levy and Myers, 2016). Several sequencing technologies have been applied for whole genome sequencing, like the First Generation Sequencing (Sanger sequencing), Next-Generation Sequencing (NGS) or Second Generations Sequencing (SGS), and Third Generation Sequencing (TGS), like Oxford Nanopore technologies (Kumar et al., 2019). Now the long-read sequencing platforms have been made available to provide more accurate assembly of the whole genome (Fig. 1, Table 1). Further, structural variation, copy number variation, haplotypes, genotyping-by-sequencing and exome sequencing can be utilized for allele mining and Single Nucleotide Polymorphism (SNP) discovery in the organisms.

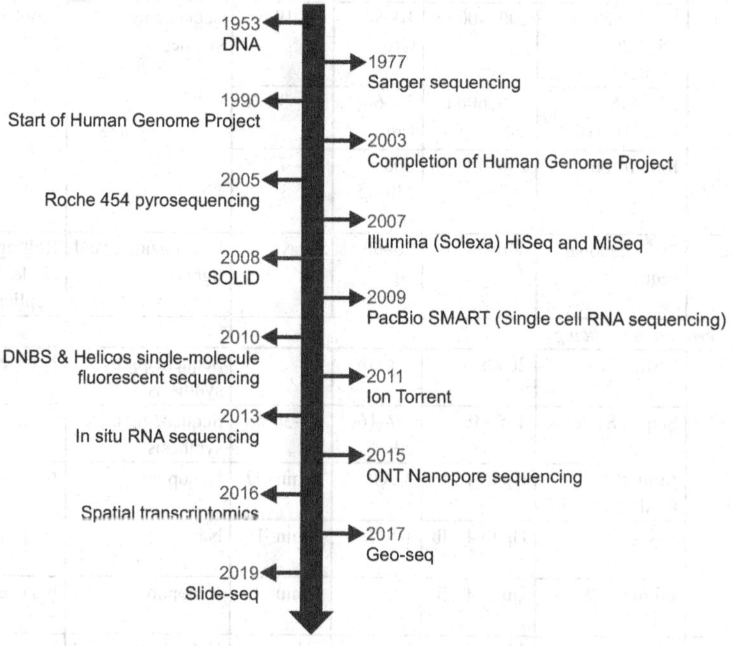

Fig. 1. Time line of the genome sequencing technologies.

Table 1. Features of genome sequencing technologies.

Company	Sequencing technology	Read length	Data output	Run time	Sequencing chemistry	Template preparation
First Generation Sequencing						
Life Sciences	Sanger	800 bp	1 read	2 h	Dideoxy-nucleoside terminator	Bacterial cloning
Next Generation Sequencing						
Roche	Roche 454 pyrosequencing	700 bp	0.7 Gb	24 h	Sequencing by synthesis, pyrosequencing	Emulsion PCR
Illumina	iSeq100	2 × 150 bp	1.2 Gb	9–10 h	Reversible terminator sequencing	Bridge PCR
	MiniSeq	2 × 150 bp	7.5 Gb	4–24 h		
	MiSeq	2 × 300 bp	15 Gb	4–55 h		
	NextSeq 550	2 × 150 bp	120 Gb	12–30 h		
	NextSeq 1000 & 2000	2 × 150 bp	330 Gb	11–48 h		
	NovaSeq 6000	2 × 250 bp	6000 Gb	13–44 h		
	HiSeq 3000/4000/ X Series	100 bp	1500–1800 Gb	3–4 days		
ABI	SOLiD	50–75 bp	30 Gb	7–14 days	Sequencing by ligation	Emulsion PCR
Ion Torrent	Ion GeneStudio S5 (+Plus/ Prime)	200–400 bp	10–50 Gb	10–19 h	Sequencing by synthesis	Emulsion PCR
	Ion PGM 314/316/318	200–600 bp	30–600 Mb	2.3–3.7 h		
	Ion Proton	200–400 bp	600 Mb–15 Gb	2.5–7.3 h		
Complete Genomics	DNA nanoball sequencing	440–500 bp	20–60 Gb	9 days	Hybridization and ligation	Rolling circle Replication
Third Generation Sequencing						
PacBio SMRT	RSII	20 kb	~ 1 Gb	4 h	Sequencing by synthesis	No need
	Sequel Systems (II/IIe)	4–5 Mb	3.7–160 Gb	20–30 h	Sequencing by synthesis	No need
Oxford NanoPore	MinION/ GridION Mk1	Up to 4 Mb	42 Gb	1 min–72 h	Nanopore	No need
	Flongle	Up to 4 Mb	1.8 Gb	1 min–16 h	Nanopore	No need
	PromethION	Up to 4 Mb	245 Gb	1 min–72 h	Nanopore	No need
Helicos bio-sequencing	Helicos sequencing	25–60 bp	21–35 Gb	8 days	Hybridization and synthesis	No need

2.2 First-generation Sequencing Technology

DNA is the basic genetic material of any organism containing gene information. Hence, deciphering the nucleotides sequences of the DNA structure is of utmost important for genome sequencing to reveal the underlying genetic information. The first DNA sequencing (chemical) method was discovered by Maxam and Gilbert (1977) and later chain termination method was invented by Sanger et al. (1977). They were the pioneer researchers who laid the foundation stone for automated DNA sequencing, called 'Sanger sequencing' (Smith et al., 1986). This first-generation Sanger sequencing was made popular by various companies, like Life Sciences/ Applied Biosystems Instruments (ABI). This is based on the Sanger dideoxynucleotide method and major platforms, like ABI capillary-based sequencer, Megabase and Beckman Coulter. Further, sequencing technology continued to advance with improvement in read length, accuracy and high throughput, opening avenues in an array of plant species for broad applications (Levy and Boone, 2019). Despite the good read length, these platforms were not suitable for large genomes because of high cost and low throughput. Additionally, Sanger sequencing technology suffers from various limitations, such as (i) handling limited length of DNA, (ii) may result in biological biasness, (iii) require high amount of template DNA, and (iv) allow limited number of samples at a time.

2.3 Next Generation Sequencing Technology

Vast progress in research has been made in genome-sequencing technologies in the fields of fluidics, imaging, detection power, accuracy, resolution and computation tools (Levy and Boone, 2019). The complete genome sequencing of the human genome project (Yamey, 2000) was the start-up of DNA sequencing. The first publication on the human genome sequence was made by Venter et al. (2001). With further research and development in advanced sequencing strategies, the discovery of high-throughput DNA sequencing technology of Second Generation Sequencing (SGS), called the Next Generation Sequencing (NGS) (Varshney et al., 2009), became possible. As the time progressed, the sequencing cost got reduced with increase in high throughput data. NGS technologies have the capacity to produce several times more data of short reads length as compared to Sanger sequencing. Moreover, these NGS technologies open avenues to wider applications in plant research in the area of Chromatin Immune-Precipitation (ChIP) sequencing (ChIP-seq), ChIP DNA microarray chip (ChIP-chip), transcriptome sequencing (RNA-seq), small RNA sequencing, whole genome resequencing, de novo assembly, structural variation analysis, organelle genomes analysis, to name a few. Later, several animal and plant genomes were published in public (www.ensembl.org/info/about/species.html). The potato genome was deciphered using Roche 454 pyrosequencing, Illumina and Sanger sequencing technologies (Visser et al., 2009; The Potato Genome Sequencing Consortium, 2011).

There are several advantages of NGS, like reduced per sample cost, high throughput data and integration of various downstream genomics applications. Despite a wide range of applications of NGS platforms in plant research, various limitations, such as short read and lesser accuracy, are seen. Moreover, generation of accurate assembly with shorter reads poses a major challenge to NGS applications. The short reads cannot fully reconstruct repetitive regions, leading to fragmented assemblies and collapsed regions, thus requiring new platforms with longer reads length.

2.3.1 Roche 454 Pyrosequencing

In 2005, the NGS platform began with the first development of Roche 454 pyrosequencing by synthesis (Roche 454 GS20). This technology has increased throughput and reduced sequencing cost but with reduced read length and lesser accuracy. This technology allows detection of a large number of samples in parallel at a given light intensity, compared to Sanger sequencing. Later, in 2008, the upgraded version 454 GS FLX Titanium platform was released. The major problems of Roche platforms were the high error rates and very high reagent cost. Now this technology has been discontinued since more than six to seven years.

2.3.2 Illumina

In 2007, the most important NGS discovery was made by Illumina (Solexa) sequencing technology. This was introduced in the marker based on even short reads but greater high throughput. Illumina is also based on sequencing by synthesis chemistry. As of now, Illumina has discovered several platforms, such as NextSeq 550, HiSeq series 2500, 3000 and 4000, and HiSeq X series and so on. Also, Illumina has introduced MiSeq, a small compact sequencer with 0.3–15 Gb data output. Further, a new TrueSeq technology has been introduced for long read sequencing. Currently, Illumina is the most popular available NGS technology offering a wide range of sequencing platforms for various genomics applications at affordable price and choice based on study design.

2.3.3 SOLiD (Sequencing by Oligo Ligation and Detection)

In 2008, the SOLiD platform was first introduced by Applied Biosystems Instruments (ABI) based on sequencing by ligation chemistry with short read and high throughput. Earlier, SOLiD 5500 W series was applied for whole genome, exome and transcriptome sequencing applications. The major drawback of this platform was short read length (50–75 bp), long run time and computational tools to analyze the data. Now this technology has been discontinued since more than four to five years.

2.3.4 Ion Torrent

In 2011, the Ion Torrent technology was introduced by the 454 sequencing inventors, such as Ion Personal Genome Machine (PGM) and high throughput proton sequencers. The major problems with this technology were reading homopolymer repeats and it provides more or less a similar kind of sequencing options like Illumina.

2.3.5 Complete Genomics Technology

In 2010, the Complete Genomics Technology was discovered by Beijing Genomics Institute (BGI) and MGI Tech Co. Ltd (MGI). The BGI/MGI sequencing technology offered several sequencing applications, like whole genome, transcriptome, etc. It is comparable with other sequencers, including Illumina.

2.4 Third Generation Sequencing Technology

In recent years, the single-molecule based newer sequencing technologies of longer reads, called Third Generation Sequencing (TGS), has gained great interest and has become popular (Li et al., 2017). The TGS technologies directly target single DNA molecules and perform real time sequencing. TGS was first introduced by the Pacific Biosciences Inc. and later followed with a portable Oxford Nanopore technology. TGS follows single molecule sequencing without PCR amplification of DNA in less time for sample preparation and sequencing with high accuracy and reduced chances of errors. The long read sequences produced by the TGS platforms make easier and more accurate genome assembly and thus, unlike NGS, avoid time-consuming computational tools. TGS having long reads also allows easier identification of structural variants in the genome than in short reads of NGS. This technology is more efficient in terms of high-throughput, less cost per sample in less time.

2.4.1 PacBio

The first successful Single Molecule Real Time (SMRT) sequencing technology with highly accurate long reads (up to 20 Kb) was first introduced in 2009 by Pacific Biosciences (PacBio, http://www.pacb.com/). This technology has been applied to whole genome sequencing, transcriptome sequencing, epigenetic and many more studies. PacBio workflows are available for SNP, InDels and structural variants using long read sequences. With modern reagents and sequencing kits, the typical throughput of PacBio RS II system produces 0.5–1 Gb per SMRT cell, with a mean read length of roughly 10 kilobases (kb). Nonetheless, PacBio reads have a significantly higher error rate (10–15%) than in NGS-generated reads (> 2%) (Larsen et al., 2014). Fortunately, these sequencing errors are randomly distributed; the rates can therefore be greatly reduced through the use of circular consensus sequencing, where a single molecule template and its complement strand are sequenced multiple times to generate a unique consensus.

2.4.2 Oxford Nanopore

Oxford Nanopore Technology (ONT) was released in 2014 and is the newest TGS platform (Laver et al., 2015; Kanzi et al., 2020). The ONT platform offers affordable and portable features to enhance flexibility for a small laboratory setup. It yields ultra-long read sequences, particularly for large size genomes. Unlike NGS which requires tedious and time-consuming library preparation and sequencing protocols, ONT technology provides an easy and rapid methodology, even with automation. The MinION sequencer is a portable real-time device for genome and transcriptome sequencing which is integrated with an early access program (The MinION Access Program, MAP). This device can be connected with laptop and works even under limited electric power availability and therefore, can be used anywhere without any labs, such as jungle, mountains, space, etc. No additional computing infrastructure is required. Moreover, these disadvantageous areas would benefit from the real time nature of this sequencing platform. Thus, it opens new opportunities for resource-limited countries.

The MinION generates data in real time which allows simultaneous data analysis during the experiment. Each consumable flow cell can now generate 10–20 Gb of DNA sequence data. Ultra-long read lengths are possible (100s of kb) as we can choose the fragment length. It weighs under 100 g and plugs into a desktop or laptop, using a high-speed USB 3.0 cable. The read length ONT device is very similar to that of PacBio, with a maximum length up to 100 thousands base pairs. However, ONT reads have error rates higher than PacBio reads, with accuracy ranging between 65–88%. In addition, the throughput per MinION flow cell run is not very stable at the moment, varying from below 0.1–1 Gb of raw sequence data (Ip et al., 2015; Laver et al., 2015). Due to its smaller size and low cost, the MinION machine is becoming popular, particularly in pathogen surveillance and clinical diagnostic applications, and is yet to be explored in plant sciences. The MinION is being used for a number of biological analysis techniques, including *de novo* sequencing, targeted sequencing, metagenomics, epigenetics and more. The ever-increasing progress in sequencing technology would certainly lead to discovery of newer technologies in modern science to identify alleles/markers for agronomic traits in potato.

Other TGS technologies that are still in early stages of development, such as Helicos fluorescent sequencing, is available for true single molecule sequencing (http://seqll.com/). Likewise, several others are available, such as Complete Genomics Technology in ligation technique used by SOLiD (Supported Oligonucletide Ligation and Detection) and GnuBIO technology by BioRad (CA, USA) and so on. Moreover, efforts are continuing to improve sequencing in terms of ease of application, data output, quality, read length, short sequencing protocols, reduced cost and simple data analysis.

2.4.3 Mapping of Long Reads to Optical Maps

NGS technologies have advanced the crop genomes sequencing. However, they have limitations of less accuracy in sequencing assembly of repetitive elements of crop species, which often lead to incomplete or erroneous assemblies. These challenges were overcome by the introduction of long-read sequencing platforms, as discussed above. These long-reads data can be used in optical mapping to produce high-quality assemblies for complex genomes.

Optical mapping is a light microscope-based technique that captures images of restriction sites to produce fingerprints of DNA sequences. Earlier the technique was used to assemble the small genome of microbes but with increase in throughput and semi-automation, it was used to assemble many complex genomes. Optical mapping has proven to be useful for genome scaffolding and structure variation analysis, and provides valuable physical linkage information without reliance on recombination and genetic maps. In most studies to date, optical mapping was mainly used to validate assemblies and few studies used optical mapping as the principal method of genome scaffolding or structural variation detection (Zhou et al., 2004). Optical mapping competes with other physical mapping approaches, such as Hi-C and the Dovetail ChicagoTM method and because both continue to advance, it is unclear which is likely to dominate in future. Long-read sequencing and optical mapping technologies offer new solutions to accurately assemble and compare large and complex crop reference genomes. Accurate reference assemblies facilitate the identification of candidate genes for agronomic traits—information which can be applied for molecular breeding. As long-read sequencing and optical mapping technologies continue to develop and are applied for crop genomes, scientists will gain a better understanding of crop genomic diversity, evolution and gene function to accelerate plant breeding.

Advantages of TGS:

* *Long reads*: TGS technology provides long reads of more than 200 kb, covering the repetitive region of the genome and thus facilitating more accurate genome assembly.
* *Less sequencing time*: Reduces sample sequencing time from days to hours or minutes for real-time application is the greatest advantage of TGS technology.
* *Easy sample preparation*: Allows easy sample preparation for sequencing and not much time and skill are required.
* *Less per sample cost*: Incurs less per sample cost than NGS platforms.
* *Reduced sequencing biasness*: Reduction or elimination of sequencing biasness is another important advantage.
* *Real time analysis*: Enables real time sequencing, where reads are available for analysis as soon as they have passed through the sequencer. It allows immediate access to the data and downstream application thereof.
* *Easy multiplexing*: Permits multiplexing of different samples with easy barcoding system to run simultaneously at a time on one-flow cell.
* *Accessibility*: Has easy access with limited capital cost, easy installation and less lab infrastructure requirements.

Disadvantages of TGS:

* Higher error rates (e.g., PacBio: 10–15% than NGS: > 2%).
* Low throughput (e.g., PacBio RS II: 0.5–1 Gb; MinION: 0.1 to 1 Gb).
* High initial costs of equipments.

2.5 Fourth Generation Sequencing

Fourth Generation Sequencing (FGS) has come to be known very recently in humans for *in situ* sequencing of mRNA (Mignardi and Nilsson, 2014). In the *in situ* sequencing method, spatial distribution and subcellular resolution of reads can be visualized. Additionally, this technology is high throughput in terms of simultaneous analysis of the number of cells. Single-cell RNA sequencing method is a robust technology which uses just a few picograms of tissues, and thus increases the rate of single-cell sequencing at reduced cost. This method has the potential to visualize, at single nucleotide level, point mutation, allelic variants, add molecular information to microscopic view and tissue expression profiling. However, it is computationally challenging to analyze single-cell-based information where tissues are composed of thousands of cells. Till now, this technology was applied in human clinical diagnostics and its future prospects in plants is yet to be seen. Taken together, third and fourth generation sequences are designed for 'single molecule sequencing' and have a tremendous potential in a wide range of applications in plant sciences (Suzuki, 2020).

3. The Potato Genome Sequence

To undertake the challenge of complex potato genome sequence, the Potato Genome Sequencing Consortium (PGSC), comprising 26 international institutes belonging to 14 countries, was constituted. The PGSC deciphered the potato genome sequence in 2011, which was published in the high-impact factor journal *Nature* (The Potato Genome Sequencing Consortium, 2011). This was the first genome of a plant belonging to Asterid clade of eudicot that represents 25% of flowering plant species. A total of 39,031 protein-coding genes were predicted in the potato genome of 840 Mb size. The potato genome provides new insights into eudicot genome evolution. A combination of data from the unique and homozygous doubled-monoploid potato clone DM, and the vigorous, heterozygous diploid clone RH was used to address the form and extent of heterozygosity and complexities of inbreeding depression in potato. The salient features of the potato genome are:

- A unique and homozygous doubled-monoploid potato clone (*S. tuberosum* Group Phureja DM1-3 516 R44, referred as DM (CIP801092)) was employed for genome sequencing to overcome the problems associated with genome assembly due to high levels of heterozygosity in the cultivated tetraploid potato.
- Additionally, a heterozygous diploid cone, *S. tuberosum* Group Tuberosum RH89-039-16 (referred as RH), was also used in the sequencing project.
- Approximately 78,000 clones of Bacterial Artificial Chromosome (BAC) libraries were aligned to nearly 7,000 physical map contigs, of which, nearly 30,000 BACs were anchored to the Ultra High Density genetic map of potato comprising 10,000 AFLP markers.
- Out of this integrated genetic-physical map, about 50–150 seed BACs were identified for every potato chromosome and was basic material to BAC-by-BAC sequencing strategy.
- Fluorescent *in situ* hybridization of selected BAC clones confirmed the anchor points in each chromosome.
- Nuclear and organelle genomes of DM were sequenced using whole-genome shotgun sequencing approach, i.e., Illumina Genome Analyser and Roche Pyrosequencing, as well as conventional Sanger sequencing technologies.
- Combined data of both genotypes for whole genome sequence and transcriptome sequence were generated to decipher the potato genome sequence.
- The potato genome consists of 12 haploid chromosomes. Approximately 96.6 Gb raw genome sequence data was generated. Finally, the genome assembly amounted to 727 Mb (86%) of the potato genome size (844 Mb), based on the flow cytometry analysis.

- A total of 39,031 protein coding genes were annotated in the potato genome. There is also evidence of alternative transcripts for as many as 60% of these genes.
- The potato genome presents evidence for at least two genome duplication events indicative of a palaeopolyploid origin.
- The genome is highly duplicated and contains many short segments showing conserved synteny with other plants, such as rice and grape.
- Potato is the first genome sequence of an asteroid and reveals 2,642 genes specific to this large angiosperm clade.
- A total of 917 superscaffolds were anchored, ordered and assembled into pseudomolecules corresponding to the 12 haploid potato chromosomes.
- The DM Whole Genome Shotgun sequencing project has been deposited at DDBJ/EMBL/GenBank under the accession AEWC00000000.
- The potato genome sequence is available freely at the Spud DB, Potato Genomics Resources, Michigan State University, USA.
- Potato genome website: http://solanaceae.plantbiology.msu.edu/pgsc_download.shtml.

4. Genome Resequencing of Wild/Cultivated Potatoes (*Solanum* spp.)

4.1 Impact of Post-genome Sequencing

The potato genome sequence has opened new vistas for genomics-aided next-generation potato breeding. The potato collections can now be more efficiently mined for novel alleles and beneficial traits of economical and industrial importance. The integrated genome sequence and genetic reference map will allow traits phenotyping loci or quantitative trait loci (QTL) defined by sequence-based markers to be linked to specific genetic and physical regions of the genome. Such regions can be used to define markers for fine-scale mapping, or candidate genes search directly from the genome sequence and associated annotation data. This step change, facilitating sequence-based genomics and aiding molecular breeding in potato, would accelerate gene discovery and gene isolation for numerous traits. This would further shorten the time to breed new varieties and also significantly improve parental genotypic assessment. Genome-tagged molecular marker studies will be more meaningful and enable more accurate estimates of population genetic and Linkage Disequilibrium (LD) parameters. The shift towards sequence-based polymorphism rather than fragment based, will virtually replace centimorgan (cM) position by sequence co-ordinates and greatly increase the information output and accuracy of mapping procedures. The integrated potato genetic and physical reference map forms an important resource for linking all current and future genetic mapping efforts by the potato community and help to alleviate many of the complicating aspects of potato as a genetic system. With the release of the genome of the other economically important Solanaceous plant—tomato, comparative linkage mapping and in-depth

Table 2. Genome sequencing/resequencing and sequence-based recent studies in potato.

S.No.	*Solanum* species	Key findings	References
Genome sequencing/resequencing			
1.	*S. tuberosum* Group Phureja DM1-3 516 R44 (DM), and *S. tuberosum* Group Tuberosum RH89-039-16 (RH)	First potato genome sequence by the PGSC Genome: 840 Mb Genes: 39,031	Visser et al. (2009); The Potato Genome Sequencing Consortium (2011)
2.	*S. commersonii* (PI 243503)	Genome: 830 Mb Genes: 39,290	Aversano et al. (2015)
3.	67 *Solanum* species	Diversity, SNP and domestication analysis	Hardigan et al. (2017)
4.	*S. chacoense* (M6)	Genome: 825.76 Mb Genes: 37,740	Leisner et al. (2018)
5.	202 wild and cultivated diploid potatoes	Elucidated the phylogeny of Solanum section Petota based on plastid genomes	Huang et al. (2019)
6.	95 dihaploid lines (Superior × IVP101, a haploid inducer *S. phureja* called *in vitro* pollinator: IVP)	Detected somatic translocation events and other chromosome-scale abnormalities in during dihaploid production in potato	Pham et al. (2019)
7.	6 *Solanum* species	Structural variation and CNV analysis using long read (TGS) technology	Kyriakidou et al. (2020a)
8.	12 *Solanum* species	SNP and CNV analysis	Kyriakidou et al. (2020b)
9.	167 dihaploid lines (cv. Alca Tarma × haploid inducer *S. phureja* pollens from IVP101 or PL4)	Structural variation in 167 dihaploid indicated that dihaploid are free of haploid inducer DNA	Amundson et al. (2020)
10.	DM1–3 516 R44 (DM: a doubled monoploid clone of *S. tuberosum* Group Phureja, as used originally by the PGSC	Generation of a high-quality, long-read, chromosome-scale, new (v6.1) genome assembly and improved annotation dataset for the reference potato genotype (DM)	Pham et al. (2020)
11.	Diploid breeding line 'Solyntus' (Diploid potato × self-compatible *S. chacoense*)	Deciphered the *de novo* sequenced genome of Solyntus as the next standard reference in potato genome in future. Solyntus plant produces good tuber numbers and yields with round shape and creamy flesh	van Lieshout et al. (2020)
12.	RH89-039-16 (2*x*) (RH, heterozygous diploid potato *S. tuberosum* group Tuberosum)	The haplotype-resolved genome assembly using a combination of multiple sequencing strategies, including circular consensus sequencing provides a holistic view of the complex genome organization of a heterozygous diploid potato	Zhou et al. (2020)

Table 2 contd. ...

...*Table 2 contd.*

S.No.	*Solanum* species	Key findings	References
13.	4 different genotypes: • Interspecific somatic hybrid P8 (C-13 + *S. pinnatisectum*) (4*x*) • *S. tuberosum* dihaploid 'C-13' (2*x*) • Wild species: *S. pinnatisectum* (2*x*) • Hybrid progeny 'MSH/14-112' (P8 × cv. Kufri Jyoti) (4*x*)	• P8 (Genome 725.01 Mb and 39260 genes), • *S. pinnatisectum* (Genome 724.95 Mb and 25711 genes), • MSH/14-112 (Genome 725.01 Mb and 39730 genes), • C-13 (Genome 809.59 Mb and 30241 genes)	Tiwari et al. (2021)
14.	6 potato varieties (4*x*): • Fresh market: Colomba, Spunta, • Chip processing: Atlantic • Frozen processing: Castle Russet • Starch: Altus, Avenger	Phased, chromosome-scale genome assemblies of six tetraploid potato cultivars revealed extensive allelic diversity	Hoopes et al. (2022)
15.	3 potato cultivars (4*x*) (Otava, Hera and Stieglitz)	Reported chromosome-scale and haplotype-resolved genome assembly, based on high-quality long reads, single-cell sequencing of 717 pollen genomes and Hi-C data of a tetraploid potato cultivar Otava	Sun et al. (2022)

sequence-based synteny analysis in *Solanaceae* family will be feasible. Given the biological and economic importance of Solanaceous species and the diversity of their phenotypes/products (agriculturally useful parts—tubers, berries, etc., growth habits, wide geographical growing range, clonal propagation, regeneration), comparative Solanaceous genomics will provide a fundamental framework for tackling both applied and basic questions. Over a decade after potato genome sequencing, a few more wild and cultivated potato genotypes have been sequenced worldwide, as described in later sections (Table 2).

4.2 Genome Sequence of Wild Species S. commersonii

S. commersonii is a tuber-bearing, diploid and 1 EBN (Endosperm Balance Number) wild potato species and native to central and South America. A cold-tolerant accession (PI 243503) was used for draft whole genome and transcriptome sequencing. In 2015, draft whole genome sequence of a cold/freezing tolerant wild potato species *S. commersonii* was completed (Aversano et al., 2015). The *S. commersonii* genome consists of ~ 830 Mb genome size compared to the reference potato (840 Mb). The *S. commersonii* genome comprises 44.5% (~ 383 Mb) transposable elements, which are comparatively lower than the potato (55%) and tomato (63%) genomes. *S. commersonii* shows reduced heterozygosity and variation in genome size due to variation in intergenic sequence region length. A total of 39,290 protein coding genes have been predicted in the genome, of which 126 are specific to cold stress.

Table 3. Number of disease resistance (R) genes domain proteins between *S. commersonii* and *S. tuberosum* genomes.

Domain protein	***S. commersonii***	***S. tuberosum***
CC-NBS-LRR	186	194
TIN-NBS-LRR	36	46
NBS-LRR	98	165
NBS	71	199
TIR	10	36
LRR	144	199
TIR-NBS	12	14
TIR-LRR	1	2
Receptor like kinase	252	313
Receptor like protein	180	237

Phylogenetic analysis indicates duplications and divergence and indicated nearly 2.3 million years old lineages relationship between *S. commersonii* and domesticated potatoes (Table 3).

4.3 Genome Sequence of Wild Species S. chacoense (M6)

Tetraploidy and highly heterozygous nature of the cultivated potato causes major problems in genome assembly analysis and breeding. Wild potato species are reservoirs of agronomic important traits to introgression into cultivated potato. *S. chacoense* is the diploid wild potato species and M6, an inbred clone, is self-compatible and has desirable tuber quality, disease resistance and reduced cold-induced sweetening traits. In 2018, whole genome sequence of *S. chacoense* M6 clone was completed (Leisner et al., 2018) (Table 4). The M6 genome assembly size was 825.76 Mb, of which 508 Mb anchored on 12 pseudomolecules with 37,740 genes. Comparative analysis of M6 and six other Solanaceous species revealed a core set of 158,367 genes, of which 1,897 genes are unique in three potato species. SNP analysis indicates high residual heterozygosity on the chromosomes 4, 8 and 9 than on other chromosomes. M6 genome provides important resources for genes identification and rapid potato breeding.

Table 4. Statistics of *Solanum chacoense* M6 genome assembly.

Total scaffold size	825767562 bp
No. of scaffolds	8260
N50 scaffold size	713 601 bp
Maximum scaffolds length	7385816 bp
Minimum scaffolds length	909 bp
Total anchored scaffold length	508150181 bp
Total unanchored scaffold length	317617381 bp
Total number of genes	37740

4.4 Genome-based Structural Variation Analysis in Potato Species

Genome sequences of 12 wild species and cultivated potatoes have been completed (Kyriakidou et al., 2020b). This study includes wild potato species, namely *S. stenotomum* subsp. *goniocalyx* (2*x*, 2 accessions), *S. stenotomum* subsp. *stenotomum* (2*x*), *S. phureja* (2*x*), *S. ajanhuiri* (2*x*), *S. bukasovii* (2*x*), *S. juzepczukii* (3*x*), *S. chaucha* (3*x*), *S. tuberosum* subsp. *andigenum* (4*x*, 2 accessions), *S. tuberosum* subsp. *tuberosum* (4*x*) and *S. curtilobum* (5*x*). Genome analysis revealed structural variation (CNV: copy number variation) in these potato species compared with the two published potato genomes, i.e., doubled monoploid clone DM (*S. phureja*), and a diploid inbred clone M6 (*S. chacoense*). Given that highly hererozygosity and complex genome of tetraploid, comparative genome analysis reveals CNV that is likely to contribute key roles in trait diversification and evolutionary process in these species. Comparison of genome sequences of 12 species with DM and M6 reveals a greater number of deletions than the number of duplications. However, the duplication region was found more than the deletion region. Some genome regions are highly abundant in CNVs events with 3.8–10.5 million SNPs. Most SNPs are found in the intergenic regions while the least are in exonic regions. In particular, a few genes are found to be heavily impacted by CNV events, such as abiotic stress-responsive, auxin-induced SAUR genes, disease-resistance genes, particularly NBS-LRR, methylketones, mannan endo-1,4-β-mannosidase, the 2-oxoglutarate/Fe(II)-dependent oxygenase superfamily proteins and genes of unknown function. The study clearly indicates that heterozygosity increases with increase in ploidy number of the species.

4.5 Genome-aided Domestication Study in Potato Species

A panel of 67 wild and cultivated *Solanum* species were sequenced and analyzed for diversity and domestication of potato (Hardigan et al., 2017). These potato species (ploidy and number of accession) are cultivated potato of *S. tuberosum* groups Tuberosum (4*x*, 23), Chilotanum (4*x*, 2), Andigena (4*x*, 8), Phureja (2*x*, 4), Stenotomum (2*x*, 6); and wild landraces *S. berthaultii* (2*x*, 1), *S. boliviense* (2*x*, 1), *S. megistracrolobum* (2*x*, 1; now *S. boliviense*), *S. brevicaule* (2*x*, 5; one each from earlier taxa name *S. brevicaule*, *S. gourlayi*, *S. leptophyes*, *S. sparsipilum*, *S. spegazzinii*), *S. candolleanum* (2*x*, 2; one each from previous taxa *S. bukasovii*, *S. multidissectum*), *S. chacoense* (2*x*, 1), *S. chomatophilum* (2*x*, 1), *S. commersonii* (2*x*, 1), *S. ehrenbergii* (2*x*, 1), *S. etuberosum* (2*x*, 1), *S. infundibuliforme* (2*x*, 1), *S. jamesii* (2*x*, 1), *S. kurtzianum* (2*x*, 1), *S. medians* (2*x*, 1), *S. microdontum* (2*x*, 1), *S. okadae* (2*x*, 1), *S. raphanifolium* (2*x*, 1), *S. vernei* (2*x*, 1), and *S. verrucosum* (2*x*, 1).

Genomic variations in terms of sequence and structural variation (CNV) were captured in 67 accessions representing cultivated and wild progenitors potatoes compared with the reference potato, *S. tuberosum* Group Phureja (DM). High density of CNV across the diversity panel evidences heterogeneous genome containing agronomic-important trait loci in the species. Of the total nearly 70 million SNP, 44 millions were identified within the conserved genome. The study suggested increase in heterozygosity with increase in ploidy number from diploid wild species and landraces to cultivated tetraploid potato varieties. Moreover, the asexual mode of propagation results in retention of alleles and desirable loci, thus causing successful domestication of potato for a wide range of agronomic important traits. This study uncovers new alleles to accelerate genome-enabled development of potato varieties; even reconstruction and designing of very old and popular varieties.

4.6 Genome Sequence Using Long Read Third Generation Sequencing Technology

Draft genome sequences of six wild and cultivated potatoes were analyzed recently. These species are *S. chaucha* (3*x*), *S. juzepczukii* (3x), *S. tuberosum* Group Andigena (4*x*, 2 acc.), *S. tuberosum* Group Tuberosum (4*x*) and *S. curtilobum* (5*x*) (Kyriakidou et al., 2020a). Genome assembly of tetraploid potato is very cumbersome due to high heterozygosity caused by duplication events and a high level of transposable elements. Genomes of six wild and cultivated potatoes were analysed, based on TGS technology generating long read sequences data which result in high quality and accurate genome assembly. This study suggests that a great diversity exists in these species and identified numerous CNVs, particularly for biotic/abiotic stress responses, like disease resistance. These polyploid genome data serve as a useful resource for genetic enhancement and potato breeding, particularly with reduced genome assembly problems using TGS technology.

The genome sequence of DM1–3 516 R44, a double monoploid clone of *S. tuberosum* Group Phureja, was deciphered by the PGSC in 2011, using a whole-genome shotgun sequencing approach with short-read sequence data (The Potato Genome Sequencing Consortium, 2011). Pham et al. (2020) presented an updated version of the DM1–3 516 R44 genome sequence (v6.1) using Oxford Nanopore Technologies long reads coupled with proximity-by-ligation scaffolding (Hi-C), yielding a chromosome-scale assembly. The new (v6.1) assembly represents 741.6 Mb of sequence (87.8%) of the estimated 844 Mb genome enabled annotation of 32,917 high-confidence protein-coding genes. In another study, Zhou et al. (2020) reported the 1.67-Gb haplotype-resolved assembly of an important *S. tuberosum* heterozygous diploid potato clone RH89-039-16 (RH) using a combination of multiple sequencing strategies, including circular consensus sequencing. This study offers a holistic view of the genome organization of a clonally propagated diploid species and provides insights into technological evolution in resolving complex genomes.

Very recently, Sun et al. (2022) reported the 3.1 Gb haplotype-resolved (at 99.6% precision), chromosome-scale assembly of the potato cultivar 'Otava' based on high-quality long reads, single-cell sequencing of 717 pollen genomes and Hi-C data. This work provides insights on the recent breeding history of potato, the functional organization of its tetraploid genome and has the potential to strengthen the future of genomics-assisted breeding. Hoopes et al. (2022) analyzed chromosome-scale phased genome assemblies and revealed extensive allelic diversity and structural variation distribution across the homologous chromosomes.

To undertake diploid breeding research, van Lieshout et al. (2020) deciphered the *de novo* sequenced genome of Solyntus, a diploid breeding line developed by cross between diploid potato and self-compatible *S. chacoense*, as the next standard reference in potato genome studies. A true *Solanum tuberosum* made up of 116 contigs that is also highly homozygous, diploid, vigorous and self-compatible, Solyntus provides a more direct and contiguous reference genome. It was constructed by sequencing with state-of-the-art long and short read technology. This study will strengthen the diploid hybrid breeding program.

4.7 Genome Sequence of Potato Somatic Hybrid, Parents and Progeny

Draft genome sequences of interspecific potato somatic hybrid, fusion parents and hybrid progeny were deciphered and analyzed particularly for late blight resistance genes (Tiwari et al., 2021). These four potato genotypes are (*i*) interspecific potato somatic hybrid clone P8 ($2n = 4x = 48$) (*S. tuberosum* dihaploid clone C-13 + wild species *S. pinnatisectum*), (*ii*) *S. pinnatisectum* ($2n = 2x = 24$), (*iii*) MSH/14-112 ($2n = 4x = 48$) (P8 × cv. Kufri Jyoti), and (*iv*) C-13 ($2n = 2x = 24$) (Figs. 2 and 3). Earlier, dihaploid C-13 was developed by another culture from potato cv. Kufri Chipsona-2, and then interspecific somatic hybrids (e.g., P8) were also produced successfully by protoplast fusion; and further a hybrid progeny MSH/14-112 was regenerated through traditional breeding (P8 × Kufri Jyoti) to improve agronomic traits. The genotypes P8, *S. pinnatisectum* and MSH/14-112 are highly resistant to late blight, the most devastating disease of potato, whereas C-13 is susceptible.

Draft genome sequences of above four genotypes were generated using the Illumina platform. The reference-based assemblies with the potato genome sequence (DM) reveal genomes assemblies and number of genes, such as P8 (725.01 Mb and 39260 genes), *S. pinnatisectum* (724.95 Mb and 25711 genes), MSH/14-112 (725.01 Mb and 39730 genes), and C-13 (809.59 Mb and 30241 genes). Comparative genomics identifies a total of 17,411 common genes among these genotypes, of which common late blight resistance genes are *R3a*, *RGA2*, *RGA3*, *R1B-16*,

Fig. 2. Phenotype of somatic hybrid (P8), parents (C-13 and *S. pinnatisectum*) and progeny (MSH/14-112) used for genome sequencing.

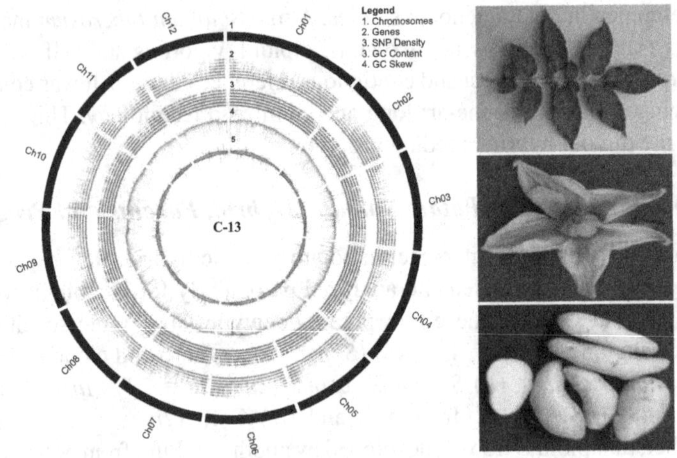

Fig. 3. Circos plot showing whole genome resequencing of potato (*Solanum tuberosum* L.) dihaploid clone 'C-13'.

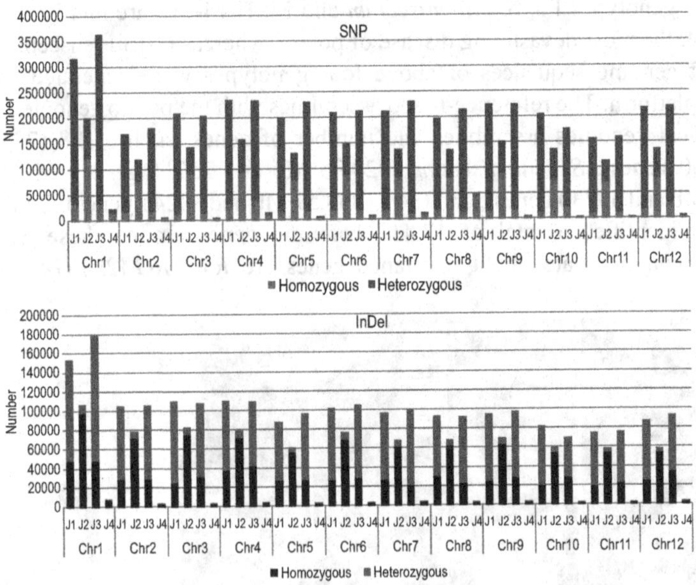

Fig. 4. Single nucleotide polymorphisms (SNP) and insertion deletion (InDels) analysis in somatic hybrid, parents and progeny.

Rpi-blb2, *Rpi* and *Rpi-vnt1*. Further, study identifies more heterozygous SNP than the homozygous one and SNP in genic region are more than inter-genic region (Figs. 4 and 5). Structural variation, particularly CNV analysis, indicates a greater number of deletions than duplications. This study provides insights on genome re-sequences, structural variation and late blight resistance genes in interspecific potato somatic hybrid, parents and progeny for future application.

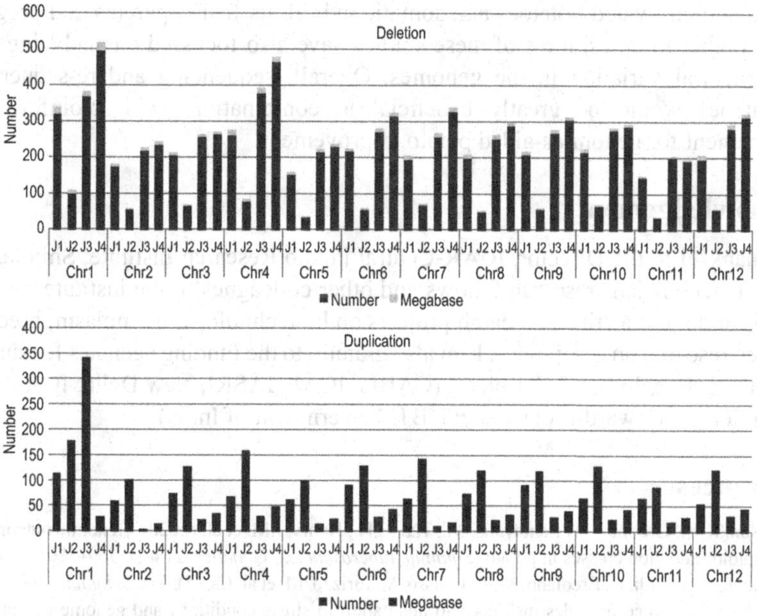

Fig. 5. Copy number variation (CNV) analysis in somatic hybrid, parents and progeny.

5. Conclusion

The rapid progress in sequencing technologies has greatly advanced the genome sequencing of crop species, including potato. With the beginning of the Sanger sequencing (first generation) to the next generation sequencing technologies of short read, like Illumina, has become most popular in genomics research so far. Now with further technological advancements in high-throughput data, accuracy, long read and reducing cost per sample sequencing have led to the discovery of third generation sequencing, like PacBio and Nanopore technologies. These third generation sequencing have great potential to generate more accurate genome assembly with wider application. Further, advancements in bioinformatics tools and simple protocols, particularly with a portable sequencer like Nanopore, have allowed easy access of these technologies for large size genomes.

Potato plays a pivotal role in food production and security and importantly, the potato genome provides very valuable resources for genomics-aided breeding. Genome sequencing of a unique doubled-monoploid potato clone DM overcomes the problems associated with genome assembly due to high heterozygosity. Finally, a high-quality draft potato genome sequence was deciphered. Many agronomic traits of quantitative and qualitative nature would be simplified for tailor-made variety development. Greater research has been conducted at the diploid level in potato and most potato cultivars are tetraploid. Hence, it would be challenging to incorporate desirable alleles in the cultivated background. With the availability of potato genome sequences, in a few recent years several wild and cultivated potato (*Solanum*) species have been sequenced, such as *S. commersonii*, *S. chacoense* (M6), nearly a hundred of wild and cultivated potatoes and somatic hybrid, its fusion parents and progeny. Besides other findings, most of these studies have also focussed on analyzing SNP and structural variation in the genomes. Overall, sequencing and re-sequencing of potatoes would be greatly beneficial in combinations with bioinformatics advancement for genomics-aided potato improvement.

Acknowledgement

I am thankful to the Director, ICAR-Central Potato Research Institute, Shimla, and scientists/technicians/research fellows and other colleagues of the institute for their support under the institute research projects on biotechnology, germplasm, breeding and seed research on aeroponics. I am also grateful to the funding agencies for support under the externally funded projects (CABin, ICAR-IASRI, New Delhi; ICAR-LBS Young Scientist Award Project, and DBT, Government of India).

References

Amundson, K.R., Ordoñez, B., Santayana, M., Tan, E.H., Henry, I.M. et al. (2020). Genomic outcomes of haploid induction crosses in potato (*Solanum tuberosum* L.). *Genetics*, 214(2): 369–380.

Aversano, R., Contaldi, F., Ercolano, M.R., Grosso, V., Iorizzo, M. et al. (2015). The *Solanum commersonii* genome sequence provides insights into adaptation to stress conditions and genome evolution of wild potato relatives. *The Plant Cell*, 27(4): 954–968.

Bennett, M.D. and Leitch, I.J. (2011). Nuclear DNA amounts in angiosperms: Targets, trends and tomorrow. *Ann. Bot.*, 107: 467–590.

FAOSTAT. (2022). https://www.fao.org/faostat/en/#data accessed on 2nd February, 2022.

Hardigan, M.A., Laimbeer, F., Newton, L., Crisovan, E., Hamilton, J.P. et al. (2017). Genome diversity of tuber-bearing *Solanum* uncovers complex evolutionary history and targets of domestication in the cultivated potato. *Proc. Natl. Acad. Sci., U.S.A.*, 114(46): E9999–E10008.

Hoopes, G., Meng, X., Hamilton, J.P., Achakkagari, S.R., de Alves Freitas Guesdes, F. et al. (2022). Phased, chromosome-scale genome assemblies of tetraploid potato reveal a complex genome, transcriptome and predicted proteome landscape underpinning genetic diversity. *Mol. Plant S1674-2052(22)00003-X*, Advance online publication. https://doi.org/10.1016/j.molp.2022.01.003; https://phytozome-next.jgi.doe.gov/accessed on 10th May, 2021.

Huang, B., Ruess, H., Liang, Q., Colleoni, C. and Spooner, D.M. (2019). Analyses of 202 plastid genomes elucidate the phylogeny of Solanum section Petota. *Sci. Rep.*, 9(1): 4454.

Ip, C., Loose, M., Tyson, J.R., de Cesare, M., Brown, B.L. et al. (MinION Analysis and Reference Consortium). (2015). MinION analysis and reference consortium: Phase 1 data release and analysis. *F1000Res.*, 4: 1075.

Kanzi, A.M., San, J.E., Chimukangara, B., Wilkinson, E., Fish, M. et al. (2020). Next generation sequencing and bioinformatics analysis of family genetic inheritance. *Front. Genet.*, 11: 544162.

Kumar, K.R., Cowley, M.J. and Davis, R.L. (2019). Next-generation sequencing and emerging technologies. *Semin. Thromb. Hemost.*, 45(7): 661–673.

Kyriakidou, M., Anglin, N.L., Ellis, D., Tai, H.H. and Strömvik, M.V. (2020a). Genome assembly of six polyploid potato genomes. *Sci. Data*, 7(1): 88.

Kyriakidou, M., Achakkagari, S.R., Gálvez López, J.H., Zhu, X., Tang, C.Y. et al. (2020b). Structural genome analysis in cultivated potato taxa. *Theor. Appl. Genet.*, 133(3): 951–966.

Larsen, P.A., Heilman, A.M. and Yoder, A.D. (2014). The utility of PacBio circular consensus sequencing for characterizing complex gene families in non-model organisms. *BMC Genomics*, 15(1): 720.

Laver, T., Harrison, J., O'Neill, P.A., Moore, K., Farbos, A. et al. (2015). Assessing the performance of the Oxford nanopore technologies MinION. *Biomol. Detect. Quantif.*, 3: 1–8.

Leisner, C.P., Hamilton, J.P., Crisovan, E., Manrique-Carpintero, N.C., Marand, A.P. et al. (2018). Genome sequence of M6, a diploid inbred clone of the high-glycoalkaloid-producing tuber-bearing potato species *Solanum chacoense*, reveals residual heterozygosity. *The Plant J.*, 94(3): 562–570.

Levy, S.E. and Myers, R.M. (2016). Advancements in next-generation sequencing. *Annu. Rev. Genomics Hum. Genet.*, 17: 95–115.

Levy, S.E. and Boone, B.E. (2019). Next-generation sequencing strategies. *Cold Spring Harb. Perspect. Med.*, 9(7): a025791.

Li, C., Lin, F., An, D., Wang, W. and Huang, R. (2017). Genome sequencing and assembly by long reads in plants. *Genes*, 9(1): 6.

Maxam, A.M. and Gilbert, W. (1977). A new method for sequencing DNA. *Proc. Natl. Acad. Sci., U.S.A.*, 74: 560–564.

Mignardi, M. and Nilsson, M. (2014). Fourth-generation sequencing in the cell and the clinic. *Genome Med.*, 6(4): 31.

Pham, G.M., Braz, G.T., Conway, M., Crisovan, E., Hamilton, J.P. et al. (2019). Genome-wide inference of somatic translocation events during potato dihaploid production. *The Plant Genome*, 12(2): 10.3835/plantgenome2018.10.0079.

Pham, G.M., Hamilton, J.P., Wood, J.C., Burke, J.T., Zhao, H. et al. (2020). Construction of a chromosome-scale long-read reference genome assembly for potato. *GigaScience*, 9(9): giaa100.

Sanger, F., Nicklen, S. and Coulson, A.R. (1977). DNA sequencing with chain-terminating inhibitors. *Proc. Natl. Acad. Sci., U.S.A.*, 74: 5463–5467.

Smith, L.M., Sanders, J.Z., Kaiser, R.J., Hughes, P., Dodd, C. et al. (1986). Fluorescence detection in automated DNA sequence analysis. *Nature*, 321(6071): 674–679.

Spooner, D.M. and Hijmans, R.J. (2001). Potato systematics and germplasm collecting, 1989–2000. *Am. J. Potato Res.*, 78: 237–268.

Sun, H., Jiao, W.B., Krause, K., Campoy, J.A., Goel, M. et al. (2022). Chromosome-scale and haplotype-resolved genome assembly of a tetraploid potato cultivar. *Nature Genetics*, 10.1038/s41588-022-01015-0.

Suzuki, Y. (2020). Advent of a new sequencing era: Long-read and on-site sequencing. *J. Hum. Genet.*, 65(1): 1.

The Potato Genome Sequencing Consortium. (2011). Genome sequence and analysis of the tuber crop potato. *Nature*, 475: 189–195.

Tiwari, J.K., Rawat, S., Luthra, S.K., Zinta, R., Sahu, S. et al. (2021). Genome sequence analysis provides insights on genomic variation and late blight resistance genes in potato somatic hybrid (parents and progeny). *Mol. Biol. Rep.*, 48(1): 623–635.

van Lieshout, N., van der Burgt, A., de Vries, M.E., Ter Maat, M., Eickholt, D. et al. (2020). Solyntus, the new highly contiguous reference genome for potato (*Solanum tuberosum*). *G3 (Bethesda)*, 10: 3489–3495.

Varshney, R.K., Nayak, S.N., May, G.D. and Jackson, S.A. (2009). Next-generation sequencing technologies and their implications for crop genetics and breeding. *Trends in Biotechnol.*, 27(9): 522–530.

Venter, J.C., Adams, M.D., Myers, E.W., Li, P.W., Mural, R.J. et al. (2001). The sequence of the human genome. *Science*, 291(5507): 1304–1351.

Visser, R.G.F., Bachem, C.W.B., de Boer, J.M., Bryan, G.J., Chakrabati, S.K. et al. (2009). Sequencing the potato genome: outline and first results to come from the elucidation of the sequence of the world's third most important crop. *Am. J. Potato Res.*, 86: 417–429.

Yamey, G. (2000). Scientists unveil first draft of human genome. *BMJ*, 321(7252): 7.

Zhou, Q., Tang, D., Huang, W., Yang, Z., Zhang, Y. et al. (2020). Haplotype-resolved genome analyses of a heterozygous diploid potato. *Nat. Genet.*, 52(10): 1018–1023.

Zhou, S., Kile, A., Bechner, M., Place, M., Kvikstad, E. et al. (2004). Single-molecule approach to bacterial genomic comparisons via optical mapping. *J. Bacteriol.*, 186(22): 7773–7782.

Chapter 3

Genomics in Potato Germplasm Management and Utilization

1. Introduction

The cultivated potato (*Solanum tuberosum* L.) belongs to the genus *Solanum* and family Solanaceae, which it is among the biggest genus in plant species. The *Solanum* genus contains over 200 wild, semi-cultivated and cultivated potato species. By the 16th century, potato was unknown to the people of Europe, Asia, Africa and North America. The cultivated potato (*Solanum tuberosum* Group Tuberosum) originated on the Andes mountain of Peru and Bolivia in South America, and more specifically, in the basin of Lake Titicaca (Hawkes, 1990). Evidences indicate that hybridization of the wild species, *S. stenotomum* with *S. sparsipilium* and subsequent chromosome doubling, produced tetraploid *S. tuberosum* Group Andigena in the central Andes, and from which cultivated potato originated. In South America, potato was the most productive source of main food for centuries in the high Andes and Chile. Potatoes were dried by Andean Indians to make *chuno*, still an important food in highlands of Peru, for use during food shortage between successive crops. Potato grows in its natural habitat in central America and Andean region of north-western Argentina, Peru and southern Bolivia. The distribution of wild tuber-bearing *Solanum* species ranges from the south-western states of the USA to Guatemala, Honduras, Costa Rica and western Panama. Some wild potato species are adapted to grow in the cold and very high Andean regions at 3,000–4,500 m (Hawkes, 1990, 1994). Due to their wide distribution and adaptation to extreme climatic conditions, wild species have become adapted to stress environments and have strong resistances to a wide range of pests and diseases (Hawkes, 1994). Potato, being a clonally propagated crop, makes conservation of potato genetic resources as the most important for true-to-type clonal identity and widening of narrow genetic base of the cultivated potato. With the availability of potato genome, characterization of germplasm and molecular genetic diversity analysis are inevitable for rapid potato improvement.

2. Taxonomy, Origin and Domestication of Potato

2.1 Taxonomic Classification

The potato and various other taxonomically related species are placed in section *Petota* under subgenus *Potatoe*. The section *Petota* contains two subsections, viz. *Estolonifera* and *Potatoe* (Hawkes, 1994). The species of subsection *Estolonifera* do not produce tubers and which include two series, namely *Etuberosa* and *Juglandifolia*. The species of *Etuberosa* series can be hybridized with the tuber-bearing species but the species of *Juglandifolia* do not hybridize and appear to be related to tomato (old genus *Lycopersicon*, now *Solanum lycopersicum*). The subsection *Potatoe* is divided into two superseries, *Stellata* and *Rotata*, characterized by star-shaped and rotate or wheel-shaped corollas, respectively. These superseries include 19 series containing tuber-bearing potato species. This subsection is of potential interest to breeders. The cultivated potato *S. tuberosum* is placed in the series *Tuberosa*. *S. tuberosum* is generally divided into two subspecies, namely subsp. *tuberosum*, the universally cultivated potato, and subsp. *andigena*, a primitive taxon cultivated to limited extent in the Andes region.

Taxonomic classification of potato (wild and cultivated) species seems difficult due to polyploidy, interspecific sexual crossing barriers and method of propagation (true potato seed and tuber). A comprehensive taxonomic classification of potatoes was reported by Hawkes (1990) who stated 235 potato species (228 cultivated and seven wild). Since then, various studies were carried out, applying molecular tools in a wide range of potato accessions and suggested for reconsideration of taxonomic classification of potato (Jacobs et al., 2008, 2011; Spooner, 2009). Later, a combination of morphological and molecular tools was carried out that led to reduced number of 111 potato species (four cultivated and 107 wild). A detailed taxonomic description of cultivated and wild potato species has been illustrated by Spooner et al. (2014).

2.2 Origin and Domestication

The cultivated potato seems to have evolved through geographical and ecological isolation. The archeological data reveal that the cultivated potato is believed to be domesticated nearly 10,000–7,000 years ago and originated in Lake Titicaca of the Andes region of southern Peru and north Bolivia, where still, potato landraces are grown (Hawkes, 1990; Ovchinnikova et al., 2011). The cultivated species were at one time confined to the Andes of South America and the lowlands of southern Chile, as they adapted to cool temperate climates of these regions, whereas, the related wild species were more widespread. Distribution of cultivated potato ranges from the upland Andes of western Venezuela to northern Argentina to the lowlands of southern-central Chile and adapting to medium to high mean sea level (3000–4000 m amsl). Potato landraces are highly diverse in colour, shape and size and about more than 3,000 landraces or native potatoes are cultivated in South America (Spooner et al., 2014; de Haan and Rodriguez, 2016). The important characters during domestication of wild potato species were large tuber size, flesh and skin color and low glycoalkaloids (Gavrilenko et al., 2013). A large number of studies were conducted on taxonomic classification of potato and its wild relatives by applying morphological, physio-biochemical and molecular markers and current study shows 111 (107 wild and four cultivated) potato species. Since, phylogenetics and geographical classifications could not clearly classify the potato species (de Haan and Rodriguez, 2016) and therefore, later orthologous gene sequences data determined the allopolyploidy origin of *S. tuberosum* (Spooner et al., 2008, 2010; Rodrıguez and Spooner, 2009; Rodríguez et al., 2009).

The domestication of cultivated potato *S. tuberosum* Gp Andigenum is believed to have originated from wild progenitor *S. brevicaule* complex (*S. bukasovii*, *S. canasense* and *S. multissectum*) in southern Peru and had ancestral relationship with the most primitive cultivated potato species (*S. stenotomum*) (Spooner et al., 2005). Following several sexual polyploidization and natural hybridization, cultivated potato was developed in different climatic conditions. The diploid cultivated species, *S. stenotomum*, grown from central Peru to central Bolivia, is believed to be the most primitive and probably derived from the diploid wild species, *S. leptophyes*. It is believed that at least four wild potato species (*S. acaule*, *S. sparsipilum*, *S. leptophyes* and *S. megistacrolobum*) were involved in the process of evolution (Hawkes, 1994). Evidence indicates that hybridization of *S. stenotomum* with the weedy species *S. sparsipilum,* either as $2n$ gamete or from normal haploid (n) gamete followed by subsequent chromosome doubling, produced the tetraploid *S. tuberosum* subsp. *andigena* in the central Andes. This tetraploid subspecies was carried into southern Chile, where it became adapted to long photoperiod to evolve into *S. tuberosum* subsp. *tuberosum*. Cultivated diploid species *S. phureja* evolved from *S. stenotomum* through artificial selection by Andean farmers in lower and warmer eastern valleys and acquired shorter dormancy. Natural hybridization of *S. stenotomum* with *S. megistacrolobum* gave rise to the diploid *S. ajanhuiri*. Similarly, a series of hybridizations between *S. stenotomum* and the wild tetraploid *S. acaule* gave rise to a highly sterile triploid, *S. juzepczukii*, and further natural cross between *S. juzepczukii* and *S. tuberosum* subsp. *andigena* produced the pentaploid *S. curtilobum*. The spread of potato was started in the 16th century from Peru, South America to Europe (Spain: 1570, and UK: 1590) and then from Europe to Asian and several other countries in 17th century, like in India from UK/Portugal in 1610 and later domesticated to all over the world (Figs. 1 and 2).

2.3 Cultivated and Wild Potato Species

2.3.1 Cultivated Potato Species

The genus *Solanum* is very large and contains over 2,000 species of which nearly 235 are tuber bearing potatoes (Hawkes, 1990). The tuber-bearing *Solanum* species form a polyploid series of diploid ($2n = 2x = 24$), triploid ($2n = 3x = 36$), tetraploid ($2n = 4x = 48$), pentaploid ($2n = 5x = 60$) and hexaploid ($2n = 6x = 72$) with a basic chromosome number of 12 ($x = 12$) (Howard, 1970). Chromosome counts have been made for only three-fourth of the species. Among the tuber-bearing *Solanum* species, about 73% are diploids, 4% triploids, 15% tetraploids and 6% hexaploids (Hawkes, 1994). The cultivated potato is an autotetraploid with total 48 chromosomes ($2n = 4x = 48$) (Swaminathan and Howard, 1953). The modern cultivated potato is generally viewed as having an autopolyploid or segmental polyploidy origin (Table 1).

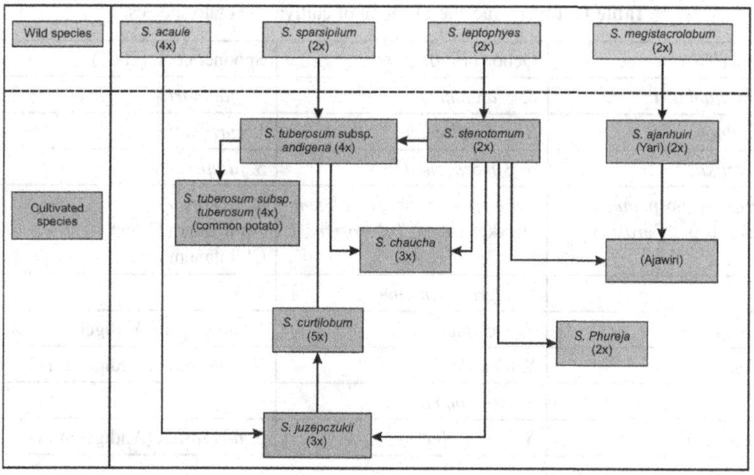

Fig. 1. Evolutionary relationships of cultivated potato and their ploidy levels (Hawkes, 1990).

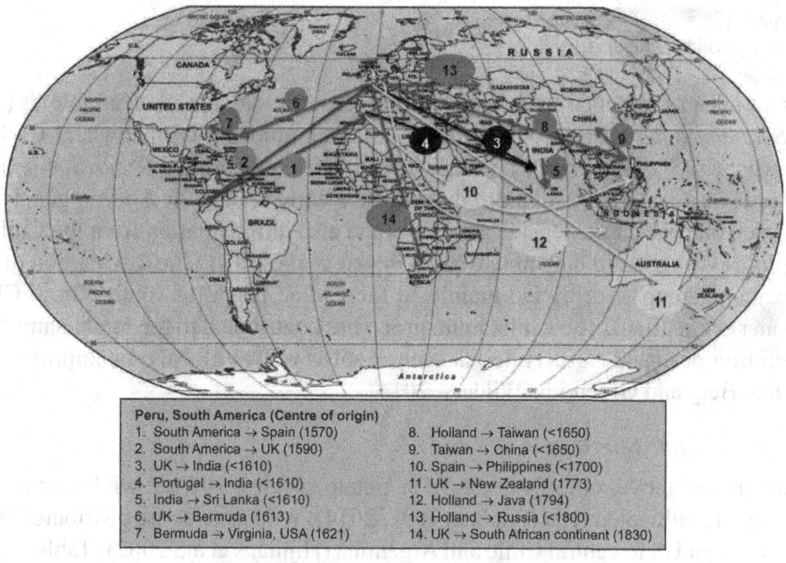

Fig. 2. The journey of potato in the world.

In addition to the cultivated *S. tuberosum*, there are other seven primitive/native potato species, viz., *S. ajanhuiri*, *S. goniocalyx*, *S. phureja* and *S. stenotomum* (all diploids); *S. chaucha* and *S. juzepczukii* (both triploids); and *S. curtilobum* (pentaploid) under cultivation. *S. hygrothermicum* listed below is now almost extinct (Ochoa, 1984). The cultivated species cross either directly or through bridging species with several species belonging to other series. *Solanum tuberosum* is the original species from which modern cultivars have been developed worldwide (Ames and Spooner, 2008). The recent classification by Spooner et al. (2007) classifies the cultivated potato species as following: (i) *Solanum tuberosum* Group Andigenum of

Table 1. Taxonomic classification of cultivated potato species.

Hawkes (1990)	Ochoa (1990, 1999)	Spooner et al. (2007)
Solanum ajanhuiri	*S.* × *ajanhuiri*	*S. ajanhuiri*
S. curtilobum	*S.* × *curtilobum*	*S. curtilobum*
S. juzpeczukii	*S.* × *juzepczukii*	*S. juzepczukii*
S. tuberosum subsp. *andigena* Hawkes subsp. *tuberosum*	*S. tuberosum* subsp. *andigena* Hawkes subsp. *Tuberosum*	*S. tuberosum* Andigenum Group Chilotanum Group
	S. hygrothermicum	
S. chaucha	*S.* × *chaucha*	*S. tuberosum* (Andigenum Group)
S. phureja	*S. phureja*	*S. tuberosum* (Andigenum Group)
	S. stenotomum	
S. stenotomum	*S. goniocalyx*	*S. tuberosum* (Andigenum Group)

Diploid ($2n = 2x = 24$): *S. ajanhuiri, S. phureja, S. stenotomum, S. goniocalyx,*
Triploid ($2n = 3x = 36$): *S.* × *chaucha, S. juzepczukii*
Tetraploid ($2n = 4x = 48$): *S. tuberosum* Group Andigenum and *S. tuberosum* Group Chilotanum
Pentaploid ($2n = 5x = 60$): *S. curtilobum*
Source: (Machida-Hirano, 2015)

upland Andean genotypes containing diploids (2*x*), triploids (3*x*) and tetraploids (4*x*); and *Solanum tuberosum* Group Chilotanum of tetraploid (4*x*) of lowland Chilean landraces; (ii) *S. ajanhuiri* (2*x*); (iii) *S. juzepczukii* (3*x*); and (iv) *S. curtilobum* (5*x*) (Tables 1, 2 and 3). It is difficult to distinguish clearly among the Andean genotypes from each other due to extensive gene flow and also differentiation from the Chilean tetraploids is slight and incomplete (Gavrilenko et al., 2010). The distribution of the Chilotanum group is restricted mainly to the Chiloe Island of south-central Chile (Spooner et al., 2010). The Chilotanum group has contributed to the establishment of current European and North American gene pool as well as global crop improvement (van den Berg and Groendijk-Wilders, 2014).

2.3.2 Wild Potato Species

As mentioned above, out of the total 111 potato species, 107 are wild species and only four are cultivated type (Spooner et al., 2014). Wild species are distributed from south-western US to central Chile and Argentina (Hijmans et al., 2002) (Tables 2 and 3). Mainly two centres of diverse wild species have been identified, such as North and Central America (centre: Mexico), and another one in South America (centre: Andes mountain ranges from Venezuela to Chile). The distribution of wild species ranges from mean sea level to high altitude of 4,500 m above mean sea level (Hijmans and Spooner, 2001). This indicates a wide range of adaptation in wild species to diverse environments of high temperature, desert, cold, forest, cultivated field, islands and many others (Hijmans et al., 2002, 2007). Because of their adaptation to several habitats of agricultural importance, they are tolerant to different biotic (late blight, viruses, potato cyst nematodes, bacterial wild, storage rot and insect-pests) and abiotic (heat, drought, cold and salinity) stresses (Hawkes, 1990, 1994; Spooner and Bamberg, 1994; Barker, 1996; Ochoa, 1999; Hijmans et al., 2003; Bamberg

Table 2. Classification of potato and its wild relatives (Hawkes, 1990).

Genus: *Solanum* L.	*Solanum* species (basic chromosome no. $x = 12$)
Subgenus: *Potatoe* (G. Don) A'Arey	
Section: *Petota* Dumortier	
Subsection: *Estlonifera* Hawkes	
Series	
I: *Etuberosa* Juzepczuk	2x: *S. brevidens, S. etuberosum*
II: *Juglandifolia* (Rydb.) Hawkes	2x: *S. lycopersicoides*
Subsection: *Potatoe* G. Don	
Superseries: *Stellata* Hawkes	
Series	
I: *Morelliformia* Hawkes	2x: *S. morelliforme*
II: *Bulbocastana* (Rydb.) Hawkes	2x: *S. clarum* 2x/3x: *S. bulbocastanum*
III: *Pinnatisecta* (Rydb.) Hawkes	2x: *S. brachistotrichum, S. pinnatisectum, S. trifidum* 2x/3x: *S. cardiophyllum, S. jamesii*
IV: *Polyadenia* Bukasov ex Correll	2x: *S. polyadenium, S. lesteri*
V: *Commersoniana* Bukasov	2x/3x: *S. commersonii*
VI: *Circaeifolia* Hawkes	2x: *S. capsicibaccatum, S. circaeifolium*
VII: *Lignicaulia* Hawkes	2x: *S. lignicaule*
VIII: *Olmosiana* Ochoa	2x: *S. olmosense*
IX: *Yungasensa* Correll	2x: *S. chacoense, S. tarijense, S. yungasense*
Superseries: *Rotata* Hawkes	
Series	
X: *Megistacroloba* Card *et* Hawkes	2x: *S. boliviense, S. megistacrolobum, S. sancta-rosae, S. toralapanum*
XI: *Cuneoalata* Hawkes	2x: *S. infundibuliformae*
XII: *Conicibaccata* Bitter	2x: *S. chomatophilum, S. santolallae, S. violaceimarmoratum* 4x: *S. agrimonifolium, S. colombianum, S. longiconium, S. oxycarpum* 5x: *S. moscopanum*
XIII: *Piurana* Hawkes	2x: *S. piurae* 4x: *S. tuquerrense*
XIV: *Ingifolia* Ochoa	2x: *S. ingifolium*
XV: *Maglia* Bitter	2x/3x: *S. maglia*
XVI: *Tuberosa* (Rudb.) Hawkes • Wild	2x: *S. alandiae, S. berthaultii, S. brevicaule, S. bukasovii, S. canasense, S. gandarillasii, S. hondelmannii, S. kurtzianum, S. leptophyes, S. marinasense, S. multidissectum, S. neocardenasii, S. sparsipilum, S. spagazzini, S. vernei, S. verrucosum* 2x/3x: *S. microdontum* 4x: *S. sucrense* 2x/4x: *S. gourlayi* 2x/4x/6x: *S. oplocense*

Table 2 contd. ...

...Table 2 contd.

Genus: *Solanum* L.	*Solanum* species (basic chromosome no. $x = 12$)
• Cultivated	2*x*: *S. x ajanhuiri, S. phureja, S. stenotomum* 3*x*: *S. x chaucha, S. juzepczuki* 4*x*: *S. tuberosum* subsp. *tuberosum, S. tuberosum* subsp. *andigena* 6*x*: *S. x curtilobum*
XVII: *Acaulia* Juzepczuk	4*x*: *S. acaule* 6*x*: *S. albicans*
XVIII: *Longipedicellata* Bukasov	3*x*: *S. x vallis-mexici* 4*x*: *S. fendleri, S. hjertingii, S. papita, S. polytrichon, S. stoloniferum*
XIX: *Demissa* Bukasov	2*x*: *S. x semidemissum, S. x edinensi* 5*x*: *S. brachycarpum, S. demissum* 6*x*: *S. guerreroense, S. hougasii, S. iopetalum, S. schenckii*

Table 3. Cultivated potato species.

Ploidy	**Species**	**Area of cultivation**	**Trait of interest**
Diploid	*S. stenotomum*	South Peru and Bolivia	Frost resistant, yellow flesh, good flavor
	S. goniocalyx	Mostly central and northern Peru	Yellow flesh, good flavor, now has been placed as a sub species of *S. stenotomum*
	S. ajanhuiri	High elevations in Peru and Bolivia	Frost resistant, bitter
	S. phureja	South America, mostly at moderate elevations	No tuber dormancy, high dry matter; genes for unreduced gametes
Triploid	*S. x chaucha*	Bolivia and Peru	Natural hybrids of *S. tuberosum* subsp. *andigena* and *S. stenotomum*
	S. x juzepczukii	High elevations of Bolivia and Peru	Natural hybrids of *S. stenotomum* and *S. acaule*, frost resistant, bitter
Tetraploid	*S. tuberosum* subsp. *andigena*	Andes of Venezuela, Colombia, Equador, Peru, Bolivia and northwest Argentina	Adaptation to short days conditions, Variability for many traits, especially disease resistance and quality
	S. tuberosum subsp. *tuberosum*	Most widespread and cultivated in Europe, Asia and elsewhere	Adaptation to long days or short days conditions in subtropics, disease resistance, good appearance
	S. hygrothermicum	Amazon Basin of Peru	Adaptation to lowland tropics, said to be almost extinct
Pentaploid	*S. curtilobum*	High elevations of Bolivia and Peru	Natural hybrid *of S. tuberosum* subsp. *andigena* and *S. juzepczukii*, frost resistant, somewhat bitter
Diploid	*S. stenotomum*	South Peru and Bolivia	Frost resistant, yellow flesh, good flavor
	S. goniocalyx	Mostly central and northern Peru	Yellow flesh, good flavor, now has been placed as a sub species of *S. stenotomum*
	S. ajanhuiri	High elevations in Peru and Bolivia	Frost resistant, bitter
	S. phureja	South America, mostly at moderate elevations	No tuber dormancy, high dry matter; genes for unreduced gametes
Triploid	*S. x chaucha*	Bolivia and Peru	Natural hybrids of *S. tuberosum* subsp. *andigena* and *S. stenotomum*
	S. x juzepczukii	High elevations of Bolivia and Peru	Natural hybrids of *S. stenotomum* and *S. acaule*, frost resistant, bitter

Table 3 contd. ...

...Table 3 contd.

Ploidy	Species	Area of cultivation	Trait of interest
Tetraploid	*S. tuberosum* subsp. *andigena*	Andes of Venezuela, Colombia, Equador, Peru, Bolivia and northwest Argentina	Adaptation to short days conditions, Variability for many traits, especially disease resistance and quality
	S. tuberosum subsp. *tuberosum*	Most widespread and cultivated in Europe, Asia and elsewhere	Adaptation to long days or short days conditions in subtropics, disease resistance, good appearance
	S. hygrothermicum	Amazon Basin of Peru	Adaptation to lowland tropics, said to be almost extinct
Pentaploid	*S. curtilobum*	High elevations of Bolivia and Peru	Natural hybrid *of S. tuberosum* subsp. *andigena* and *S. juzepczukii*, frost resistant, somewhat bitter

Source: Adapted and modified from Hoops and Plaisted (1987)

and del Rio, 2005). The wild potato species have been characterized world over by morphological and molecular tools and deployed for introgression of desirable traits in cultivated potato through conventional and molecular breeding. Genes from wild sources have been isolated and cis/transgenics have been developed for desirable traits. Likewise, a large number of examples are available in potato for utilization of wild species in cultivated background (Bamberg and del Rio, 2005).

2.4 Potato Genepool and Crossability

Potato improvement mainly relies on availability of genetic resources and their crossability to develop new genotypes. The wild species are not crossable with the cultivated potato species due to differences in ploidy level and endosperm balance number (EBN). These are the major problems in utilization of wild species for genetic enhancement and potato improvement. A number of techniques have been deployed to overcome the sexual crossing barriers and harness the potential of wild species in potato breeding and biotechnology. Harlan and de Wet's (1971) used the concept of genepool, stating degree of relatedness between species. Genepool is classified on the basis of degree of crossability among species and classified in three groups: (1) primary genepool: cultivated species that are easily crossable, (2) secondary genepool: less closely related species from which gene transfer is possible through conventional breeding, and (3) tertiary genepool: wild species from which trait introgression is impossible through conventional breeding, or requires modern tools for gene transfer from wild to cultivated species. Several attempts have been made to apply the genepool concept in potato (Veilleux and De Jong, 2007; Bradeen and Haynes, 2011). Recently, five crossability groups have been defined in potato, based on EBN and self-compatibility/self-incompatibility (Spooner et al., 2014). The primary genepool of potato includes all landraces and cultivars of the cultivated *S. tuberosum* Gp. Tuberosum, which are tetraploid ($2n = 4x = 48$) and 4 EBN (Johnston et al., 1980). The secondary genepool of potato includes related wild species. These species are very large in number with diverse genetic variations and also a rich source of various biotic and abiotic stress resistance/tolerance and

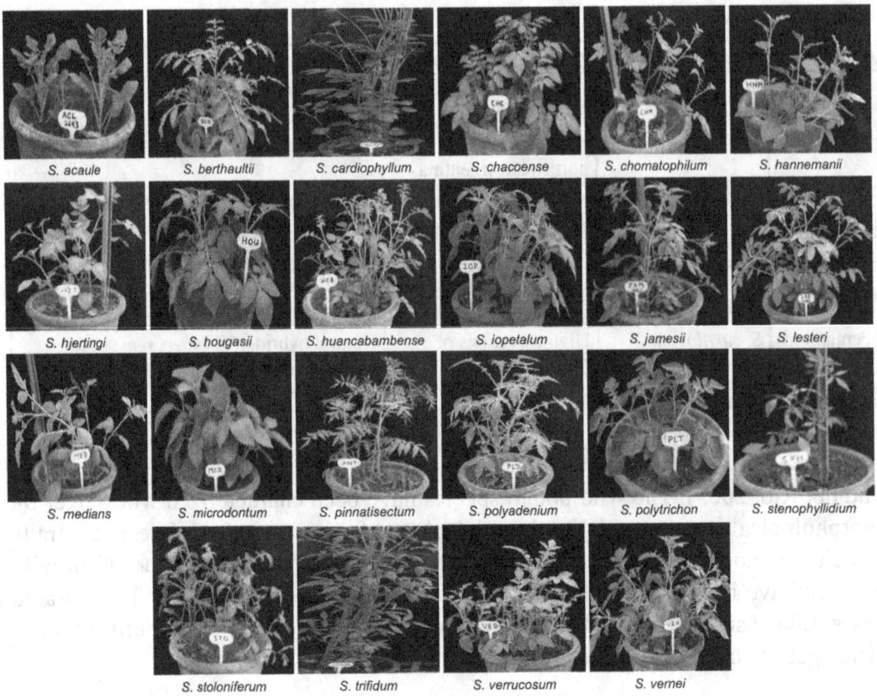

Fig. 3. Potato wild species (1EBN) with late blight resistance and non-crossable with cultivated potato (4EBN).

quality traits. Johnston et al. (1980) proposed the concept of the EBN determining the success of interspecific crosses. Potatoes are classified as 1EBN (2*x*), 2EBN (2*x*/4*x*) and 4EBN (4*x*/6*x*) (Spooner and Hijmans, 2001). Generally, hybridization within the same EBN species is successful, whereas it is unsuccessful across the species with different EBN (Fig. 3). The genepool concept and the EBN provide guidelines for successful hybridization among the potato species (Bradeen and Haynes, 2011).

Many research methodologies have been deployed to overcome sexual barriers in potato for transfer of genes of target traits from secondary or even tertiary genepool into cultivated potato. These methods are ploidy manipulation (McHale and Lauer, 1981; Hermundstad and Peloquin, 1985; Iwanaga et al., 1989; Camadro and Espinillo, 1990), bridge crosses for tertiary genepool transfer (Hermsen, 1994; Jansky and Hamernik, 2009), embryo rescue and mentor pollination (Iwanaga et al., 1991; Watanabe et al., 1995) and somatic hybridization (Fock et al., 2000; Chen et al., 2013). Transgenics have been developed for several traits, like disease and insect-pest resistance and tolerance to abiotic stresses, such as drought, heat stress and many others (Wu et al., 1995; van der Vossen et al., 2003; Missiou et al., 2004).

3. Germplasm Conservation

3.1 Potato Genebanks

The rich genetic resources of potato (*Solanum* species) offer tremendous opportunities for sustainable crop production. Hence, conservation of potato germplasm is most essential to maintain genetic integrity of genotypes for potato improvement. The potato genebanks play very important roles in conservation, characterization and distribution of germplasm to various countries for potato improvement. The centres of origin of crop species are the richest source of genetic diversity. Prof. S.M. Bukasov and his co-workers were the first to explore the centres of diversity of tuber-bearing *Solanum* species during 1925–26 and in subsequent years to collect from remote and non-accessible areas the old land races and primitive and wild species. In early 1920's, Russian scientists made intensive germplasm collection in central and South America and established the foundation stone of N.I. Vavilov Institute of Plant Industry in St. Petersburg, Russia (Ovchinnikova et al., 2011). In later years, explorations were done by scientists from other countries as well and valuable genetic material was collected. Presently, the International Potato Centre (CIP), Lima, Peru, a CGIAR institute, holds the largest potato germplasm collection. Other important collections are available with N.I. Vavilov Institute of Plant Industry, St. Petersburg, Russia; United States Potato Introduction Project (NRSP-6), Sturgeon-Bay, Wisconsin, USA; Dutch German Potato Collection (CGN), Wageningen, The Netherlands; Institute of Plant Genetics and Crop Plant Research (GLKS), Gross-Lusewitz, Germany; Commonwealth Potato Collection (CPC), Scottish Crop Research Institute (SCRI), Dundee, Scotland and Instituto Nacional de Tecnologia Agropecuaria (INTA), Balcarce, Argentina. Likewise several efforts were made in different parts of the countries and national/international genebanks were established the world over. Now, potato germplasm are being conserved in several countries in different forms like *in vitro*, tuber and true potato seed, and made available in public for potato researchers.

The Food and Agriculture Organization of the United Nations reported that globally more than 98,000 accessions are conserved *ex situ (in vitro)* and of which 80% are maintained in 30 key collections. Accessions are conserved as botanical seeds or vegetatively as tubers and *in vitro* plantlets. Latin American collections contain many native cultivars and wild relatives and the collections in Europe and North America contain modern cultivars and breeding materials, as well as wild relatives. The cultivated potatoes are conserved as tubers (live plants in field), *in vitro* plantlets and cryopreservation of plantlets, whereas wild species are conserved in the form of true potato seed (TPS) (Salas et al., 2008). Conservation through TPS has several advantages, like reduces cost, increases shelf life, saves storage space, allows easy transportation, and free from Potato Spindle Tuber Viroid (PSTVd). An integrated genebank database, named Genesys, has been launched in 2008 and is managed by the Global Crop Diversity Trust and supported by the CGIAR institutions. This online platform allows exploration of information about plant genetic resources for food and agriculture conserved in genebanks worldwide (https://www.genesys-pgr.org). In total, this database contains 4 million genebank accessions, which is about a half of

Table 4. Summary of important genomics resources and genebanks of potato (*Solanum* spp.).

Name of resource	Description	URL
Genomics resources		
Spud DB: Potato Genomics Resource	The web browser of the potato genome sequence deciphered originally by the Potato Genome Sequencing Consortium (PGSC), and the latest version of the potato genomes sequenced recently.	http://solanaceae.plantbiology.msu.edu/index.shtml
NCBI Genome *Solanum tuberosum* (potato)	The representative potato genome database of *Solanum tuberosum* (assembly SolTub_3.0).	https://www.ncbi.nlm.nih.gov/genome/400
Solanum tuberosum Genome	A resources for complete plant genomics including potato genome.	http://www.plantgdb.org/StGDB/
NCBI GEO Datasets (*Solanum tuberosum*)	A dataset of gene expression profiling for potato.	https://www.ncbi.nlm.nih.gov/gds/?term=Solanum+tuberosum
ArrayExpress (*Solanum tuberosum*)	A gene expression dataset based on arrays for *S. tuberosum*.	https://www.ebi.ac.uk/arrayexpress/search.html?query=Solanum+tuberosum
PoMaMo Database (Potato Maps and More)	A GABI primary database for potato maps, sequences, SNPs, tools, etc.	http://www.gabipd.org/projects/Pomamo/#Tools
NSF Potato Genome Project	SSR and microarray database of potato genomics resources.	http://potatogenome.berkeley.edu/nsf5/
Sol Genomics Network	A genomics resource database for Solanaceous crops (tomato, pepper, eggplant, potato and tobacco).	https://solgenomics.net/
Solanaceae Coordinated Agricultural Project (SolCAP)	Genotype and phenotype database of Solanaceous crops germplasm.	http://solcap.msu.edu/index.shtml
GoMapMan	A tool for integration, consolidation and visualization of plant gene annotations within the MapMan ontology.	http://www.gomapman.org/
Genebanks		
GENESYS database	An online platform on information about Plant Genetic Resources for Food and Agriculture (PGRFA) conserved in genebanks worldwide.	https://www.genesys-pgr.org/
Leibniz Institute for Plant Genetics and Crop Plant Research (IPK), Germany	The IPK Genebank is one of the largest collections of crop plants and their wild relatives including potatoes.	https://www.ipk-gatersleben.de/en/genebank/
International Potato Center (CIP), Peru	World collection of potato (and sweet potato) cultivated and wild species.	http://cipotato.org/genebank/
NRSP-6-United States Potato Genebank	Germplasm collection of cultivated and wild potato species.	http://www.ars-grin.gov/ars/MidWest/NR6/
U.S. National Plant Germplasm System	A collection of germplasm of plants species in the USA.	https://npgsweb.ars-grin.gov/gringlobal/search.aspx

Table 4 contd. ...

...Table 4 contd.

Name of resource	Description	URL
Centre for Genetic Resources, The Netherlands (CGN)	The potato collection at CGN since 1995 consisting of wild and Andean cultivated potato species.	http://www.wur.nl/en/Expertise-Services/Statutory-research-tasks/Centrefor-Genetic-Resources-the-Netherlands-1/Centre-for-Genetic-Resourcesthe-Netherlands-1/Expertise-areas/Plant-Genetic-Resources/CGN-cropcollections/Potato.htm http://cgngenis.wur.nl/
N.I. Vavilov Institute of Plant Genetic Resources (VIR), Russia	Wild and cultivated potato (*Solanum*) species.	http://vir.nw.ru http://vir.nw.ru/test/vir.nw/index.php?lang=en
The Association of Potato Intergenebank Collaborators (APIC)	An online intergenebank collaboration of NRSP-6 and its sister genebanks of global inventory of wild potato genetic resources.	http://germplasmdb.cip.cgiar.org/index.jsp
Canadian Potato Genetic Resources	Plant gene resources of Canada maintains potato germplasm collection in the country.	http://pgrc3.agr.gc.ca/index_e.html
Commonwealth Potato Collection	The Commonwealth Potato Collection is the UK's genebank of landrace and wild potatoes at the James Hutton Institute.	https://ics.hutton.ac.uk/germinate-cpc/#/data/germplasm https://www.hutton.ac.uk/about/facilities/commonwealth-potato-collection
Genetic Resources Center, NARO (National Agriculture and Food Research Organization), Japan	NARO Genebank is the central coordinating institute in Japan for conservation of plants, microorganisms, and animals related to agriculture in Japan.	https://www.gene.affrc.go.jp/about_en.php
Crop Trust (Svalbard Global Seed Vault)	The Seed Vault represents the world's largest collection of crop diversity.	https://www.croptrust.org/our-work/svalbard-global-seed-vault/
American Potato Varieties Database	A catalogue of US potato varieties by the Potato Association of America.	http://potatoassociation.org/industry/varieties#Breeding
Canadian Potato Varieties Database	A catalogue of Canadian potato varieties.	http://www.inspection.gc.ca/plants/potatoes/potatovarieties/eng/1299172436155/1299172577580
The European Cultivated Potato Database	A catalogue of European potato varieties.	https://www.europotato.org/menu.php
AHDB Potato Variety Database, UK	A catalogue of British potato varieties.	http://varieties.ahdb.org.uk/
European Cooperative Programme for Plant Genetic Resources	Resources of crop germplasm database of various crops including potatoes.	https://www.ecpgr.cgiar.org/resources/germplasm-databases/list-of-germplasm-databases/

the estimated total number in the world (https://www.genesys-pgr.org). For potato, 11 key international genebank collections together conserve 86% of the total number of accessions, as reported in the global strategy for the *ex situ* conservation of potato. In Genesys database, 23,047 potato accessions are listed (https://www.genesys-pgr.org/ accessed on 5th August 2021), of which 19% are landraces and 24% are wild relatives. In this Genesys database, the institutes located in Peru hold maximum germplasm followed by USA. The largest genebank collections of potato are held by the IPK Gatersleben genebank in Germany, the USDA Potato Germplasm Introduction Station in MI, USA and by the International Potato Center in Peru. Important potato genomics resources and potato genebanks are given in Table 4.

Before modernization of agriculture, peasant farmers on the Island of Chiloé cultivated 800 to 1,000 varieties of potato and now one finds only about 270 varieties. The cultivated Andean diploid species *Solanum phureja* is also reported to be vulnerable. A recent study on the effect of climate change predicts that 7–13 out of 108 wild potato species studied may become extinct. The potato collections can be divided into four types of germplasm (modern cultivars, naïve cultivars, wild species and other germplasm (FAO, 2010)). Among them, wild species are the largest group followed by native cultivars at the centre of origin in the Latin American countries. The key collections are conserved in Latin and North America (e.g., Peru and USA), Europe and a few in Asian countries.

1) Modern cultivars (and old varieties) of the common potato (*Solanum tuberosum* subsp. *tuberosum*), the most cultivated potato subspecies in the world;
2) Native cultivars, including local potato cultivars occurring in the center of diversity (seven to 12 species depending on taxonomic treatment);
3) Wild relatives, consisting of wild tuber-bearing species and a few nontuber-producing species, occurring in the center of diversity (180–200 species, depending on taxonomic treatment);
4) Other germplasm or research material; all types of genetic stocks, e.g., interspecific hybrids, breeding clones, genetically enhanced stocks, etc.

3.2 *In Vitro* Conservation

In vitro conservation is the most common method worldwide to maintain the clonal identity of potato genotypes. *In vitro* cultures can be regenerated any time, irrespective of season/environment, to be used for development of new cultivars. This method has several advantages, such as (a) maintenance of cultures round the year in diseases-free condition, (b) safety from natural calamities, (c) possibility of conserving a large number of genotypes in limited space, and (d) easy exchange in *in vitro* form. Hence, establishment of tissue culture protocols is the most crucial step in this system. A number of protocols have been standardized in the literature and practiced world over, varying from short (three months) to middle (three years) term storage (Oka and Niino, 1997). However, there could be chances of somaclonal variations in tissue culture-derived materials after repeated subculturing. Therefore, minimal growth medium is desirable for reducing the number of subculturing. Gopal and Chauhan (2010) suggested that the frequency of subculturing can be reduced by incubation under low temperature, low light intensity and varying photoperiod MS medium supplemented with growth retardants or osmotic stress (Murashige and Skoog, 1962). For example, at CIP, Lima, Peru, *in vitro* cultures are conserved under MS medium containing 40 g/l sorbitol, 20 g/l sucrose at 6–8°C temperature and 16 h light photoperiod (22 μmol/m²/s). This allows *in vitro* plantlets to be stored for nearly two years without sub-culturing (Niino and Valle Arizaga, 2015). At the ICAR-CPRI, Shimla, India, more than 3,000 accessions of *in vitro* cultures are maintained on MS medium supplemented with 40 g/l sucrose and 20 g/l mannitol at 6–8°C and 16 h light photoperiod up to 2–2.5 years without subculturing (Gopal and Ghauhan, 2010). Genetic fidelity of *in vitro* propagated potato microtubers was assessed using molecular markers (RAPD, ISSR, SSR and AFLP). Our study suggests that genetic stability of *in vitro* conserved microplants for three years on MS medium supplemented with 40 g/L sorbitol and 7 g/L agar at low temperature (7 ± 1°C) under controlled conditions is a safe method for conservation of true-to-type potato genotypes. Further, genetic and epigenetic changes (DNA methylation) were examined by molecular markers (AFLP-Amplified fragment length polymorphism and MSAP-methylation-sensitive amplified polymorphism) in the tissue-culture propagated somatic hybrids. Study showed that *in vitro* propagated somatic hybrids could be multiplied 'true-to-type' upto 30th cycles of sub-culturing from micropropagated plants for their future use in potato breeding.

3.3 *Cryo-conservation*

Cryo-conservation is the cutting-edge technology for the long-term storage of potato. This method includes conservation of plant samples at –196°C in liquid nitrogen and restricts plant growth (cell division, metabolic and biochemical processes, etc.) while maintaining viability of the tissues. Thus tissues can be stored for long term without deterioration or modification for a long period of time and avoid contamination and genetic changes in the samples. Cryo-conservation of *in vitro* shoot tips is considered to be a suitable method for long-term storage of potato (Niino and Valle Arizaga, 2015) and other clonally propagated crops as well.

In potato, cryo-conservation method is used to maintain accessions using shoot tips in liquid nitrogen. The shoot-tip method has been applied to maintain potato accessions at the international potato genebanks at CIP, Peru and IPK, Germany using DMSO droplet and droplet vitrification methods (Niino and Valle Arizaga, 2015). Moreover, two additional methods have been developed for simple, reproducible and reliable cryo-conservation, namely V cryo-plate and D cryo-plate methods (Yamamoto et al., 2011). As of now, several agronomic, physiological, biochemical and molecular studies have been executed to investigate genetic stability in cryo-conserved materials and no significant differences were observed in regenerants and mother plants (Maki et al., 2015). There is a need to establish cryo-banks in each country for conservation of potato genetic resources, which would serve as a backup plan of field and *in vitro* conservation methods (Niino et al., 2007). Nevertheless, it necessitates continuous monitoring and examination of tissue culture-derived materials to see genetic and epigenetic changes using modern genomics tools, like SNP array and sequencing technologies.

3.4 In Situ Conservation

In situ or on-farm conservation (field genebank) method is still practiced in the farmers of the crop's centre of origin and diversity in Peru and Bolivia and reports are available for 110 species with five or fewer observation records. Little is known about the *in situ* conservation status of wild potato species. A few researchers have studied the *in situ* conservation in Peru, using different methodologies (Pradel, 2013; Asociación ANDES, 2016). Now with the emerging modern genomics tools, an integration of farmers' knowledge along with *in situ* conservation with rich genetic diversity is essential for development of new varieties. World potato germplasm collections are summarized in Tables 5, 6, 7 and 8.

Table 5. Summary of potato collections in the different countries in the Genesys database.

Country of holding institute	Accessions
Peru	7,467
United States of America	4,051
Ukraine	2,795
Czech Republic	2,192
Poland	1,765
Netherlands	1,471
Romania	829
Estonia	803
Ecuador	450
Republic of Ireland	386
Other	838

Source: https://www.genesys-pgr.org

Table 6. Brief summary of species wise potato collection in the Genesys database.

Solanum spp.	Accessions
S. tuberosum	14,599
S. acaule	957
Solanum sp.	771
S. stoloniferum	602
S. brevicaule	576
S. stenotomum	423
S. bukasovii	375
S. chacoense	326
S. demissum	321
S. boliviense	251
Other	3,846

Source: https://www.genesys-pgr.org

Table 7. Type of potato germplasm storage in the Genesys database.

Type of storage	Accessions
In vitro collection	9,874
Long-term seed collection	8,459
Field collection	5,735
Seed collection	3,690
Medium-term seed collection	1,544
Others	302
Cryopreserved collection	167
Short-term seed collection	13
Not specified	598

Source: https://www.genesys-pgr.org

4. Germplasm Characterization

4.1 Genetic Diversity by Molecular Markers

Various types of markers have been employed to study genetic diversity in potato species. Classical tools for such investigations are morphological markers. They are easily identifiable simply on the phenotype of an organism. Later, biochemical markers (or isozymes) were discovered; they are based on the relative mobility of enzyme isoforms. However, both markers are influenced by the environment and plant-growth stage. A quantum jump towards genome mapping was made possible after introducing DNA markers, which are based on the DNA sequence variation and are least affected by the environment and growth stages. A wide range of molecular markers have been developed over the past four decades. Different types of molecular markers have been developed. The first molecular-marker technique was RFLP, followed by PCR-based marker systems, such as RAPD, SSR, AFLP, SSCP and CAPS. Sequencing technologies allowed the detection of Single Nucleotide Polymorphisms (SNPs) markers (Machida-Hirano, 2015). The progress of Next Generation Sequencing (NGS) technologies and the decreasing prices for sequence runs have led to a number of novel techniques for the detection of polymorphic markers. Some recent examples are genotyping by sequencing (GBS) and SNP array chip.

Genetic diversity analysis, germplasm characterization and DNA fingerprinting (genotyping) are important applications of molecular markers. Unlike morphological descriptors, profiles created by using molecular data are independent of environmental effect. Therefore, the International Union for the Protection of (New) Plant Varieties (UPOV) has constituted a working group to critically examine the feasibility of using Biochemical and Molecular Techniques (BMT) for variety identification. DNA fingerprints can be used to establish distinctness and check the uniformity and stability of a particular variety. A wide range of genetic diversity studies have been carried out in potato the world over. Consequently, genetic relationships have been established in cultivated and wild potato species. Various molecular markers, such as RAPD, ISSR, SSR, AFLP, DArT and SNP have been extensively used for diversity analysis

Table 8. Potato germplasm collections (*ex situ* conservation) in the world's major genebanks.

Genebank	Country	Accessions		Type of accession (%)				
		No.	%	Wild species	Landraces/ old cultivars	Breeding lines/ research materials	Advanced cultivars	Others*
INRA-RENNES	France	10461	11	6	2	84	8	-
VIR	Russian Federation	8889	9	-	46	3	26	25
CIP	Peru	7450	8	2	69	2	< 1	27
IPK	Germany	5392	5	18	37	7	32	6
NR6	United States	5277	5	65	21	9	5	< 1
NIAS	Japan	3408	3	3	1	31	-	65
CORPOICA	Colombia	3043	3	-	-	-	-	100
CPRI	India	2710	3	-	-	-	-	100
BNGTRA-PROINPA	Bolivia	2393	2	26	74	-	-	-
HBROD	Czech Republic	2207	2	5	1	29	52	13
BAL	Argentina	1739	2	85	15	-	-	-
CNPH	Brazil	1735	2	-	-	-	-	100
SASA	United Kingdom	1671	2	-	-	-	-	100
ROPTA	Netherlands	1610	2	3	1	-	1	95
PNP-INIFAP	Mexico	1500	2	-	-	-	-	100
TARI	Taiwan	1282	1	-	-	-	-	100
SamAI	Uzbekistan	1223	1	-	-	-	-	100
IPRBON	Poland	1182	1	-	-	8	92	-
RIPV	Kazakhstan	1117	1	26	2	15	57	-
SVKLOMNICA	Slovakia	1080	1	1	2	47	41	9
Others (154)		32916	33	19	15	3	16	46
Total		**98285**	**100**	**15**	**20**	**16**	**14**	**35**

* Others: the type is unknown or a mixture of two or more types

INRA-RENNES: Institut national de la recherche agronomique/Station d'Amélioration des Plantes (France); **VIR** N.I. Vavilov All-Russian Scientific Research Institute of Plant Industry (Russian Federation); **CIP** Centro Internacional de la Papa, Peru; **CNPH** Embrapa Hortaliças (Brazil); **CORPOICA** Centro de Investigación La Selva, Corporación Colombiana de Investigación Agropecuaria (Colombia); CPRI: Central Potato Research Institute (India); **BAL** Banco Activo de Germoplasma de Papa, Forrajeras y Girasol Silvestre (Argentina); **BNGTRA-PROINPA** Banco Nacional de Germoplasma de Tubérculos y Raíces Andinas, Fundación para la Promoción e Investigación de Productos Andinos (Bolivia, Plurinational State of); **HBROD** Potato Research Institute Havlickuv Brod Ltd. (Czech Republic); **IPK (DEU271)** External Branch North of the Department Genebank, Leibniz Institute of Plant Genetics and Crop Plant Research, Oil Plants and Fodder Crops in Malchow (Germany); **IPRBON** Institute for Potato Research, Bonin (Poland); **NIAS** National Institute of Agrobiological Sciences (Japan); **NR6** Potato Germplasm Introduction Station, United States Department of Agriculture, Agricultural Research Services (USA); **PNP-INIFAP** Programa Nacional de la Papa, Instituto Nacional de Investigaciones Forestales, Agrícolas y Pecuarias (Mexico); **RIPV** Research Institute of Potato and Vegetables (Kazakhstan); **ROPTA** Plant Breeding Station Ropta (Netherlands); **SamAI** Samarkand Agricultural Institute named F. Khodjaev (Uzbekistan); **SASA** Science and Advice for Scottish Agriculture, Scottish Government (United Kingdom); **SVKLOMNICA** Potato Research and Breeding Institute (Slovakia); **TARI** Taiwan Agricultural Research Institute.

in potato germplasm and varieties. Isozyme, RAPD and AFLP markers were used for diversity analysis and to test the genetic integrity of potato after micropropagation and long-term conservation. SSR has been used to analyze genetic diversity within wild species showing noteworthy resistances to *R. solanacearum* and aphids. Scientists at the International Potato Centre (CIP), Peru developed and characterized a set of 24 highly informative microsatellite markers with size standards for inter-laboratory standardization of fragment lengths, called Potato Genome Identification (PGI) Kit (Ghislain et al., 2009) which distinguished over 93% of 742 potato accessions maintained by CIP. This kit is useful for both estimating population structure of clones and for verifying accession identity. This new set of 24 highly informative SSR markers (two from each linkage group) has been applied worldwide, including in India for characterization of potato varieties and wild species (Tiwari et al., 2018). This set has been used by various researchers across countries for structure analysis in different potato populations. Recently, a set of two SSR markers (STIIKA and STU6SNRN) has been developed for varietal identification, genetic fidelity testing, DUS testing and molecular characterizations in Indian potato varieties (Tiwari et al., 2018). Further, SSR-based molecular characterization of wild potato species is accession specific and development of an allelic dataset for all the accessions would strengthen their utilization in potato. Molecular diversity was validated in the core collection of 77 Andigena accession using 24 SSR markers. Molecular diversity of the Andigena core collection, based on the microsatellite data, appears to have quite distinct genotypes. The core collection of 77 accessions was prepared, based on 21 morpho-agronomic traits of total 740 accessions. Further, PGI kit has also been used to identify the structure among the tetraploid cultivated Indian potato and indigenous cultivars. The role of cytoplasmic markers (T/β, W/α, W/γ and A/ε) has also been studied in potato, using plastome- and chondriome-specific markers (Tiwari et al., 2014). Likewise numerous works are available in literature on characterization of potato germplasm the world over. In recent years, a wide range of SNP arrays have recently been developed and applied in potato to characterize germplasm and gene discovery.

4.2 Evaluation for Agronomic Traits and Biotic and Abiotic Stresses

Efforts are continuing for evaluation of germplasm for various traits, such as agronomic and tuber traits, resistance to biotic stresses like diseases and insect-pests, and tolerance to abiotic stresses, such as heat, drought, salinity, cold temperature and climate-change resilience. To make the best use of germplasm in breeding programmes, it is necessary to have sufficient information on desirable as well as undesirable traits in a gene pool, either form own evaluation trials or secondary passport data. Breeders prefer to use the well-adapted and cultivated (*Solanum tuberosum* subsp. *tuberosum*) germplasm or similar breeding materials for development of new varieties. A large number of exotic germplasm has been exploited in breeding and biotechnology in potato improvement for numerous traits. A substantial amount of materials, ranging from 23 to 7630 accessions per year, have been exchanged by the international genebanks to national genebanks; however, limited data on evaluation of these germplasm is available with the donor countries.

Nevertheless, a large number of varieties have been developed the world over, using the cultivated and wild potato species.

5. Application of Genomics in Germplasm Characterization

5.1 Germplasm Collection and Genotyping

A highly complex genetic and heterozygosity of autotetraploid nature of the crop hampers potato improvement. Morphological, cellular, biochemical and molecular technologies have been applied to elucidate taxonomy and phylogenetics of potato species (Spooner et al., 2007; Jacobs et al., 2008; Watanabe, 2015). Further, advancement in genome sequencing/resequencing technologies is likely to advance our understanding of the genetics and genomics of potato. A summary of potato germplasm characterization and SNP genotyping is outlined in Tables 9 and 10, respectively.

As described in the earlier chapter, the original potato genome sequence was deciphered using a homozygous Doubled Monoploid (DM) of cultivated potato Phureja group (The Potato Genome Sequencing Consortium, 2011). Recently, Gálvez et al. (2017) and Hirsch et al. (2014, 2016) reviewed the potato genome database. Till now, a large number of publications have come out, using the genome sequences and also several wild and cultivated potato species have been sequenced to discover new markers and genes. Genomics and transcriptomics resources have increased substantially, applying modern genome sequencing technologies in wild and cultivated potatoes (Massa et al., 2011; The Potato Genome Sequencing Consortium, 2011; Sharma et al., 2013; Hardigan et al., 2016). An update on molecular markers in potato linked to various traits, such as disease resistance (Ramakrishnan et al., 2015) and stress tolerance (Kikuchi et al., 2015) have been reviewed for marker-assisted breeding.

Since then, in recent years, genome sequence of wild potato species, such as *S. commersonii* have definitely increased our understanding on phylogenetics and taxonomic relationship of wild and cultivated species (Aversano et al., 2015). Moreover, availability of tomato genome sequence and a few more wild tomato species would also provide study across the species and comparative genomics in Solanaceae family. In this way, identification of genes for trait of interest would be possible for breeding application. Germplasm collection, conservation and distribution are the major functions of potato genebanks (Bamberg and del Rio, 2007). The reduction in sequencing cost and more affordable high-throughput genotyping facilities allow scientists to utilize a large set of germplasm collections for massive scale genotyping. This further allows discovering of genes and markers for a number of target traits in potato. Moreover, sharing information on genotyping in combination with phenotype data would be a great help for the potential users of potato researchers community in the world. For example, CIP has already genotyped its germplasm collection, using 12 K SNP array chip (Ellis, 2018). Similarly in other crops, like wheat and maize also, SNP genotyping of germplasm collections have been completed alongwith phenotype data and being exploited by the researchers.

Table 9. Cultivated and wild potato (*Solanum*) species, ploidy, endosperm balance number (EBN) and reported useful traits.

SN	*Solanum* species (section *Petota*) and synonyms in bullets	Country	Ploidy (EBN)	Fungus resistance			Bacterial resistance		Virus resistance				Insect resistance		Nematode resistance		Physiological characters			
				Late blight	Wart	Common scab	Bacterial wilt	Soft rot: black leg	Potato virus X	Potato virus Y	Potato leaf roll virus	Spindle tuber viroid	Colorado beetle	Myzus persicae, Macrosiphum euphorbiae (aphids)	Globodera rostochiensis, G. pallida (Potato cyst nematode)	Meloidgyne incognita (root knot nematode)	Frost	Heat	Drought	No tuber blackening
Cultivated species																				
1.	*S. tuberosum* L. Chilotanum group • *S. tuberosum* subsp. *tuberosum*	Chilean landraces	4*x* (4EBN)	+	+				+	+	+			+	+			+	+	
2.	*S. tuberosum* Andigenum group • *S. chaucha* Juz. & Bukasov; • *S. phureja* Juz. & Bukasov • *S. stenotomum* Juz. & Bukasov • *S. stenotomum* Juz. & Bukasov subsp. *goniocalyx* (Juz. & Bukasov) Hawkes • *S. tuberosum* subsp. *andigenum* Hawkes	Venezuela south to Argentina	2*x* (2EBN), 3*x*, 4*x* (4EBN) 2*x* (2EBN) 2x (2EBN)	+	+	+	+	+	++	+	+				+	+	+	+	+	
3.	*S. ajanhuiri* Juz. & Bukasov	BOL, PER	2*x* (2EBN)														+		+	
4.	*S. curtilobum* Juz. & Bukasov	BOL, PER	5*x*						+							+	+		+	
5.	*S. juzepczukii* Juz.	ARG, BOL, PER	3*x*						+								+			

Table 9 contd. ...

...Table 9 contd.

SN	*Solanum* species (section *Petota*) and synonyms in bullets	Country	Ploidy (EBN)	Fungus resistance			Bacterial resistance		Virus resistance				Insect resistance		Nematode resistance		Physiological characters			
Wild species																				
6.	*S. acaule* Bitter • *S. acaule* f. *incuyo* Ochoa • *S. acaule* var. *punae* (Juz.) Hawkes	ARG, BOL, PER	4*x* (2EBN), 6*x*		++		+		++	+	++	+	+		+	+	++	+	+	
7.	*S. acroglossum* Juz.	PER	2*x* (2EBN)																	
8.	*S. acroscopicum* Ochoa • *S. lopez-camarenae* Ochoa	PER	2*x*																	
9.	*S.* × *aemulans* Bitter & Wittm. • *S. acaule* subsp. *aemulans* (Bitter & Wittm.) Hawkes & Hjert. • *S.* × *indunii* K.A. Okada & A.M. Clausen	ARG	3*x*, 4*x* (2EBN)																	
10.	*S. agrimonifolium* Rydb.	GUA, HON, MEX	4*x* (2EBN)																	
11.	*S. albicans* (Ochoa) Ochoa • *S. acaule* subsp. *palmirense* Kardolus	ECU, PER	6*x* (4EBN)																	
12.	*S. albornozii* Correll	ECU	2*x* (2EBN)																	
13.	*S. amayanum* Ochoa	PER	2*x* (2EBN)																	
14.	*S. anamatophilum* Ochoa • *S. peloquinianum* Ochoa	PER	2*x* (2EBN)																	

15.	*S. andreanum* Baker • *S. burtonii* Ochoa • *S. correllii* Ochoa • *S. cyanophyllum* Correll • *S. paucijugum* Bitter • *S. regularifolium* Correll • *S. serratoris* Ochoa. • *S. solisii* Hawkes • *S. suffrutescens* Correll • *S. tuquerrense* Hawkes	COL, ECU	2*x* (2EBN), 4*x* (4EBN)											+						
16.	*S. augustii* Ochoa	PER	2*x* (1EBN)																	
17.	*S. ayacuchense* Ochoa	PER	2*x* (2EBN)																	
18.	*S. berthaultii* Hawkes • *S. flavoviridens* Ochoa • *S. tarijense* Hawkes • *S.* ×*litusinum* Ochoa • *S.* ×*trigalense* Cárdenas • *S.* ×*zudaniense* Cárdenas	ARG, BOL	2*x* (2EBN), 3*x*	+	+				+			+	+	+	+	+				
19.	*S.* × *blanco-galdosii* Ochoa	PER	2*x* (2EBN)																	
20.	*S. boliviense* Dunal in DC. • *S. astleyi* Hawkes & Hjert. • *S. megistacrolobum* Bitter • *S. megistacrolobum* f. *purpureum* Ochoa • *S. sanctae-rosae* Hawkes • *S. toralapanum* Cárdenas & Hawkes	ARG, BOL, PER	2*x* (2EBN)	+	+			+			+		+	+	+	+	++	+	+	
21.	*S. bombycinum* Ochoa	BOL	4*x*																	

Table 9 contd. ...

...Table 9 contd.

SN	*Solanum* species (section *Petota*) and synonyms in bullets	Country	Ploidy (EBN)	Fungus resistance			Bacterial resistance		Virus resistance				Insect resistance		Nematode resistance		Physiological characters			
22.	*S. brevicaule* Bitter • *S. alandiae* Cárdenas • *S. avilesii* Hawkes & Hjert. • *S. gourlayi* Hawkes • *S. gourlayi* subsp. *pachytrichum* (Hawkes) Hawkes & Hjert. • *S. gourlayi* subsp. *saltense* A.M. Clausen & K.A. Okada • *S. gourlayi* subsp. *vidaurrei* (Cárdenas) Hawkes & Hjert. • *S. hondelmannii* Hawkes & Hjert. • *S. hoopesii* Hawkes & K.A. Okada • *S. incamayoense* K.A. Okada & A.M. Clausen • *S. leptophyes* Bitter • *S. oplocense* Hawkes • *S. setulosistylum* Bitter • *S. sparsipilum* (Bitter) Juz. & Bukasov • *S. spegazzinii* Bitter • *S. sucrense* Hawkes • *S. ugentii* Hawkes & K.A. Okada • *S. virgultorum* (Bitter) Cárdenas & Hawkes • *S.* × *subandigena* Hawkes	ARG, BOL, PER	2*x* (2EBN) 4*x* (4EBN) 6*x* (4EBN)		+						+		+	+	+	+	+	+	+	
23.	• *S.* × *brucheri* Correll • *S.* × *viirsoii* K.A. Okada & A.M. Clausen	ARG	3*x*																	
24.	*S. buesii* Vargas	PER	2*x* (2EBN)																	

25.	*S. bulbocastanum* Dunal in Poir • *S. bulbocastanum* subsp. *dolichophyllum* (Bitter) Hawkes • *S. bulbocastanum* subsp. partitum (Correll) Hawkes	GUA, HON, MEX	2*x* (1EBN), 3*x*	++	+		+	+						++	+	+		+	+	
26.	*S. burkartii* Ochoa • *S. irosinum* Ochoa • *S. irosinum* forma *tarrosum* Ochoa	PER	2*x*																	
27.	*S. cajamarquense* Ochoa	PER	2*x* (1EBN)																	
28.	*S. candolleanum* Berthault • *S. abancayense* Ochoa • *S. achacachense* Cárdenas • *S. ambosinum* Ochoa • *S. ancoripae* Ochoa • *S. antacochense* Ochoa • *S. aymaraesense* Ochoa • *S. bill-hookeri* Ochoa • *S. bukasovii* Juz. • *S. bukasovii* var. *multidissectum* (Hawkes) Ochoa • *S. bukasovii* forma *multidissectum* (Hawkes) Ochoa • *S. canasense* Hawkes • *S. canasense* var. *xerophilum* (Vargas) Hawkes • *S. chillonanum* Ochoa • *S. coelestispetalum* Vargas • *S. hapalosum* Ochoa • *S. huancavelicae* Ochoa • *S. longiusculus* Ochoa • *S. marinasense* Vargas • *S. multidissectum* Hawkes • *S. orophilum* Correll	PER	2*x* (2EBN), 3*x*		+		+		+	+			+	+	+	+	+	+	+	

Table 9 contd. ...

...Table 9 contd.

SN	*Solanum* species (section *Petota*) and synonyms in bullets	Country	Ploidy (EBN)	Fungus resistance			Bacterial resistance		Virus resistance				Insect resistance		Nematode resistance		Physiological characters			
	• *S. ortegae* Ochoa • *S. pampasense* Hawkes • *S. puchupuchense* Ochoa • *S. sarasarae* Ochoa • *S. sawyeri* Ochoa • *S. saxatile* Ochoa • *S. sicuanum* Hawkes • *S. sparsipilum* subsp. *calcense* (Hawkes) Hawkes • *S. tapojense* Ochoa • *S. tarapatanum* Ochoa • *S. × mollepujroense* Cárdenas & Hawkes																			
29.	*S. cantense* Ochoa	PER	2*x* (2EBN)																	
30.	*S. cardiophyllum* Lindl. • *S. cardiophyllum* subsp. *lanceolatum* (Berthault) Bitter	MEX	2*x* (1EBN), 3*x*	++	+					+			+	+	+	+				
31.	*S. chacoense* Bitter • *S. arnezii* Cárdenas • *S. calvescens* Bitter • *S. chacoense* subsp. *chacoense* • *S. chacoense* subsp. *muelleri* (Bitter) Hawkes • *S. tuberosum* subsp. *yanacochense* Ochoa; (=*S. yanacochense* (Ochoa) Gorbatenko) • *S. yungasense* Hawkes	ARG, BOL, BRA, PAR, PER, URU	2*x* (2EBN), 3*x*	+	+	+	+	+	+	++			+			+	+	+	+	
32.	*S. chilliasense* Ochoa	ECU	2*x* (2EBN)																	

33.	*S. chiquidenum* Ochoa • *S. ariduphilum* Ochoa • *S. chiquidenum* forma *amazonense* Ochoa • *S. chiquidenum* var. gracile Ochoa • *S. chiquidenum* var. *robustum* Ochoa	PER	2*x* (2EBN)																	
34.	*S. chomatophilum* Bitter • *S. chomatophilum* forma *sausianense* Ochoa • *S. chomatophilum* var. *subnivale* Ochoa • *S. huarochiriense* Ochoa • *S. jalcae* Ochoa • *S. pascoense* Ochoa • *S. taulisense* Ochoa	ECU, PER	2*x* (2EBN)										+	+			++			
35.	*S. clarum* Correll	GUA, MEX	2*x*				+						+	+	+					
36.	*S. colombianum* Dunal • *S. cacetanum* Ochoa • *S. calacalinum* Ochoa • *S. jaenense* Ochoa • *S. moscopanum* Hawkes • *S. nemorosum* Ochoa • *S. orocense* Ochoa • *S. otites* Dunal • *S. pamplonense* L.E. López • *S. subpanduratum* Ochoa • *S. paramoense* Bitter • *S. sucubunense* Ochoa	COL, ECU, PER, VEN	4*x* (2EBN)	+																
37.	*S. commersonii* Dunal	ARG, BRA, URU	2*x* (1EBN), 3*x*	+	+	+		+					+			+	++			
38.	*S. contumazaense* Ochoa	PER	2*x* (2EBN)																	

Table 9 contd. ...

...Table 9 contd.

SN	*Solanum* species (section *Petota*) and synonyms in bullets	Country	Ploidy (EBN)	Fungus resistance			Bacterial resistance		Virus resistance				Insect resistance		Nematode resistance		Physiological characters			
39.	*S. demissum* Lindl. • *S.* × *semidemissum* Juz.	GUA, MEX	6*x* (4EBN)	++	+			+	+	++	+		+			+	+		+	
40.	*S.* × *doddsii* Correll	BOL	2*x* (2EBN)																	
41.	*S. dolichocremastrum* Bitter • *S. chavinense* Correll • *S. huanuchense* Ochoa	PER	2*x* (1EBN)																	
42.	*S.* × *edinense* Berthault • *S.* × *edinense* subsp. *salamanii* (Hawkes) Hawkes	MEX	5*x*	+									+							
43.	*S. ehrenbergii* (Bitter) Rydb • *S. cardiophyllum* subsp. *ehrenbergii* Bitter	MEX	2*x* (1EBN)																	
44.	*S. flahaultii* Bitter • *S. neovalenzuelae* L.E.López	COL	4*x*																	
45.	*S. gandarillasii* Cárdenas	BOL	2*x* (2EBN)																	
46.	*S. garcia-barrigae* Ochoa • *S. donachui* (Ochoa) Ochoa	COL	4*x*																	
47.	*S. gracilifrons* Bitter	PER	2*x*																	
48.	*S. guerreroense* Correll	MEX	6*x* (4EBN)									+								
49.	*S. hastiforme* Correll	PER	2*x* (2EBN)																	
50.	*S. hintonii* Correll	MEX	2*x*																	
51.	*S. hjertingii* Hawkes • *S. hjertingii* var. *physaloides* (Correll) Hawkes • *S. leptosepalum* Correll5 • *S. matehualae* Hjert. & T.R. Tarn	MEX	4*x* (2EBN)					+												+
52.	*S. hougasii* Correll	MEX	6*x* (4EBN)																	
53.	*S. huancabambense* Ochoa	PER	2*x* (2EBN)																	
54.	*S. humectophilum* Ochoa	PER	2*x* (1EBN)																	
55.	*S. hypacrarthrum* Bitter • *S. guzmanguense* Whalen & Sagást.	PER	2*x* (1EBN)																	

56.	*S. immite* Dunal • *S. yamobambense* Ochoa	PER	2*x* (1EBN), 3*x*																	
57.	*S. incasicum* Ochoa	PER	2*x* (2EBN)																	
58.	*S. infundibuliforme* Phil	ARG, BOL	2*x* (2EBN)		+						+		+	+			+			
59.	*S. iopetalum* (Bitter) Hawkes • *S. brachycarpum* (Correll) Correll	MEX	6*x* (4EBN)		+				+							+	+	+	+	
60.	*S. jamesii* Torr.	MEX, USA	2*x* (1EBN)		+						+		+			+				
61.	*S. kurtzianum* Bitter & Wittm • *S. ruiz-lealii* Brücher	ARG	2*x* (2EBN)		+				+		+		+		+	+	+	+	+	
62.	*S. laxissimum* Bitter • *S. neovargasii* Ochoa • *S. santolallae* Vargas	PER	2*x* (2EBN)																	
63.	*S. lesteri* Hawkes & Hjert	MEX	2*x*																	
64.	*S. lignicaule* Vargas	PER	2*x* (1EBN)											+						
65.	*S. limbaniense* Ochoa	PER	2*x* (2EBN)																	
66.	*S. lobbianum* Bitter	COL	4*x* (2EBN)																	
67.	*S. longiconicum* Bitter	CRI, PAN	4*x*																	
68.	*S. maglia* Schltdl	ARG, CHL	2*x*, 3*x*																	
69.	*S. malmeanum* Bitter	ARG, BRA, PAR, URU	2*x* (1EBN), 3*x*																	
70.	*S. medians* Bitter • *S. arahuayum* Ochoa • *S. sandemanii* Hawkes • *S. tacnaense* Ochoa • *S. weberbaueri* Bitter	CHL, PER	2*x* (2EBN), 3*x*											+						
71.	*S.* × *michoacanum* (Bitter) Rydb	MEX	2*x*																	
72.	*S. microdontum* Bitter • *S. microdontum* subsp *gigantophyllum* (Bitter) Hawkes & Hjert. • *S. microdontum* var. *montepuncoense* Ochoa	ARG, BOL	2*x* (2EBN), 3*x*	+	+		+	+			+			+		+		+	+	
73.	*S. minutifoliolum* Correll	ECU	2*x* (1EBN)																	

Table 9 contd. ...

...Table 9 contd.

SN	*Solanum* species (section *Petota*) and synonyms in bullets	Country	Ploidy (EBN)	Fungus resistance			Bacterial resistance		Virus resistance				Insect resistance		Nematode resistance		Physiological characters			
74.	*S. mochiquense* Ochoa • *S. chancayense* Ochoa • *S. incahuasinum* Ochoa	PER	2*x* (1EBN)																	
75.	*S. morelliforme* Bitter & Muench	BOL, GUA, MEX, HON	2*x*	+									+				+			
76.	*S. multiinterruptum* Bitter • *S. chrysoflorum* Ochoa • *S. moniliforme* Correll • *S. multiinterruptum* forma *albiflorum* Ochoa • *S. multiinterruptum* forma *longipilosum* Correll • *S. multiinterruptum* var. *machaytambinum* Ochoa	PER	2*x* (2EBN), 3*x*																	
77.	*S. neocardenasii* Hawkes & Hjert.	BOL	2*x*																	
78.	*S. neorossii* Hawkes & Hjert.	ARG	2*x*																	
79.	*S. neovavilovii* Ochoa	BOL	2*x* (2EBN)																	
80.	*S.* × *neoweberbaueri* Wittm.	PER	3*x*																	
81.	*S. nubicola* Ochoa	PER	4*x* (2EBN)																	
82.	*S. okadae* Hawkes & Hjert.	BOL	2*x*																	
83.	*S. olmosense* Ochoa	ECU, PER	2*x* (2EBN)																	
84.	*S. oxycarpum* Schiede	MEX	4*x* (2EBN)	+									+							
85.	*S. paucissectum* Ochoa	PER	2*x* (2EBN)																	
86.	*S. pillahuatense* Vargas	PER	2*x* (2EBN)																	
87.	*S. pinnatisectum* Dunal	MEX	2*x* (1EBN)	++	+		+	+		+			+			+		+	+	
88.	*S. piurae* Bitter	PER	2*x* (2EBN)																	
89.	*S. polyadenium* Greenm.	MEX	2*x*	+	+		+				+		+	+	+	+	+	+	+	
90.	*S. raphanifolium* Cárdenas & Hawkes • *S. hawkesii* Cárdenas	PER	2*x* (2EBN)		+						+					+	+	+	+	

91.	*S. raquialatum* Ochoa • *S. ingaefolium* Ochoa	PER	2*x* (1EBN)																	
92.	*S.* × *rechei* Hawkes & Hjert.	ARG	2*x*, 3*x*																	
93.	*S. rhomboideilanceolatum* Ochoa	PER	2*x* (2EBN)																	
94.	*S. salasianum* Ochoa	PER	2*x*																	
95.	*S.* × *sambucinum* Rydb.	MEX	2*x*																	
96.	*S. scabrifolium* Ochoa	PER	2*x*																	
97.	*S. schenckii* Bitter	MEX	6*x* (4EBN)																	
98.	*S. simplicissimum* Ochoa	PER	2*x* (1EBN)																	
99.	*S. sogarandinum* Ochoa	PER	2*x* (1EBN), 3*x*																	
100.	*S. stenophyllidium* Bitter • *S. brachistotrichium* (Bitter) Rydb. • *S. nayaritense* (Bitter) Rydb.	MEX	2*x* (1EBN)														+			
101.	*S. stipuloideum* Rusby • *S. circaeifolium* Bitter • *S. circaeifolium* subsp. *quimense* Hawkes & Hjert. • *S. capsicibaccatum* Cárdenas • *S. soestii* Hawkes & Hjert.	BOL	2*x* (1EBN)	+	+								+	+	+	+				
102.	*S. stoloniferum* Schltdl. • *S. fendleri* A. Gray • *S. fendleri* subsp. *arizonicum* Hawkes • *S. papita* Rydb. • *S. polytrichon* Rydb. • *S. stoloniferum* subsp. *moreliae* Hawkes	MEX, USA	4*x* (2EBN)	++	+			+		++	+		+	+	+	+		+	+	
103.	*S. tarnii* Hawkes & Hjert.	MEX	2*x*																	
104.	*S. trifidum* Correll	MEX	2*x* (1EBN)	+	+								+	+			+			
105.	*S. trinitense* Ochoa	MEX	2*x* (1EBN)																	
106.	*S.* × *vallis-mexici* Juz.	MEX	3*x*																	
107.	*S. venturii* Hawkes & Hjert.	ARG	2*x* (2EBN)																	

Table 9 contd. ...

...Table 9 contd.

SN	*Solanum* species (section *Petota*) and synonyms in bullets	Country	Ploidy (EBN)	Fungus resistance			Bacterial resistance		Virus resistance				Insect resistance		Nematode resistance		Physiological characters			
108.	*S. vernei* Bitter & Wittm. • *S. vernei* subsp. *ballsii* (Hawkes) Hawkes & Hjert.	ARG	2*x* (2EBN)	+			+						+		+	+	++	+	+	
109.	*S. verrucosum* Schltdl. • *S. macropilosum* Correll	MEX	2*x* (2EBN), 3*x*, 4*x*	+					+				+			+		+	+	
110.	*S. violaceimarmoratum* Bitter • *S. multiflorum* Vargas • *S. neovavilovii* Ochoa • *S. urubambae* Juz. • *S. villuspetalum* Vargas	BOL, PER	2*x* (2EBN)										+							
111.	*S. wittmackii* Bitter	PER	2*x* (1EBN)																	
112.	*S. woodsonii* Correll	PAN	4*x*																	

Adapted from Spooner et al. (2014) and Machida-Hirano (2015). ++Indicates species possesses immunity/high resistance/tolerance or high quality; +species includes resistance/tolerance, or good quality. County code: ARG (Argentina), BOL (Bolivia), BRA (Brazil), CHL (Chile), COL (Colombia), CRI (Costa Rica), ECU (Ecuador), GUA (Guatemala), HON (Honduras), MEX (Mexico), PAN (Panama), PAR (Paraguay), PER (Peru), URU (Uruguay), USA (United States of America), VEN (Venezuela).

Table 10. Application of some SNP array markers in characterization of potato germplasm.

SN	Genotype	Genotyping method	Objectives	References
1.	Tetraploid varieties (214)	22 K SNP Potato Array (Illumina)	Population structure, genetic diversity and construction of core collection of potato germplasm, breeding lines and varieties (commercial and reference).	Pandey et al. (2021)
2.	Tetraploid varieties/ progenitor clones (537)	20 K SNP Potato Array (Illumina)	Population structure, linkage disequilibrium decay, SNP genotyping, QTL discovery and haplotype-specific SNPs discovery.	Vos et al. (2017)
3.	Germplasm accessions (250)	12 K SNP Potato Array (Illumina)	Genetic diversity and population structure analysis of potato germplasm collection at International Potato Centre (CIP), using SNP genotyping.	Ellis et al. (2018)
4.	Diploid and tetraploid lines (144)	8.3 K SNP Potato Array (Illumina)	Genetic diversity, population structure, linkage disequilibrium and identification of duplicates in potato germplasm and breeding collection.	Berdugo-Cely et al. (2021)
5.	Diploids and tetraploids cultivars, landrace and wild species (67)	Genome sequence	Genomic diversity analysis potato germplasm collection (cultivars, wild and landraces) and impact on domestication of key loci associated with traits.	Hardigan et al. (2017)
6.	Commercial varieties (330)	20 K SNP Potato Array	Genome-wide association study (GWAS) for potato wart disease and identification of haplotype-specific SNP markers for genomics-assisted breeding in potato.	Prodhomme et al. (2020)
7.	Breeding clones and varieties (143)	12 K SNP Potato Array	GWAS for common scab in potato and identification of QTLs-associated SNP markers.	Yuan et al. (2020)
8.	Tetraploid F_1 mapping population (160)	8.3 K SNP Potato Array	QTL mapping in tetraploid population, using SNP genotyping for internal heat necrosis resistance and identification of SNP markers.	Schumann et al. (2017)
9.	Tetraploid F_1 mapping population (133)	8.3 K SNP Potato Array and 195 SSR markers	Mapping of wart resistance genes in a biparental progeny of F_1 progenies and identification of linked SNP markers.	Obidiegwu et al. (2015)
10.	Tetraploid cultivars (184)	8.3 K SNP Potato Array and candidate gene-based SNPs (total 9000 SNPs)	GWAS for plant maturity correlated late blight resistance in potato, using SNPs and identification of trait associated SNPs for breeding.	Mosquera et al. (2016)
11.	F_1 populations (MCD: 92 and PAM: 71)	Genotyping-by-sequencing (GBS)	Genome wide mapping of late blight resistance QTL in F_1 populations of two wild species *S. microdontum* and *S. pampasense* and identification of associated SNP markers.	Meade et al. (2020)

Table 10 contd. ...

...Table 10 contd.

SN	Genotype	Genotyping method	Objectives	References
12.	Potato varieties (90)	22 K SNP Potato Array	Genome-wide association study (GWAS) for tuber starch properties and SNP identification for breeding uses.	Khlestkin et al. (2020)
13.	Potato varieties (90)	22 K SNP Potato Array	GWAS for SNPs discovery to develop diagnostic markers to accelerate breeding for starch phorphorous content in potato.	Khlestkin et al. (2019)
14.	Potato varieties (277)	20 K SNP Potato Array	GWAS for QTL identification and candidate gene for enhancing protein content in potato for genomics-aided breeding.	Klaassen et al. (2019)
15.	*S. boliviense* x *S. tuberosum*	12 K SNP Potato Array	GWAS analysis for SNP identification for increasing folate content in potato.	Bali et al. (2018)
16.	Tetraploid potato clones (237)	12 K SNP Potato Array	Population structure, diversity and GWAS analysis for tuber yield and quality traits.	Zia et al. (2020)
17.	Tetraploid F_1 progenies (162)	8.3 K SNP Potato Array	GWAS analysis by SNP genotyping and QTLs identification for traits, like tuber sugar concentration, processing quality, vine maturity and other agronomic traits in potato.	Massa et al. (2018)
18.	Tetraploid varieties and breeding clones (448)	8.3 K SNP Potato Array and RADseq	GWAS analysis and SNP markers identification for tuber yield and starch content in potato.	Schönhals et al. (2017)

5.2 Core Collection

Core collection is a small set of lines (~ 10%) representing maximum genetic diversity in whole germplasm collection (Brown, 1989). Core collection is a very important resource for traits and genes discovery in limited resources. Genomics information can be exploited in the use of germplasm collection, mainly exploiting core collection to obtain maximum impact (van Treuren and van Hintum, 2014). The core collection can be developed for target traits and both genomic and phenotyping data could be made available in public for maximum benefits to the research community (van Treuren and van Hintum, 2014). Further, such core collection can be maintained as reference collection for use in other studies, like comparative genomics.

In potato, many core collections have been developed for tetraploid cultivated potatoes based on morphological and disease-pest resistance and diploid wild species, mainly on molecular markers (Bamberg and del Rio, 2014; Bamberg et al., 2016). The major problems faced in potato are ploidy variation in *Solanum* species wth intricate genotyping data analysis (Ghislain et al., 2006). Further, correct taxonomic classification on homogenous lines of wild potato species is another big challenge in construction of core collection. Heterogeneity affect considerably the classification of wild and cultivated species, and heterogeneity within accession of a particular species would be more problematic in construction of core collection and therefore hampers utilization in breeding and biotechnology. Thus, development of accurate core collection and their usefulness is highly dependent upon the quality of materials and correct phenotype and genotype data (Janskey et al., 2015). Therefore, international cooperation is required in terms of exchange of phenotype and molecular data across the laboratories in different countries to strengthen germplasm database for efficient utilization of germplasm collection.

Application of modern sequencing technologies (second and third generations) would be highly useful tools to study an array of studies, like taxonomy, evolution, domestication and conservation of potato germplasm collection (van Treuren and van Hintum, 2014; Janskey et al., 2015). Although environment affects comparison of phenotyping of lines conserved *in vitro* and field genebanks, application of SNP markers-based genotyping would be a highly useful technology for germplasm curators to manage the potato collection (Bastien et al., 2018). Taken together, genomics data is highly beneficial to potato researchers and germplasm managers to analyze genetic diversity in total collection (*in situ*/*ex situ*) or scientists utilizing the germplasm for desirable traits in breeding and biotechnology in the post-genomics era (Bethke et al., 2019).

6. Conclusion

Over 98,000 accessions of potato germplasm collection are maintained in the various national/international genebanks. These collections are conserved mostly through *ex situ* (*in vitro*) form and a few under *in situ* (on farm) methods. A very tiny fraction of germplasm collection, particularly of wild species, have been exploited in potato breeding and biotechnology. Most breeders prefer the tetraploid cultivated potato in breeding method for desirable traits, whereas wild species have undesirable traits as well and are not preferred while developing new varieties. The available germplasm needs to be extensively evaluated for most desirable traits, such as agronomic, processing, quality, nutrition, diseases and pests resistance, abiotic stress (heat, drought, cold, salinity and nutrient use efficiency), and climate change. Under-utilization of cultivated as well as wild species in varietal development is primarily due to want of desirable traits in parental breeding materials. Genebank collections, therefore, need to be transformed in usable form for maximum utilization of collections in breeding programs. The modern genome sequencing technologies offer tremendous opportunities in characterization and utilization of wild and cultivated potatoes. SNP markers allow genotyping of a whole set of germplasm collection and cost less and take less time. Both genotyping and phenotyping data would be highly beneficial for potato research community in management and efficient utilization of germplasm collection. Genomics resources allow development of potato core collection representing maximum genetic diversity and will be valuable for potato improvement in future.

Potato germplasm conservation and utilization face several big challenges, such as gaps in conservation of wild species. A study identified under-representation of wild species in genebank, for example *ex situ* conservation of 73 species revealed that 32 species were not represented (Castañeda-Álvarez et al., 2015). They also noticed that four wild species (*S. ayacuchense*, *S. neovavilovii*, *S. olmosense* and *S. salasianum*) were not available in the international genebanks. Secondly, accurate taxonomic classification and characterization of collection is necessary and needs to be validated across the genebanks. As per the current taxonomic classification, *S. phureja* Juz. & Bukasov (2*x*) has been classified into *S. tuberosum* subsp. *andigenum* (Juz. & Bukasov) Hawkes (= *S. tuberosum* Group Andigenum) (4*x*) (Spooner et al., 2007). Moreover, this has been updated in the CIP database as well. Therefore, such knowledge is very important for proper utilization of genetic resources. Third, adequate conservation of germplasm and maintenance of enough genetic diversity in genebank is important for effective utilization in breeding programs. Lastly, enough phenotypic variation is a must for conservation and utilization. Morphological variation could be a major concern particularly in wild potato species where they are maintained through TPS. Maintenance of clonal identity through segregating generations of true seeds in diploid species is an issue of germplasm conservation, given that natural crossing and domestication under natural habitat maintenance of original material is a challenging task in potato. Therefore, application of genomics resources is important for efficient germplasm conservation, characterization and utilization for potato improvement.

Acknowledgement

I am thankful to the Director, ICAR-Central Potato Research Institute, Shimla, and scientists/technicians/research fellows and other colleagues of the institute for their support under the institute research projects on biotechnology, germplasm, breeding and seed research on aeroponics. I am also grateful to the funding agencies for support under the externally funded projects (CABin, ICAR-IASRI, New Delhi; ICAR-LBS Young Scientist Award Project, and DBT, Government of India).

References

Ames, M. and Spooner, D.M. (2008). DNA from herbarium specimens settles a controversy about origins of the European potato. *Am. J. Bot.*, 95: 252–257.

Asociación ANDES. (2016). Resilient farming systems in times of uncertainty: Biocultural innovations in the potato park, Peru. *IIED*, London.

Aversano, R., Contaldi, F., Ercolano, M.R., Grosso, V., Iorizzo, M. et al. (2015). The *Solanum commersonii* genome sequence provides insights into adaptation to stress conditions and genome evolution of wild potato relatives. *The Plant Cell*, 27(4): 954–968.

Bali, S., Robinson, B.R., Sathuvalli, V., Bamberg, J. and Goyer, A. (2018). Single Nucleotide Polymorphism (SNP) markers associated with high folate content in wild potato species. *PLoS ONE*, 13(2): e0193415.

Bamberg, J.B. and del Rio, A.H. (2005). Conservation and potato genetic resources. pp. 476. *In*: Razdan, M.K. and Mattoo, A.K (eds.). *Genetic Improvement and Solanaceaous Crops*. vol. I: Potato. Science Publishers, Inc. Plymouth.

Bamberg, J.B. and del Rio, A.H. (2007). The canon of potato science—50 topics in potato science that every potato scientist should know: (1) Genetic diversity and gene banks. *Potato Res.*, 50: 207–210.

Bamberg, J.B. and del Rio, A.H. (2014). Selection and validation of an aflp marker core collection for the wild potato *Solanum microdontum*. *Am. J. Potato Res.*, 91: 368–375.

Bamberg, J.B., del Rio, A.H., Kinder, D., Louderback, L., Pavlik, B. and Fernandez, C. (2016). Core collections of potato (*Solanum*) species native to the USA. *Am. J. Potato Res.*, 93: 564–571.

Barker, H. (1996). Inheritance of resistance to potato viruses Y and A in progeny obtained from potato cultivars containing gene *Ry*: Evidence for a new gene for extreme resistance to PVA. *Theor. Appl. Genet.*, 93: 710–716.

Bastien, M., Boudhrioua, C., Fortin, G. and Belzile, F. (2018). Exploring the potential and limitations of genotyping-by-sequencing for SNP discovery and genotyping in tetraploid potato. *Genome*, 61(6): 449–456.

Berdugo-Cely, J.A., Martínez-Moncayo, C. and Lagos-Burbano, T.C. (2021). Genetic analysis of a potato (*Solanum tuberosum* L.) breeding collection for southern Colombia using single nucleotide polymorphism (SNP) markers. *PLoS ONE*, 16(3): e0248787.

Bethke, P.C., Halterman, D.A. and Jansky, S.H. (2019). Potato germplasm enhancement enters the genomics era. *Agronomy*, 9: 575.

Bradeen, J.M. and Haynes, K.G. (2011). Introduction to potato. pp 1–19. *In*: Bradeen, J. and Kole, C. (eds.). *Genetics, Genomics, and Breeding of Potato*. CRC Press, Boca Raton, Fl.

Brown, A.H.D. (1989). Core collections: a practical approach to genetic resources management. *Genome*, 31: 818–824.

Camadro, E.L. and Espinillo, J.C. (1990). Germplasm transfer from the wild tetraploid species *Solanum acaule* Bitt. to the cultivated potato *S. tuberosum* L. using 2n eggs. *Am. Potato J.*, 67: 737–749.

Castañeda-Álvarez, N.P., de Haan, S., Juárez, H., Khoury, C.K., Achicanoy, H.A., Sosa, C.C., Bernau, V., Salas, A., Heider, B., Simon, R., Maxted, N. and Spooner, D.M. (2015). *Ex situ* conservation priorities for the wild relatives of potato (*Solanum* L. Section petota). *PloS One*, 10(4): e0122599.

Chen, L., Guo, X., Xie, C., He, L., Cai, X. et al. (2013). Nuclear and cytoplasmic genome components of *Solanum tuberosum* + *S. chacoense* somatic hybrids and three SSR alleles related to bacterial wilt resistance. *Theor. Appl. Genet.*, 126: 1861–1872.

de Haan, S. and Rodriguez, F. (2016). Potato origin and production. pp 1–32. *In*: Singh, J. and Kaur, L. (eds.). *Advances in Potato Chemistry and Technology*. Elsevier Inc., London, GB.

Ellis, D., Chavez, O., Coombs, J., Soto, J., Gomez, R. et al. (2018). Genetic identity in genebanks: Application of the SolCAP 12K SNP array in fingerprinting and diversity analysis in the global in trust potato collection. *Genome*, (7): 523–537.

FAO. (2010). *The Second Report on the State of the World's Plant Genetic Resources for Food and Agriculture*. Rome, Italy, 370 p.

Fock, I., Collonnier, C., Purwito, A., Luisetti, J., Souvannavong, V. et al. (2000). Resistance to bacterial wilt in somatic hybrids between *Solanum tuberosum* and *Solanum phureja*. *Plant Sci.*, 160: 165–176.

Gálvez, J.H., Tai, H.H., Barkley, N.A., Gardner, K., Ellis, D. and Strömvik, M.V. (2017). Understanding potato with the help of genomics. *AIMS Agriculture and Food*, 2: 16–39.

Gavrilenko, T., Antonova, O., Ovchinnikova, A., Novikova, L., Krilova, E. et al. (2010). A microsatellite and morphological assessment of the Russian national cultivated potato collection. *Genet. Resour. Crop Evol.*, 57: 1151–1164.

Gavrilenko, T., Antonova, O., Shuvalova, A., Krylova, E., Alpatyeva, N. et al. (2013). Genetic diversity and origin of cultivated potatoes based on plastid microsatellite polymorphism. *Genet. Resour. Crop Evol.*, 60: 1997–2015.

Ghislain, M., Andrade, D., Rodríguez, F., Hijmans, R.J. and Spooner, D.M. (2006). Genetic analysis of the cultivated potato *Solanum tuberosum* L. Phureja Group using RAPDs and nuclear SSRs. *Theor. Appl. Genet.*, 113: 1515–1527.

Ghislain, M., Núñez, J., del Rosario Herrera, M., Pignataro, J., Guzman, F. and Spooner, D.M. (2009). Robust and highly informative microsatellite-based genetic identity kit for potato. *Mol. Breed.*, 23: 377–388.

Gopal, J. and Chauhan, N.S. (2010). Slow growth *in vitro* conservation of potato germplasm at low temperature. *Potato Res.*, 53: 141–149.

Hardigan, M.A., Crisovan, E., Hamilton, J.P., Kim, J., Laimbeer, P. et al. (2016). Genome reduction uncovers a large dispensable genome and adaptive role for copy number variation in asexually propagated *Solanum tuberosum*. *The Plant Cell*, 28(2): 388–405.

Hardigan, M.A., Laimbeer, F., Newton, L., Crisovan, E., Hamilton, J.P. et al. (2017). Genome diversity of tuber-bearin-+-g Solanum uncovers complex evolutionary history and targets of domestication in the cultivated potato. *Proc. Natl. Acad. Sci., U.S.A.*, 114(46): E9999–E10008.

Harlan, J.R. and DeWet, J.M.J. (1971). Toward a rational classification of cultivated plants. *Taxon*, 20: 509–517.

Hawkes, J.G. (1990). *The Potato: Evolution, Biodiversity and Genetic Resources*. Smithsonian Institution Press, Washington, D.C.

Hawkes, J. (1994). Origins of cultivated potatoes and species relationships. pp. 3–42. *In*: Bradshaw, J.E. and Mackay, G.R. (eds.). *Potato Genetics*. CAB International, Wallingford.

Hermsen, J.G.H. (1994). Introgression of genes from wild species, including molecular and cellular approaches. pp. 515–538. *In*: Bradshaw, J.E. and Mackay, G.R. (eds.). *Potato Genetics*. CAB International, Wallingford, U.K.

Hermundstad, S. and Peloquin, S.J. (1985). Germplasm enhancement with potato haploids. *J. Heredity*, 76: 463–467.

Hijmans, J.R., Jacobs, M., Bamberg, J.B. and Spooner, D.M. (2003). Frost tolerance in wild potato species: assessing the predictivity of taxonomic, geographic and ecologic factors. *Euphytica*, 130: 47–59.

Hijmans, R.J., Spooner, D.M., Salas, A.R., Guarino, L. and de la Cruz, J. (2002). Atlas of wild potato, *Systematic and Ecogeographic Studies on Crop Genepools*. International Plant Genetic Resources Institute, Rome, p. 130.

Hijmans, R.J., Gavrilenko, T., Stephenson, S., Bamberg, J., Salas, A. and Spooner, D.M. (2007). Geographic and environmental range expansion through polyploidy in wild potatoes (*Solanum section* Petota). *Global Ecol. Biogeo*., 16: 485–495.

Hirsch, C.D., Hamilton, J.P., Childs, K.L., Cepela, J., Crisovan, E. et al. (2014). Spud DB: A resource for mining sequences, genotypes, and phenotypes to accelerate potato breeding. *Plant Genome*, 7(1): 1–12. Doi: 10.3835/plantgenome2013.12.0042.

Hirsch, C.D., Buell, C. and Hirsch, C.N. (2016). A toolbox of potato genetic and genomic resources. *Am. J. Potato Res*., 93: 21–32.

Hoops, R.W. and Plaisted, R.L. (1987). Potato. pp. 385–435. *In*: Fehr, W.R. (ed.). *Principles of Cultivar Development*. vol. 2, *Crop Species*, MacMillon Publishing Company, New York.

Howard, H.W. (1970). *Genetics of the Potato*. Springer Verlag, New York.

Iwanaga, M., Jatala, P., Ortiz, R. and Guevara, E. (1989). Use of FDR 2n pollen to transfer resistance to root-knot nematodes into cultivated 4x potatoes. *J. Am. Soc. Hortic. Sci*., 114: 1008–1013.

Iwanaga, M., Ortiz, R., Cipar, M.S. and Peloquin, S.J. (1991). A restorer gene for genetic-cytoplasmic male sterility in cultivated potatoes. *Am. Potato J*., 68: 19–28.

Jacobs, M.M.J., van den Berg, R.G., Vleeshouwers, V.G.A.A., Visser, M., Mank, R. et al. (2008). AFLP analysis reveals a lack of phylogenetic structure within Solanum section Petota. *BMC Evol. Biol*., 8: 145.

Jacobs, M.M.J., Smulders, M.J.M., van den Berg, R.G. and Vosman, B. (2011). What's in a name; Genetic structure in Solanum section Petota studied using population-genetic tools. *BMC Evol. Biol*., 11: 42.

Janskey, S.H. and Hamernik, A.J. (2009). The introgression of 2x 1EBN Solanum species into the cultivated potato using *Solanum verrucosum* as a bridge. *Genet. Resour. Crop Evol*., 56: 1107–1115.

Janskey, S.H., Dawson, J. and Spooner, D.M. (2015). How do we address the disconnect between genetic and morphological diversity in germplasm collections? *Am. J. Bot*., 102: 1213–1215.

Johnston, S.A., den Nijs, T.P., Peloquin, S.J. and Hanneman, R.E. Jr. (1980). The significance of genic balance to endosperm development in interspecific crosses. *Theor. Appl. Genet*., 57(1): 5–9.

Khlestkin, V.K., Rozanova, I.V., Efimov, V.M. and Khlestkina, E.K. (2019). Starch phosphorylation associated SNPs found by genome-wide association studies in the potato (*Solanum tuberosum* L.). *BMC Genet*., 20(Suppl 1): 29.

Khlestkin, V.K., Erst, T.V., Rozanova, I.V., Efimov, V.M. and Khlestkina, E.K. (2020). Genetic loci determining potato starch yield and granule morphology revealed by genome-wide association study (GWAS). *Peer J*., 8: e10286.

Kikuchi, A., Huynh, H.D., Endo, T. and Watanabe, K. (2015). Review of recent transgenic studies on abiotic stress tolerance and future molecular breeding in potato. *Breed Sci*., 65: 85–102.

Klaassen, M.T., Willemsen, J.H., Vos, P.G., Visser, R.G.F., van Eck, H.J. et al. (2019). Genome-wide association analysis in tetraploid potato reveals four QTLs for protein content. *Mol. Breed*., 39: 151.

Machida-Hirano, R. (2015). Diversity of potato genetic resources. *Breed. Sci*., 65(1): 26–40.

Maki, S., Hirai, Y., Niino, T. and Matsumoto, T. (2015). Assessment of molecular genetic stability between long-term cryopreserved and tissue cultured wasabi (*Wasabia japonica*) plants. *CryoLett*., 36: 318–324.

Massa, A.N., Childs, K.L., Lin, H. et al. (2011). The transcriptome of the reference potato genome *Solanum tuberosum* Group *Phureja* clone DM1-3 516R44. *PLoS ONE*, 6(10): e26801.

Massa, A.N., Manrique-Carpintero, N.C., Coombs, J., Haynes, K.G., Bethke, P.C. et al. (2018). Linkage analysis and QTL mapping in a tetraploid russet mapping population of potato. *BMC Genetics*, 19(1): 87.

McHale, N.A. and Lauer, F.I. (1981). Inheritance of tuber traits from *Phureja* in diploid *Phureja-Tuberosum* hybrids. *Am. Potato J.*, 58: 93–102.

Meade, F., Hutten, R., Wagener, S., Prigge, V., Dalton, E. et al. (2020). Detection of novel qtls for late blight resistance derived from the wild potato species *Solanum microdontum* and *Solanum pampasense*. *Genes*, 11: 732.

Missiou, A., Kalantidis, K., Boutla, A., Tzortzakaki, S., Tabler, M. and Tsagris, M. (2004). Generation of transgenic potato plants highly resistant to potato virus Y (PVY) through RNA silencing. *Mol. Breed.*, 14: 185–197.

Mosquera, T., Alvarez, M.F., Jiménez-Gómez, J.M., Muktar, M.S., Paulo, M.J. et al. (2016). Targeted and untargeted approaches unravel novel candidate genes and diagnostic snps for quantitative resistance of the potato (*Solanum tuberosum* L.) to *Phytophthora infestans* causing the late blight disease. *PLoS ONE*, 11(6): e0156254.

Murashige, T. and Skoog, F. (1962). A revised medium for rapid growth and bio assays with tobacco tissue cultures. *Physiol. Plant.*, 15: 473–497.

Niino, T., Tanaka, D., Tantely, R.R., Fukui, K. and Shirata, K. (2007). Cryopreservation of basal stem buds of *in vitro*-grown mat rush (*Juncus* spp.) by vitrification. *CryoLett.*, 28: 197–206.

Niino, T. and Arizaga, M.V. (2015). Cryopreservation for preservation of potato genetic resources. *Breed. Sci.*, 65: 41–52.

Obidiegwu, J.E., Sanetomo, R., Flath, K., Tacke, E., Hofferbert, H.R. et al. (2015). Genomic architecture of potato resistance to *Synchytrium endobioticum* disentangled using SSR markers and the 8.3k SolCAP SNP genotyping array. *BMC Genetics*, 16: 38.

Ochoa, C. (1984). *S. hygrothermicum*, new potato species cultivated in the lowlands of Peru. *Econ. Bot.*, 38: 128–133.

Ochoa, C.M. (1990). [actual release date 13 June 1991]. *The Potatoes of South America: Bolivia*. Cambridge University Press, Cambridge.

Ochoa, C.M. (1999). *Las papas de Sudamerica: Peru (parte I)*. International Potato Center, Lima, Peru.

Oka, S. and Niino, T. (1997). Long-term storage of pear (*Pyrus* spp.) shoot cultures *in vitro* by minimal growth method. *Japan Agric. Res. Quar.*, 31: 1–7.

Ovchinnikova, A., Krylova, E., Gavrilenko, T., Smekalova, T., Zhuk, M. et al. (2011). Taxonomy of cultivated potatoes (*Solanum* section *Petota*: *Solanaceae*). *Bot. J. Linn. Soc.*, 165: 107–155.

Pandey, J., Scheuring, D.C., Koym, J.W., Coombs, J., Novy, R.G. et al. (2021). Genetic diversity and population structure of advanced clones selected over forty years by a potato breeding program in the USA. *Sci. Rep.*, 11(1): 8344.

Pradel, W. (2013). Importance of *in-situ* conservation initiatives in conservation of native potato varieties in the Andes of Peru, *No. 201355*, *Working Papers*, Latin American and Caribbean Environmental Economics Program.

Prodhomme, C., Vos, P.G., Paulo, M.J., Tammes, J.E., Visser, R. et al. (2020). Distribution of P1 (D1) wart disease resistance in potato germplasm and GWAS identification of haplotype-specific SNP markers. *Theor. Appl. Genet.*, 133(6): 1859–1871.

Ramakrishnan, A.P., Ritland, C.E., Sevillano, R.H.B. and Riseman, A. (2015). Review of potato molecular markers to enhance trait selection. *Am. J. Potato Res.*, 92: 455–472.

Rodríguez, F. and Spooner, D.M. (2009). Nitrate reductase phylogeny of potato (*Solanum sect* Petota) genomes with emphasis on the origins of the polyploid species. *System. Bot.*, 34: 207–219.

Rodríguez, F., Wu, F., Ané, C., Tanksley, S. and Spooner, D.M. (2009). Do potatoes and tomatoes have a single evolutionary history, and what proportion of the genome supports this history? *BMC Evol. Biol.*, 9: 191.

Salas, A., Gaspar, O., Rodríguez, W., Vargas, M., Centeno, R. and Tay, D. (2008). Regeneration guidelines: Wild potato. *In*: Dulloo, M.E., Thormann, I, Jorge, M.A. and Hanson, J. (eds.). *Crop Specific Regeneration Guidelines*. CGIAR System-wide Genetic Resource Programme, Rome, Italy, 8 p.

Schönhals, E.M., Ding, J., Ritter, E., Paulo, M.J. and Cara, N. (2017). Physical mapping of QTL for tuber yield, starch content and starch yield in tetraploid potato (*Solanum tuberosum* L.) by means of genome wide genotyping by sequencing and the 8.3 K SolCAP SNP array. *BMC Genomics*, 18(1): 642.

Schumann, M.J., Zeng, Z.B., Clough, M.E. and Yencho, G.C. (2017). Linkage map construction and QTL analysis for internal heat necrosis in autotetraploid potato. *Theor. Appl. Genet.*, 130(10): 2045–2056.

Sharma, S.K., Bolser, D., de Boer, J., Sønderkær, M., Amoros, W. et al. (2013). Construction of reference chromosome-scale pseudomolecules for potato: Integrating the potato genome with genetic and physical maps. *G3 (Bethesda)*, 3(11): 2031–2047.

Spooner, D.M. and Bamberg, J.B. (1994). Potato genetic resources: Sources of resistance and systematics. *Am. Potato J.*, 71: 325–337.

Spooner, D.M. and Hijmans, R.J. (2001). Potato systematics and germplasm collecting, 1989–2000. *Am. J. Potato Res.*, 78: 237–268.

Spooner, D.M., McLean, K., Ramsay, G., Waugh, R. and Bryan, G.J. (2005). A single domestication for potato based on mutilocus amplified fragment length polymorphism genotyping. *Proc. Natl. Acad. Sci., U.S.A.*, 102: 14694–14699.

Spooner, D.M., Núñez, J., Trujillo, G., del Rosario Herrera, M., Guzmán, F. and Ghislain, M. (2007). Extensive simple sequence repeat genotyping of potato landraces supports a major reevaluation of their gene pool structure and classification. *Proc. Natl. Acad. Sci., U.S.A.*, 104: 19398–19403.

Spooner, D.M., Fajardo, D. and Salas, A. (2008). Revision of the *Solanum medians* complex (*Solanum sect.* Petota). *System. Bot.*, 33: 579–588.

Spooner, D.M. (2009). DNA barcoding will frequently fail in complicated groups: An example in wild potatoes. *Am. J. Bot.*, 96: 1177–1189.

Spooner, D.M., Gavrilenko, T., Jansky, S.H., Ovchinnikova, A., Krylova, E. et al. (2010). Ecogeography of ploidy variation in cultivated potato (*Solanum sect.* Petota). *Am. J. Bot.*, 97: 2049–2060.

Spooner, D.M., Ghislain, M., Simon, R., Jansky, S.H. and Gavrilenko, T. (2014). Systematics, diversity, genetics, and evolution of wild and cultivated potatoes. *Bot. Rev.*, 80: 283–383.

Swaminathan, M.S. and Howard, H.W. (1953). The cytology and genetics of potato (*S. tuberosum* L.) and related species. *Bibilographia Genetica*, 16: 1–192.

The Potato Genome Sequencing Consortium. (2011). Genome sequence and analysis of the tuber crop potato. *Nature*, 475: 189–195.

Tiwari, J.K., Chandel, P., Singh, B.P. and Bhardwaj, V. (2014). Analysis of plastome and chondriome genome types in potato somatic hybrids from *Solanum tuberosum* x *Solanum etuberosum*. *Genome*, 57: 29–35.

Tiwari, J.K., Ali, N., Devi, S., Kumar, V., Zinta, R. and Chakrabarti, S.K. (2018). Development of microsatellite markers set for identification of Indian potato varieties. *Sci. Hort.*, 231: 22–30.

van den Berg, R. and Groendijk-Wilders, N. (2014). Taxonomy. pp. 12–28. *In*: Navarre, R. and Pavek, M. (eds.). *The Potato: Botany, Production and Uses*. CAB International, U.K.

van der Vossen, E., Sikkema, A., Hekkert, B.L., Gros, J., Stevens, P. et al. (2003). An ancient *R* gene from the wild potato species *Solanum bulbocastanum* confers broad-spectrum resistance to *Phytophthora infestans* in cultivated potato and tomato. *Plant J.*, 36: 867–882.

van Treuren, R. and van Hintum, T.J.L. (2014). Next-generation genebanking: plant genetic resources management and utilization in the sequencing era. *Plant Genet. Resour. Character. Utiliz.*, 12(3): 298–307.

Veilleux, R.E. and De Jong, H. (2007). Potato. pp. 17–58. *In*: Singh, R.J. (ed.). *Genetic Resources, Chromosome Engineering, and Crop Improvement*. vol. 3, CRC Press, Boca Raton.

Vos, P.G., Paulo, M.J., Voorrips, R.E., Visser, R.G., van Eck, H.J. and van Eeuwijk, F.A. (2017). Evaluation of LD decay and various LD-decay estimators in simulated and SNP-array data of tetraploid potato. *Theor. Appl. Genet.*, 130(1): 123–135.

Watanabe, K. (2015). Potato genetics, genomics, and applications. *Breed. Sci.*, 65: 53–68.

Watanabe, K.N., Orrillo, M., Vega, S., Hurtado, A., Valkonen, J.P.T. et al. (1995). Overcoming crossing barriers between non-tuber bearing and tuber-bearing *Solanum* species: Towards potato germplasm enhancement with a broad spectrum of solanaceous genetic resources. *Genome*, 38: 27–35.

Wu, G., Shortt, B.J., Lawrence, E.B., Levine, E.B., Fitzsimmons, K.C. and Shah, D.M. (1995). Disease resistance conferred by expression of a gene encoding H_2O_2-generating glucose oxidase in transgenic potato plants. *The Plant Cell*, 7: 1357–1368.

Yamamoto, S., Rafique, T., Priyantha, W.S., Fukui, K., Matsumoto, T. and Niino, T. (2011). Development of a cryopreservation procedure using aluminium cryo-plates. *CryoLett.*, 32: 256–265.

Yuan, J., Bizimungu, B., Koeyer, D.D., Rosyara, U., Wen, Z. and Lagüe, M. (2020). Genome-wide association study of resistance to potato common scab. *Potato Res.*, 63: 253–266.

Zia, M., Demirel, U., Nadeem, M.A. and Çaliskan, M.E. (2020). Genome-wide association study identifies various loci underlying agronomic and morphological traits in diversified potato panel. *Physiol. Mol. Biol. Plants*, 26(5): 1003–1020.

Chapter 4

Molecular Markers, Mapping and Genome-wide Characterization

1. Introduction

A morphological marker is simply a mutation in a particular gene, which imparts a discrete and easily identifiable phenotype to an organism. However, rarity of morphological markers was a limiting factor for their use in genome mapping. Moreover, they were applied but they are much influenced by environmental factors and are dependent on plant growth stage and show less reproducibility. It was, therefore, thought that biochemical markers that can be separated on the basis of relative mobility of enzyme isoforms would be more helpful in genome mapping. However, biochemical markers also suffered from similar limitations like morphological markers. Also, biochemical markers (isozyme/amino acid/protein-based) are limited to isozymes, amino acids or protein level variations. These biochemical markers have unusual occurrence and are subjected to post-translation changes in plants. These above markers are labour intensive also and thus limit their application for whole genome application. Moreover, in both the above cases, the markers were influenced by environment, stage of growth and tissues/organs used for analysis. A quantum leap towards genome mapping was only possible after introduction of DNA markers that are plenty in number and least affected by environment and stage of growth. The relative position of any marker in a chromosome is determined by the conventional principle of linkage analysis. Molecular markers are genetic and heritable entities associated with particular traits and differentiate the organisms, based on the marker-trait associations. Unlike above two, molecular or DNA markers show variation site at the DNA level and are heritable and reproducible in nature and not influenced by plant growth stages, seasons and environments (Table 1). Polymorphic markers can differentiate the individuals in species or genera or any taxa. Therefore, molecular markers have been popular among the geneticists and molecular biologists to map the gene and identify tightly-linked markers with a particular trait for use in marker-assisted breeding, particularly in potato (Slater et al., 2013; Tiwari et al., 2013; Ramakrishnan et al., 2015). Nevertheless, the potato crop suffers from various biotic and abiotic stresses under the climate change scenario (Raymundo et al., 2018;

Table 1. Comparison of morphological and molecular markers in plants.

Parameters	Morphological	Molecular
Environmental effect, labour intensive and unusual phenotype occurrence	Very high	No
Heritable	Not fixed	Very high
Application in genetic analysis	Very limited	Very high
Identification of genetic loci at whole plant/cell/tissue level	Limited	Yes
Possibility of identification of naturally occurring alleles	Very less	Very high
Undesirable phenotype effects of traits associated with alleles	Yes	No
Analysis of alleles of possible types of segregating progenies/ genotypes	No	Yes

Dahal et al., 2019). Hence, there is a need of molecular marker and genome mapping for marker-assisted breeding for rapid crop improvement.

2. Molecular Markers

A number of molecular markers have been described in the literature with various nomenclatures. Some of the commonly used markers for plant genome analysis are RFLP, RAPD, AFLP, SSR, ISSR, CAPS, SCAR, DArT and SNP. These can be dominant, such as RAPD, ISSR and AFLP which cannot differentiate allelic differences of a gene in heterozygous conditions and co-dominant (e.g., RFLP, SSR, SNPs) which are able to differentiate allelic differences. Markers can be dominant or co-dominant distinguishing heterozygote and homozygote genotypes. Co-dominant markers have several alleles, whereas dominant has only two alleles. Among them, only RFLP is the non-PCR-based marker, whereas others are PCR-based markers, of which, SSR markers have been used extensively in genetic diversity analysis of wild and cultivated potato species. During the last two decades, there has been rapid resurgence in development of genome sequence-based SNP marker, which is being used widely for high-throughput genotyping and various other molecular-based applications (Adhikari et al., 2017). Advances in nucleotide sequence polymorphism have led to various applications in plant science, including potato germplasm enhancement (Anisimova et al., 2019; Bethke et al., 2019). Applications of molecular markers, including biotechnology and genomics, were reviewed in potato (Barrell et al., 2013). The common molecular markers are compared in Table 2 and a few are listed below:

- Amplified fragment length polymorphism (AFLP).
- Cleaved amplified polymorphic sequences (CAPS).
- Diversity array technology (DArT).
- Inter simple sequence repeat (ISSR).
- Microsatellites or simple sequence repeats (SSR).
- Random amplified polymorphic DNA (RAPD).
- Restriction fragment length polymorphism (RFLP).

- Sequence characterized amplified regions (SCAR).
- Single nucleotide polymorphism (SNPs).

The following properties are required for ideal molecular markers:

- *Moderate to high polymorphism*: Ideal molecular markers should show moderate to high polymorphic to differentiate individuals in organisms.
- *Genomic abundance and frequent occurrence*: Markers should be more abundant and frequently found in genome to represent a versatile marker system.
- *Co-dominant inheritance and unambiguous allele differentiation*: Co-dominant markers, able to distinguish dominance and recessive alleles, so that heterozygous genotypes can be distinguished from homozygous genotypes. Preferably, co-dominant markers are located on single locus in the genome. Markers differentiate alleles unambiguously.
- *No environmental effect*: Markers are neither influenced by environmental conditions not plant growth stage.
- *Neutral genetic variation*: Molecular markers should show variation at DNA level at neutral sites without any effect to physiology of the organism.

Table 2. Comparison of five commonly used molecular markers in plants.

Feature	**RAPD**	**RFLP**	**AFLP**	**SSR**	**ISSR**
Occurrence in genome	Very high	High	Very high	Moderate	Medium
Genome covered	Whole genome	Low copy coding regions	Whole genome	Whole genome	Whole genome
Inheritance	Dominant	Co-dominant	Dominant	Co-dominant	Dominant
Polymorphism	High	Moderate	Very high	High	High
Allele detection	No	Yes	No	Yes	No
Reproducibility	Intermediate	High	High	High	Moderate-high
DNA quantity required	Low	High	Moderate	Low	Low
Automation	Moderate	Low	Medium	High	Moderate
Probes/primers	~ 10 bp random nucleotides	Low copy genomic DNA/cDNA	Specific sequence	Specific repeat DNA sequence	Specific repeat DNA sequence
Cloning/sequencing	No	Yes	No	Yes	No
Radioactivity	No	Usually yes	Yes/no	No	No
Easy to use	Easy	Labour intensive	Difficult (initially)	Easy	Easy
Effective multiplex ratio	Moderate	Low	High	Moderate	Moderate
Marker index	Moderate	Low	High	Moderate	Moderate
Development cost	Low	High	Moderate	High	Moderate

Source: Modified adapted from Adhikari et al. (2017)

- *Ease in detection, automation and low cost*: Markers should facilitate high-throughput genotyping due to ease in detection, automation, scoring and reliably, and low cost.
- *High reproducibility*: Markers should be highly reproducible and easy exchange of data is possible between laboratories.

3. Genomic Markers for High-throughput Genotyping

3.1 Single Nucleotide Polymorphism (SNP)

SNP is a single nucleotide difference in the gene sequence or any part of the genome sequence. SNP markers rely upon the sequence variation in the genome. SNPs are very common and most abundant polymorphic markers in polyploids (You et al., 2018), including potato and distributed in entire genome. For example, 1 SNP per 60–120 bp is found in maize, whereas 1 SNP per 1000 bp in humans have been reported. SNPs are found in both transcribed and non-transcribed regions of a gene and are associated with phenotypic variation in the plant. SNP markers are robust, show high polymorphism, suitable for high-throughput genotyping and various other applications. SNP marker system allows ease of automation in a wide range of applications, such as creating crop signature for varietal identification, construction of high-density genetic maps and genotyping. But SNP development may incur very high cost and require sequence information. With the available sequencing technologies and reducing cost of sequencing, it is expected to develop more and more SNP arrays for wider application in potato. Enrichment sequencing technology has been applied in disease resistance breeding in potato (Armstrong et al., 2019). Recently, haplotype-based genome analysis has been done in diploid potato, which further advances our knowledge on SNP-based potato improvement (Zhou et al., 2020).

3.2 Genotyping by Sequencing (GBS)

Genotyping by sequencing is one of the high-throughput genotyping technologies used to discover SNP and genotyping of several plant species (He et al., 2014). With the reducing cost of sequencing technology, a huge amount of data is being generated worldwide and allows the discovery of a large number of SNP. GBS has been designed for several studies, like genetic analysis, population structure, molecular characterization and SNP discovery. GBS is being studied at single to whole genome level. This is a quick and cost-effective method for high-throughput genotyping of breeding population, linkage analysis, diversity studies, molecular markers and genomic selection. To breed varieties, knowledge about genes and environment and their interaction is important for using the GBS method to select advance breeding lines with desirable traits. SNP chip-based markers are also another HTG platform available in potato for genotyping, using various platforms, such as 20 K SNPs Affymetrix Axiom (SolSTW array), Infinium 12 K V2 Potato Array (Illumina platform) and 8 K SolCAP SNP array (Illumina Infinium BeadChip). This technology has several advantages like it is cost-effective, generates adequate data at whole genome level and identifies thousands of SNPs markers for various applications, like genome-wide association mapping, population structure and genomic selection.

3.3 Diversity Arrays Technology (DArT)

DArT is also a microarray, like hybridization-based technology, allowing high-throughput genotyping of loci in the entire genome. This system includes construction of diversity arrays of DNA from a diverse range of genotypes. Briefly, protocols include isolation of genomic DNA, digested by restriction enzymes, ligation of enzyme-specific adaptors with fragment DNA and reduction of genome complexity through PCR using selective primers. After that, cloning of DNA fragments, amplification of inserts by vector-specific primers, purification and detection of amplified fragments on DNA chip are done. Finally, polymorphic clones (DArT markers) are detected, based on hybridization and signal intensities of the genotypes. This technology offers advantages, like high throughput, quick, reproducible and robust marker with high genome coverage. But it has limitations, like involves several steps, costly input in laboratory setup, skilled manpower and id limited to a few species.

4. Genome Mapping and Gene/QTL Discovery

4.1 Genetic and Physical Maps

Gene mapping is one of the important applications of molecular markers in plants. Linkage mapping is the genetic association of traits with segregating alleles of molecular markers in a defined mapping population. Linkage mapping detects genomic regions that explain phenotypic variations in a trait of interest and subsequently identifies genes/QTLs in that region. QTL mapping in potato is mainly carried out at the diploid level due to the potato's highly heterozygous nature. Many QTLs for resistance to biotic stresses, like *P. infestans* and root cyst nematodes, are known. The cultivated potato (*S. tuberosum* Group Tuberosum) varieties are tetraploid ($2n = 4x = 48$). The Solanum species contains a huge genetic diversity for its improvement (Huang et al., 2018). The complex genetic tetrasomic inheritance and high heterozygosity of potato complicate its genetic mapping and therefore diploids were used in most mapping studies. However, self-incompatability nature of the diploids prevents development of pure lines. Therefore, a number of common mapping approaches based on homozygous lines cannot be applied to potato as in other crops. The first potato genetic map was reported in 1988, based on sexual recombination frequencies and linkage groups (chromosomes) identification, using tomato RFLP markers (Bonierbale et al., 1988). Two linkage maps were obtained from a cross between a diploid clone of *S. tuberosum* Group Phureja and a diploid hybrid line from *S. tuberosum* Group Tuberosum × (*S. tuberosum* Group Phureja × *S. chacoense*). The alignment of the RFLP loci shows a high level of similarity to the tomato map and the major differences were paracentric inversions on three chromosomes. The use of PCR-based markers to select for desirable traits followed soon after. Bacterial Artificial Chromosome (BAC) libraries have become the main vehicle for performing map-based gene cloning and physical mapping in potato. Several BAC libraries were constructed, such as cultivated potato, wild species like *S. bulbocastanum* (Song et al., 2000). These libraries represent a potentially useful resource for the study of comparative genome organization and evolution in potato and other Solanaceous crops. A BAC library was used to make the Ultra High Density (UHD) genetic and physical map of potato with 10,000 AFLP loci (van Os et al., 2006). In addition, BAC libraries were used for fluorescence *in situ* hybridization (FISH) to develop chromosome-specific cytogenetic DNA markers for chromosome identification in potato (Dong et al., 2000). DArT (diversity array technology) also benefits potato genetics (Wenzl et al., 2004). Later, these 10,000 AFLP markers were exploited by the Potato Genome Sequencing Consortium (PGSC) in the potato genome sequence. Such new sets of markers have been applied with advances in the potato genome sequencing. Table 3 outlines linkage maps and molecular markers used in potato (Watanabe, 2015).

Potato genome was sequenced in 2011 and then efforts were made to generate genetic and physical maps and improve markers order of the genome assembly. Sharma et al. (2013) developed integrated maps based on the potato reference genome (DM) with a new sequence-tagged site marker-based linkage map and other physical and genetic maps of potato and the closely related species, tomato. The primary anchoring of the DM potato genome assembly was accomplished with a diploid segregating population (180 individuals). Samples were genotyped with molecular markers (DArT, AFLP, SSR, SNP) and constructed ~ 936 cM linkage map containing 2,469 marker loci. Further, *in silico* anchoring approaches utilized genetic and physical maps of the diploid potato genotype RH89-039-16 (RH) and tomato. Finally, pseudomolecules corresponding to the 12 potato chromosomes revealed 674 Mb (~ 93%) of the 723 Mb genome assembly and 37,482 (~ 96%) of the 39,031 predicted genes. A comparative study between marker distribution and physical location showed variation in chromosomal regions and segregation distortion. This study presented a great improvement in the ordering of the potato reference genome superscaffolds into pseudomolecules of the 12 chromosomes with anchor markers.

4.2 Gene/QTL Mapping

A number of techniques, like bulked segregant analysis, have been extensively used to identify markers. Many genes/QTLs for biotic and abiotic stress resistance/tolerance have been mapped in potato. Particular focus has been given to simply inherited genes, such as those conferring resistance to late blight, viruses and nematodes. Besides, many Quantitative Trait Loci (QTL) for yield, agronomic and quality traits have been identified in potato (Bradshaw et al., 2008). The QTL mapping study for abiotic stress tolerance is at infant stage. Quite a small number of QTL have been identified for drought stress tolerance (Anithakumari et al., 2012). Abiotic stress, such as drought tolerance trait, shows low level of heritability and is controlled by several genes and their epistatic in nature restricts molecular breeding in potato. Anithakumari et al. (2011) have identified 23 QTLs (13 QTLs under well-watered conditions, seven under drought stress condition and three recovery QTLs) in a diploid mapping population; and the genes underlying these QTLs were related to root to shoot ratio, plant height, shoot fresh weight, shoot dry weight, fresh root weight, root dry weight, root length, fresh biomass and dry biomass. Interestingly, the study also found the co-localization of SNPs with root to shoot

Table 3. Summary of linkage maps and molecular markers used in mapping of potato.

Sr. No.	Marker type	Number of markers	Total map length (cM)	Key findings	References
1.	RFLP (restriction fragment length polymorphism)	135	1189	First potato map	Bonierbale et al. (1988)
2.	RFLP	141	690	High heterozygosity observed in diploid genetics lines	Gebhardt et al. (1989)
3.	RFLP	304	1034	Increase of mapped loci	Gebhardt et al. (1991)
4.	RFLP	1030 (avg. 1.2 cM interval)	1276	Highly saturated map for comparison with tomato	Tanksley et al. (1992)
5.	RFLP, transposons, isozymes	175	1120	Integration of known loci and new markers	Jacobs et al. (1995)
6.	AFLP (amplified fragment length polymorphism)	770	-	AFLP used on potatoes and high resolution is possible	Vos et al. (1995) van Eck et al. (1995) Rouppe van der Voort et al. (1998)
7.	RGL (resistance gene-like fragment)	Concept generation for landmark	-	Applicability for disease and pest resistances	Leister et al. (1996)
8.	SSR	89	-	Fingerprinting	Milbourne et al. (1998)
9.	ISSR (inter simple sequence repeat)	4	-	Validation for fingerprinting	Prevost and Wilkinson (1999)
10.	RAPD (random amplified polymorphism), RFLP	100	606	An example of target-trait-specific rapid generation of mapping	Hosaka and Hanneman (1998)
11.	COS (Conserved Ortholog Set)	1025	-	Further validation of tomato-potato orthology	Fulton et al. (2002)
12.	SSR	156	-	15 SSR marker sets selected as highly informative	Ghislain et al. (2004)
13.	SSR (simple sequence repeat, microsatellite)	61	-	EST-based SSR	Feingold et al. (2005)
14.	Multiple gene family homologues (P450)	15 primer pairs produced 27 loci	-	Simple approach to make a marker and map	Yamanaka et al. (2003, 2005)

Table 3 contd. ...

...Table 3 contd.

Sr. No.	Marker type	Number of markers	Total map length (cM)	Key findings	References
15.	AFLP	10,305 (10,365 with markers such on CAPS, SCAR)	751 maternal/ 773 paternal	Small number of progeny individuals with 130 and 381 AFLP primer combinations made over 10,000 markers	Van Os et al. (2006)
16.	NB-LRR disease resistance gene homologues	738 RGLs	47 *R* genes physically mapped	BAC-based physical map of *R* genes	Bakker et al. (2011)
17.	NB-LRR *R* genes	438	370 *R* genes physically mapped	-	Jupe et al. (2012)
18.	BAC physical map	2800 contigs	1.64 times the coverage of the genome	Foundation for genome sequencing	de Boer et al. (2011)
19.	SNP (single nucleotide polymorphism), SSR, AFLP, DArT (diversity array technology)	2469	936	Integration of genetic and physical maps	Sharma et al. (2013)

Source: Modified adapted from Watanabe (2015)

ratio QTL, thus proposing their applicability in MAS for drought tolerance in potato. QTLs associated with root length allow the selection of plants with desirable root characteristics in a non-invasive method as root traits are of immense importance in tolerating drought. In addition, QTLs associated with carbon radioisotope discrimination, chlorophyll content and chlorophyll fluorescence were also identified (Anithakumari et al., 2012), as these serve as good selection criteria since they are easy to measure, fast and allow little or no sample destruction. Pathogen resistance genes, chromosome location and linked molecular markers in potato are listed in Table 4.

Single nucleotide polymorphism (SNP) has been used widely for marker generation in many species and many markers have been generated also in potatoes (Table 5). Discovery of novel genetic markers in potato will be facilitated by recent advances in genomic research. A panel of over 14,530 SNPs for potato has been developed by SolCAP, called SolSTW, and was recently used to examine genome-wide associations of the potato genome to late blight (Lindqvist-Kreuze et al., 2014). Similarly, another panel of 8,303 SNPs was developed, and called Infenium 8303 Potato Array, was used to examine genome-wide association in potato (Galvez et al., 2017). Potato marker discovery relies heavily on parallel research in tomato, but widely cultivated potatoes are tetraploid while tomatoes are diploid, so novel methods of SNP genotyping must be used. The potato genome sequence published in 2011 (The Potato Genome Sequencing Consortium, 2011) has helped to speed marker discovery. Sharma et al. (2013) have constructed a dense genetic and

Table 4. Pathogen resistance genes, chromosomal location and linked molecular markers in potato.

Gene	Chr.	Resistance source	Marker name (type)	Pathogen	References
Late blight (*Phytophthora infestans*)					
R1	5	*S. tuberosum* Group Tuberosum	AFLP1 and AFLP2 (AFLP); GP21 (RFLP); BA47f2, R11400, R11800 (SCAR)	*P. infestans*	Leonards-Schippers et al. (1992), Meksem et al. (1995), Ballvora et al. (2002), Gebhardt et al. (2004), Kuhl (2011), Mori et al. (2011)
R2	4	*S. tuberosum* Group Tuberosum	ACC/CAT-535, ACT/CAC-189, AGC/CCA-369 (AFLP); R2-800 (SCAR)	*P. infestans*	Li et al. (1998), Mori et al. (2011)
R3	11	*S. tuberosum* Group Tuberosum	TG105a, GP185, GP250(a) (RFLP)	*P. infestans*	El-Kharbotly et al. (1994)
R6, R7	11	*S. tuberosum* Group Tuberosum	185(a), GP250(a) (RFLP)	*P. infestans*	El-Kharbotly et al. (1996)
RB	8	*S. tuberosum* (+) *S. bulbocastanum* (somatic hybrids, back cross)	RB (SCAR)	*P. infestans*	Colton et al. (2006)
Rpi-blb3	4	*S. bulbocastanum*	Th21 (SCAR)	*P. infestans*	Park et al. (2005), Jacobs et al. (2010)
Rpi-smira1	11	*S. tuberosum* Group Tuberosum	45/X1(SCAR)	*P. infestans*	Tomczyńska et al. (2014)
Rpi-ber, Rpi-ber1, Rpi-ber2	10	*S. berthaultii*, *S. tuberosum* × *S. berthaultii* (back cross to *S. tuberosum*)	mCT240 (ESTS); CT214 (SCAR); TG63 (RFLP)	*P. infestans*	Ewing et al. (2000), Rauscher et al. (2006), Park et al. (2009)
Rpi-moc1	9	*S. mochiquense*	TG328 (CAPS)	*P. infestans*	Smilde et al. (2005)
Rpi1	7	*S. pinnatisectum* × *S. cardiophyllum*, (back cross to *S. cardiophyllum*)	TG20A (RFLP)	*P. infestans*	Kuhl et al. (2001)
Rpi-phu1	9	*S. tuberosum* Group Tuberosum	GP94 (PCR)	*P. infestans*	Śliwka et al. (2006, 2008, 2010)
NA	6	*S. tuberosum* Group Tuberosum, *S. vernei*	GP76 (CAPS)	*P. infestans*	Oberhagemann et al. (1999), Gebhardt et al. (2004)
NA	NA	*S. tuberosum* Group Tuberosum	NBS2_8 NBS, NBS5a6_10 NBS (SCAR)	*P. infestans*	Malosetti et al. (2007), Gebhardt (2011)

Table 4 contd. ...

...Table 4 contd.

Gene	Chr.	Resistance source	Marker name (type)	Pathogen	References
Resistance hot spot	5	*S. tuberosum* Group Tuberosum	GP179, CosA (SCAR)	*P. infestans*	Meksem et al. (1995), Gebhardt et al. (2004)
Potato viruses					
Ry_{adg}	11	*S. tuberosum* Group Andigena	RYSC3, ADG1, ADG2 (SCAR); ADG2 (CAPS); TG508 (RFLP)	PVY^O/PVY^N	Hämäläinen et al. (1997, 1998), Sorri et al. (1999), Kasai et al. (2000), Gebhardt et al. (2006), Ottoman et al. (2009), Whitworth et al. (2009), Ortega and Lopez-Vizcon (2012), Lopez-Pardo et al. (2013)
Ry_{sto}	12	*S. stoloniferum*	$SCAR_{YSTO4}$ (SCAR); YES3-3A, YES3-3B (ESTS)	$PVY^O/PVY^C/$ $PVY^N/$ PVY^{NTN}	Song et al. (2005), Cernák et al. (2008), Song and Schwarzfischer (2008)
Ry_{sto}	11	*S. stoloniferum*	M45, M5 (AFLP)	PVY^N	Brigneti et al. (1997)
$Ry\text{-}f_{sto}$	12	*S. stoloniferum*	$GP122_{718}$, $GP122_{564}$ (CAPS)	$PVY^O/PVY^C/$ $PVY^N/$ PVY^{NTN}	Flis et al. (2005), Witek et al. (2006), Gebhardt (2011), Lopez-Pardo et al. (2013)
Ry_{chc}	9	*S. chacoense*	RY186 (SCAR); CT220 (RFLP)	PVY^O/PVY^N	Sato et al. (2006), Mori et al. (2011)
Ny_{tbr}	4	*S. tuberosum* × *S. berthaultii* back cross	TG506 (RFLP)	PVY^O	Celebi-Toprak et al. (2002)
Ny-1	9	*S. tuberosum* Group Tuberosum	$SC895_{1139}$ (SCAR)	$PVY^O/PVY^N/$ PVY^{NTN}	Szajko et al. (2008)
Rl_{adg}	5	*S. tuberosum* Group Andigena	RGASC850 (SCAR); E35M48.192 (AFLP)	PLRV	Velásquez et al. (2007), Mihovilovich et al. (2014)
Nb	5	*S. tuberosum* Group Tuberosum	SPUD237 CAPS	PVX	De Jong et al. (1997)
Rx1	12	*S. tuberosum* Group Andigena	CP60 (RFLP), CP60 (CAPS)	PVX	Ritter et al. (1991), Bendahmane et al. (1997), Gebhardt et al. (2006), Mori et al. (2011)
Rx2	5	*S. acaule*	GP21 (RFLP)	PVX	Ritter et al. (1991)

Table 4 contd. ...

...*Table 4 contd.*

Gene	Chr.	Resistance source	Marker name (type)	Pathogen	References
Ns	8	*S. tuberosum* Group Andigena	$SCG17_{321}$ (SCAR), $SC811_{260}$ and CP16 (CAPS)	PVS	Marczewski et al. (2002), Witek et al. (2006)
Potato cyst nematode (PCN)					
GpaIVadg	4	*S. tuberosum* Group Tuberosum	STM3016-122/177 (SSR), C237(119)/TaqI (CAPS)	*G. pallida*	Milbourne et al. (1998), Bryan et al. (2004), Moloney et al. (2010)
H1	5	*S. tuberosum* Group Andigena	TG689, N146, N195, CP113, TG689, TG689indel12 (SCAR); 239E4left (CAPS); EM15, CMI (AFLP); CD78, CP113 (RFLP); TG689 (HRM)	*G. rostochiensis*	Pineda et al. (1993), Gebhardt et al. (1993), Niewöhner et al. (1995), Skupinová et al. (2002), Bakker et al. (2004), Pajerowska-Mukhtar et al. (2009), Milczarek et al. (2011), Mori et al. (2011), Galek et al. (2011), De Koeyer et al. (2011), Lopez-Pardo et al. (2013)
GroVI	5	*S. tuberosum* Group Tuberosum	SCAR-U14, SCAR-X02 (SCAR)	*G. rostochiensis*	Jacobs et al. (1996), Milczarek et al. (2011)
RGp5-vrnHC	5	*S. tuberosum* Group Tuberosum	HC (snp212-T/ snp444-G) SNP	*G. pallida*	Lopez-Pardo et al. (2013), Sattarzadeh et al. (2006)
Grp1	5	*S. tuberosum* Group Tuberosum	TG432 (CAPS)	*G. pallida*, *G. rostochiensis*	Finkers-Tomczak et al. (2009), Milczarek et al. (2011)
NA	5	*S. vernei*	SPUD1636 (SCAR)	*G. pallida*	Bryan et al. (2002), Milczarek et al. (2011)
Gro1	7	*S. tuberosum* Group Tuberosum	Gro1-4 (SCAR)	*G. rostochiensis*	Gebhardt et al. (2004), Paal et al. (2004), Milczarek et al. (2011)
Gro1	7	*S. spegazzinii*	CP56, CP51(c), GP516(c) (SCAR)	*G. rostochiensis*	Ballvora et al. (1995), Kuhl (2011)
Rmc1	11	*S. bulbocastanum*, *S. bulbocastanum* × *S. tuberosum* (BC_3)	M39b, CT182 (CAPS)	*M. chitwoodi*, *M. fallax*, *M. Hapla*	Rouppe van der Voort et al. (1999)

Table 4 contd. ...

...Table 4 contd.

Gene	Chr.	Resistance source	Marker name (type)	Pathogen	References
Rmc1	11	*S. bulbocastanum*	19319, 56F6, 39E18, 524F16, 406L19 (ESTS)	*M. chitwoodi*	Zhang et al. (2007)
Gpa2	12	*S. tuberosum* Group Andigena	77R (CAPS), GP34 (CAPS)	*G. pallida*	Milczarek et al. (2011), Rouppe van der Voort et al. (1999)
Black leg					
Eca1A	1	*S. tuberosum* × *S. chacoense* & *S. gungasense*	EM4-44 (AFLP)	*Erwinia carotovora* ssp. *atroseptica*	Zimnoch-Guzowska et al. (2000)
Eca2A	2	*S. tuberosum* × *S. chacoense* & *S. gungasense*	HM4-14 (AFLP)	*Erwinia carotovora* ssp. *atroseptica*	Zimnoch-Guzowska et al. (2000)
Eca6A	6	*S. tuberosum* × *S. chacoense* & *S. gungasense*	HM5-17 (AFLP)	*Erwinia carotovora* ssp. *atroseptica*	Zimnoch-Guzowska et al. (2000)
Eca11A, *Eca11B*	11	*S. tuberosum* × *S. chacoense* & *S. gungasense*	EM1-17, HM1-28, EM3-32 (AFLP)	*Erwinia carotovora* ssp. *atroseptica*	Zimnoch-Guzowska et al. (2000)
Potato wart					
Sen1	11	*S. tuberosum*	NL25 (SCAR)	*Synchytrium endobioticum*	Bormann et al. (2004), Gebhardt et al. (2006)

Source: Modified adapted from Ramakrishnan et al., 2015; Machida-Hirano, 2015

physical map for diploid backcross progeny of potato, using 2469 markers, including SSR, diversity array technology (DArT) and SNPs by using the same genotypic data of these markers. Khan et al. (2015) constructed maternal and paternal maps to carry out the first QTL study for drought tolerance. The study identified 45 genomic regions associated with nine traits in well-watered and terminal drought treatments and 26 QTLs associated with drought stress. These QTLs will promisingly be used in the breeding of potato for durable tolerance to drought, using conventional as well as genomics-assisted breeding approaches.

Internal Heat Necrosis (IHN) is an important disorder of potato tubers. Earlier studies identified AFLP markers linked to IHN susceptibility in the tetraploid potato mapping population (B2721) consisting of cross between susceptible (Atlantic) and resistant (B1829-5) genotypes. Schumann et al. (2017) constructed linkage map for both the parents based on SNP genotyping, using the 8.3 K SNP potato array (Infinium Illumina) and identified QTLs for susceptibility in the population. The linkage map yielded 1397.68 cM total length of chromosome containing 3427 SNPs. QTLs for IHN were detected on chromosomes 1, 5, 9 and 12 across multiple years of evaluation. The detected SNP markers explained 28.21% genetic variation for incidence while 25.3% variation for severity. This study strengthens genetic information on IHN in potato and identified the IHN-resistant QTL that requires confirmation on other breeding population for breeding against IHN.

Table 5. Summary of applications of SNP genotyping in genome mapping and QTL/gene discovery in potato.

SN	Genotype	Genotyping method	Objectives	References
Germplasm diversity and population structure				
1.	Tetraploid varieties (214)	22 K SNP Potato Array (Illumina)	Population structure, genetic diversity and construction of core collection of potato germplasm, breeding lines and varieties (commercial and reference)	Pandey et al. (2021)
2.	Tetraploid varieties and progenitor clones (537)	20 K SNP Potato Array (Illumina)	Population structure, linkage disequilibrium decay, SNP genotyping, QTL discovery and haplotype-specific SNPs discovery	Vos et al. (2017)
3.	Germplasm accessions (250)	12 K SNP Potato Array (Illumina)	Genetic diversity and population structure analysis of potato germplasm collection at International Potato Centre (CIP) using SNP genotyping	Ellis et al. (2018)
4.	Diploid and tetraploid lines (144)	8.3 K SNP Potato Array (Illumina)	Genetic diversity, population structure, linkage disequilibrium and identification of duplicates in potato germplasm and breeding collection	Berdugo-Cely et al. (2021)
5.	Diploids and tetraploids cultivars, landrace and wild species (67)	Genome sequence	Genomic diversity analysis potato germplasm collection (cultivars, wild and landraces) and impact on domestication of key loci associated with traits	Hardigan et al. (2017)
Disease-pest resistance				
6.	Commercial varieties (330)	20 K SNP Potato Array	Genome-wide association study (GWAS) for potato wart disease and identification of haplotype-specific SNP markers for genomics-assisted breeding in potato	Prodhomme et al. (2020)
7.	Breeding clones and varieties (143)	12 K SNP Potato Array	GWAS for common scab in potato and identification of QTLs-associated SNP markers	Yuan et al. (2020)
8.	Tetraploid F_1 mapping population (160)	8.3 K SNP Potato Array	QTL mapping in tetraploid population, using SNP genotyping for internal heat necrosis resistance and identification of SNP markers	Schumann et al. (2017)
9.	Tetraploid F_1 mapping population (133)	8.3 K SNP Potato Array and 195 SSR markers	Mapping of wart resistance genes in a biparental progeny of F_1 progenies and identification of linked SNP markers	Obidiegwu et al. (2015)
10.	Tetraploid cultivars (184)	8.3K SNP Potato Array and candidate gene-based SNPs (total 9000 SNPs)	GWAS for plant maturity correlated late blight resistance in potato using SNPs and identification of trait-associated SNPs for breeding	Mosquera et al. (2016)

Table 5 contd. ...

...Table 5 contd.

SN	Genotype	Genotyping method	Objectives	References
11.	F_1 populations (MCD: 92 and PAM: 71)	Genotyping-by-sequencing (GBS)	Genome-wide mapping of late blight resistance QTL in F_1 populations of two wild species—*S. microdontum* and *S. pampasense* and identification of associated SNP markers	Meade et al. (2020)
12.	Tetraploid F_1 clones (192)	DNA capture (RenSeq and GenSeq)	Mapping of nematode resistance gene *H2* in tetraploid potato clones applying DNA capture technology and development of allele specific KASP markers for breeding	Strachan et al. (2019)
13.	BC_1 progenies (152)	DNA capture (RenSeq and GenSeq)	SNP-based mapping of late blight resistant genes from *Solanum verrucosum* and identification of associated SNP markers using DNA capture technology for breeding	Chen et al. (2018)
14.	BC_1 progenies (84)	DArT and SSR markers (1583)	Late blight resistance gene mapping on chromosome 7 in wild potato species, *S. pinnatisectum*	Nachtigall et al. (2018)
15.	Potato differential plant Ma*R8*	BAC library sequencing	Fine mapping of late blight resistance gene *R8* derived from wild potato species *S. demissum*	Vossen et al. (2016)
16.	18 varieties and 189 tetraploid clones	SNP (100) and SSR (19) markers	Association mapping for potato cyst nematode resistance gene/QTL and identification of diagnostic SNP markers for breeding	Achenbach et al. (2009)
Agronomic and processing/nutritive traits				
17.	Potato varieties (90)	22 K SNP Potato Array	Genome-wide association study for tuber starch properties and SNP identification for breeding uses	Khlestkin et al. (2020)
18.	Potato varieties (90)	22 K SNP Potato Array	GWAS for SNPs discovery to develop diagnostic markers to accelerate breeding for starch phosphorous content in potato	Khlestkin et al. (2019)
19.	Potato varieties (277)	20 K SNP Potato Array	GWAS for QTL identification and candidate gene for enhancing protein content in potato for genomics-aided breeding	Klaassen et al. (2019)
20.	*S. boliviense* × *S. tuberosum*	12 K SNP Potato Array	GWAS analysis for SNP identification for increasing folate content in potato	Bali et al. (2018)
21.	Tetraploid potato clones (237)	12 K SNP Potato Array	Population structure, diversity and GWAS analysis for tuber yield and quality traits	Zia et al. (2020)
22.	Tetraploid F_1 progenies (162)	8.3 K SNP Potato Array	GWAS analysis by SNP genotyping and QTLs identification for traits like tuber sugar concentration, processing quality, vine maturity and other agronomic traits in potato	Massa et al. (2018)
23.	Tetraploid varieties and breeding clones (448)	8.3 K SNP Potato Array and RADseq	GWAS analysis and SNP markers identification for tuber yield and starch content in potato	Schönhals et al. (2017)

Table 5 contd. ...

...Table 5 contd.

SN	Genotype	Genotyping method	Objectives	References
24.	Potato clones	GBS	GWAS and identification of QTL for feasibility of genomic selection for fry color and resistance to cold-induced sweetening after storage in potato	Byrne et al. (2020)
25.	Tetraploid clones (318)	GBS	GBS analysis and identification of SNP markers associated with flower color in potato	Bastien et al. (2018)
26.	Tetraploid cultivars (181)	GBS-transcriptomics	High-throughput transcriptome sequencing for SNP discovery for genome-wide selection in potato	Caruana et al. (2019)
27.	Diploid wild and cultivated accessions (201)	DNA sequencing	Genome-wide genetic diversity and population structure analysis in potato wild and cultivated species. Identification of genes/QTL under selection and SNP associated with tuberization and loss of bitterness in tuber	Li et al. (2018)
28.	Elite potato varieties (6) and germplasm lines (248)	cDNA and SNP genotyping	Development of cDNA sequences in elite potato cultivars, SNP genotyping of a germplasm collection and identification of agronomic traits-specific SNP for molecular breeding	Hamilton et al. (2011)
High-density map and allele doses in tetraploid				
29.	Tetraploid F_1 population (237)	20 K SNP potato array	High density linkage mapping in tetraploid potato using SNP genotyping	Bourke et al. (2016)
30.	Tetraploid and diploid clones (639)	20 K SNP Potato Array	Population structure analysis in potato clones of various classes (market, breeding history and origin) using the 'fitTetra' software designed for autotetraploids and identification of trait-specific SNP markers	Vos et al. (2015)
31.	Tetraploid F_1 mapping population (237)	20K SNP Potato Array	Creation of high-density linkage map in tetraploid potato and identify double reduction in a biparental mapping population	Bourke et al. (2015)
32.	Diploid population (186)	8.3 K SNP Potato Array	Construction of high-density linkage map of a highly heterozygous full-sib diploid potato population and identification of QTL on chromosomes 2 and 10 for tuber shape and eye-depth	Prashar et al. (2014)
33.	Three F_1 populations	8.3 K SNP Potato Array	Development of allele dosages called software ClustalCall in tetraploid potato, using SNP array and identification of associated SNP for breeding	Carley et al. (2017)
34.	Tetraploid offspring and parents (190)	8.3 K SNP Potato Array	Development of high-density linkage maps and QTL mapping in tetraploid potato mapping population, using SNP genotyping	Hackett et al. (2013)

Table 5 contd. ...

...Table 5 contd.

SN	Genotype	Genotyping method	Objectives	References
35.	Tetraploid cultivars (83)	GBS	SNP genotyping of potato varieties for plant maturity and flesh color and identification of nucleotide diversity and genes/alleles under selection	Uitdewilligen et al. (2013)
36.	*S. tuebrosum* dihaploids (95)	Sequencing	Construction of high-density genetic maps and QTL mapping	Manrique-Carpintero et al. (2018)
37.	Diploid population (180)	DArT, AFLP, SSR and SNP markers (2469)	Creation of genetic and physical map with improvement in the ordering of the potato reference genome superscaffolds into pseudomolcules of the 12 chromosomes with anchor markers	Sharma et al. (2013)

4.3 High-density Genome Maps Using SNP Markers

4.3.1 Phenotypes and Tuber Traits

The cultivated potato (*Solanum tuberosum*) is tetraploid and highly heterozygous which leads to complex potato genome and hinders rapid development of new varieties. To understand the genetic nature of the crop, a gynogenic dihaploid was developed, followed by construction of a high-density genetic map using 12,753 SNPs (Manrique-Carpintero et al., 2018). The common QTLs were identified for tuber traits and plant phenotype (vigor and height) on chromosomes 2, 4, 7 and 10, whereas specific QTLs were observed for a number of inflorescence per plant and tuber shape on chromosomes 4, 6, 10 and 11. The study found that simplex rather than duplex were mainly associated with traits. The study observed that the Q allele detected in one or two homologous chromosomes suggested importance of allelic dose effects and the presence of undesirable alleles in QTL region. A low level of heterozygosity is associated with lower agronomic fitness in potato. The study also constructed homologous chromosome haplotypes, revealing heterogeneity in the potato genome. They also suggested that deleterious alleles are associated in tetraploid potato which concur epistasis associations in markers and phenotype.

Development of high-density linkage map of tetraploid potato is the major challenge. Prashar et al. (2014) generated a dense genetic map of a highly heterozygous full-sib diploid potato population of 186 individuals, using 8.3 K SNP potato array. The diploid population was developed by cross between two highly heterozygous diploid potato clones [HB171(13) × 99FT1b5]; both are a result of cross between diploids of *Solanum tuberosum* Group Tuberosum and *Solanum tuberosum* Group Phureja. The map contains 1,355 distinct loci and 2,157 SNPs; 802 of which co-segregate with other markers. The diploid populations were phenotyped for tuber shape and eye depth over two seasons and QTLs were analyzed. The major QTL for tuber shape was detected on chromosomes 2 and 10, whereas major QTL for tuber eye depth was co-located with tuber shape on chromosome 10. Additionally, minor-effect QTLs were observed on three other chromosomes. The study concluded that tuber shape is determined very early in the tuber development process (Prashar et al., 2014) (Table 5).

4.3.2 SNP in Tetraploid Allelic Doses and Double Reduction Analysis

Increasing number of modern SNP genotyping platforms allow high-density linkage mapping in tetraploid potato. A large number of breeding population is required to achieve accurate markers associated with traits. Bourke et al. (2016) created linkage maps of 235 F_1 genotypes of tetraploid potato based on 20 K SNP potato array. With the assumption of bivalent pairing, recombination frequencies between all markers segregation types were estimated, which were used in clustering of markers and linkage groups creation. A total of 96 homologous groups were created with a total map length of 1061 cM of 12 chromosomes using 6910 SNP markers. This study unravelled the complexities in linkage map creation tetraploid potato and also other polyploid crops (Bourke et al., 2016).

The creation of high-density linkage maps of polyploid crop is a major issue. In general, an autopolyploid crop like potato shows tetrasomic inheritance and double reduction. In normal random bivalent pairing during meiosis results in straightforward estimation of recombination frequencies through which linkage maps are created. Bourke et al. (2015) constructed a high-density linkage map of tetraploid potato and used to identify double reduction in a biparental mapping population. The frequency of multivalents required to produce double reduction was determined through simulation approach. They also determined the effect of multivalents or preferential pairing between homologous chromosomes on linkage maps. The study suggests use of highly informative markers in map creation without any effect of multivalent or preferential pairing. Double reduction in potato increases with distance from the centromeres.

Modern sequencing technologies, SNP discovery and computation tools have enabled scientists to perform linkage analysis and construct high density linkage maps in autotetraploid potato. In the advanced methodologies, Hackett et al. (2013) applied allelic doses information for linkage analysis and QTL mapping in tetraploid potato mapping population of 190 offspring and parents, using 8.3 K SNP potato array. The allelic doses information was inferred from allelic intensity and high-density linkage maps were constructed for each 12 chromosomes. Finally, positions of 3,839 SNP from a total of 5,378 SNPs could be assigned on the chromosomes. They also suggest that this methodology could be applied in autotetraploid and other autopolyploids.

Carley et al. (2017) developed a new software to genotype tetraploid potato using SNP array data to establish relationship between signal intensity and allele dosage for each marker and calibrated in F1 populations. SNP genotyping and increasing allele dosages calling software have strengthened precision breeding in potato. They have developed a computation tool ‘ClusterCall’ in R package. Genotyping of three potato families with 8 K SNP array showed 5,729 SNP markers with ClustalCall compared to 5,325 with the software fitTetra. Further, ClustalCall predicted 5,218 SNPs in SolCAP diversity panel compared to 3,521 SNPs in another study. The study concludes that ClustalCall is an efficient and accurate method for autotetraploid genotype allele calling and facilitates identification of trait-associated SNP for potato breeding.

4.3.3 Processing Quality and Other Traits

Understanding of allelic doses information underlying QTLs associated with complex agronomic traits is essential in autotetraploid potato crop via linkage analysis and QTL mapping approaches. Massa et al. (2018) investigated GWAS, deploying 8.3 K SNP potato array (Infinium, Illumina) to explore the highly heritable and complex traits, like disease resistance. A F_1 tetraploid mapping population of 162 individuals was assessed over two years at two locations and QTLs were identified for traits like tuber sugar concentration, processing quality, vine maturity and other agronomic traits. Significant QTLs were detected for tuber glucose concentration and fry color on potato chromosomes 4, 5, 6, 10 and 11 explaining 24% and 46%, phenotypic variations, respectively. Interestingly, a major QTL on chromosome 10 associated with apoplastic invertase has been identified for fry color. Additionally, on chromosome 5 QTLs, minor effects have been identified with above two traits and various others, like vine maturity, growth habit, tuber shape, early blight resistance and Verticillium wilt resistance. Finally, this study identified favorable alleles and candidate SNPs for low reducing sugar accumulation for processing potatoes. Also, discovery of minor-effect QTLs suggests an independent genetic control of the traits in potato (Massa et al., 2018).

4.3.4 Disease Resistance

Nachtigall et al. (2018) mapped a novel major late blight (*Phytophthora infestans*) resistance gene in the Mexican diploid wild species, *S. pinnatisectum*, on chromosome 7. The BC_1 progenies [F1 = resistant x susceptible; BC_1 = F1 × susceptible parent] of 84 individuals of *S. pinnatisectum* were mapped, using 1,583 DArT/SSR markers. The segregation ratio of 1:1 of the BC1 progenies indicated single dominant resistance gene. The mapping covered 12 linkage groups with a total length of 1793.5 cM of the 12 chromosomes at an average length of 1.1 cM. The resistance locus was mapped on chromosome 7. Unlike previous identification of resistance genes (*Rpi*1 and *Rpi*2), the new locus was identified on the opposite arm of chromosome 7. This study suggests development of locus-specific markers for molecular breeding for late blight resistance in potato.

Broad spectrum of late blight resistance gene R8 from wild species *S. demissum* was cloned, based on previous mapping on the lower arm of chromosome 9. The gene R8 is homologous to the NB-LRR protein Sw-5 from potato. Fine mapping in a recombinant population and Bacterial Artificial Chromosome (BAC) library screening showed a BAC contig spanning 170 kb of the R8 haplotype and sequencing revealed a cluster of at least 10 *R* gene analogues (RGAs). Of the total, only one RGA provided late blight resistance in R8 and recognises Avr8 effector. The study showed that R8 is localised clearly at distinct clade along with the Sw-5 tospovirus R protein from tomato (Vossen et al., 2016).

A broad spectrum of resistance to late blight was characterized in the Mexican diploid wild species, *S. verrucosum*, applying gene enrichment technologies with the reference potato genome, namely target resistance genes (RenSeq) and single/low-copy number genes (Generic-mapping enrichment Sequencing; GenSeq) for SNP-based mapping through bulk segregant analysis. These approaches identified resistance gene *Rpiver1* to the distal end of chromosome 9 in potato and 64 informative SNPs were also identified, of which 61 were found on chromosome 9, covering 27 candidate genes located in the position between 45.9 to 60.9 Mb of the reference potato genome. These RenSeq and GenSeq-originated allele specific KASP markers were designed for the gene at 4.3 Mb interval on chromosome 9 for marker-assisted breeding in potato (Chen et al., 2018).

5. Genome-Wide Association Studies (GWAS)

5.1 Phenotypes and Yield-contributing Traits

5.1.1 Agronomic and Tuber Traits

Unlike the above, association mapping, also known as linkage disequilibrium mapping, was developed to study genetic disorders in humans (D'Hoop et al., 2008). Association mapping is a general approach to detect correlations between genotypic and phenotypic variations in a population based on the linkage disequilibrium principles. As compared to the classical linkage mapping, which requires the use of highly related individuals, such as full sibs, association mapping can exploit the properties of more complex populations with various degrees of relatedness. Association mapping is a method to identify genes or QTLs associated with phenotypic variation in natural populations. The method takes advantage of historical meiotic recombination events related by descent and Linkage Disequilibrium (LD) value (Flint-Garcia et al., 2003) in a population consisting of diverse germplasm, including cultivars, breeding clones and landraces. Association mapping uses association of markers and QTLs within LD and is linked in the ideal case. In this method, population structure is determined, based on the genetic markers using software like structure to determine Marker Trait Association (MTA). This method provides much higher levels of resolution for the genetic dissection of quantitative traits. It has several advantages, such as no need to develop segregating generations; a collection of various cultivars and breeding lines can be utilized for mapping studies and higher mapping population may be reached with many more meiotic recombination.

In potato, tetraploid or diploid potatoes have been utilized for association mapping for desirable agronomic traits and biotic stresses. Gebhardt et al. (2004), for the first time, used association mapping in tetraploid potato germplasm to identify markers for late blight resistance and maturity traits in 600 potato cultivars. Further, association mapping was applied, based on candidate genes for resistance against *Verticillium dahliae* (Simko et al., 2004) and *Phytophthora infestans* (Pajerowska-Mukhtar et al., 2009). Genome Wide Association Mapping (GWAS) was deployed to identify genes/QTLs at whole genome level in potato by D'hoop et al. (2008). Genotyping by sequencing (GBS) is effectively used for SNP discovery and trait association mapping for multiple traits including biotic stresses (Uitdewilligen et al., 2013; Sharma et al., 2018). In view of the advantages and applications of association mapping, it can be applied in potato to develop molecular markers for drought tolerance. The feasibility of association mapping in tetraploid potato is to identify QTLs for agronomic important traits, like plant maturity and quality traits (D'Hoop et al., 2008), allele diversity (Kloosterman et al., 2013), and fry color (Byrne et al., 2020) using diverse potato populations. Berdugo-Cely et al. (2021) analyzed 809 andigenum group accessions from the Colombian Central Collections (CCC), using SNP markers. They revealed that CCC is a highly diverse germplasm collection genetically and phenotypically and useful to implement association mapping in order to identify genes related to traits of interest and to assist future potato genetic breeding programs. Genome-wide association studies are usually applied to large collections of theoretically unrelated individuals; the genetic diversity is supposed to be high and new alleles can be discovered. Furthermore, the high number of ancestral meiosis that occurred in the GWAS population can allow a precise QTL mapping. GWAS are greatly useful in diverse germplasm which offer new perspectives towards the discovery of new genes and alleles, especially for complex traits like abiotic stresses in plants. GWAS has been applied in potato for late blight resistance, maturity, *Verticilium* resistance, yield, tuber quality, chips, starch, tuber bruising, tuber shape, etc. GWAS has advantages like better mapping due to a large number of meiosis events and alleles per locus can be found in diverse population, unlike linkage mapping where recombination is defined in a single mapping population.

Vos et al. (2017) determined the number of SNP required for QTL discovery, depending on the genetic distance and LD decay in a population. In this study, LD and its decay was analyzed in tetraploid potato population of 537 genotypes by genotyping with 20 K SNP potato array. They explored the number of haplotypes and haplotype-specific SNPs in the population. Many estimators, such as average/median/percentile (r^2), were used for LD, whereas the estimator LD½, 90 was chosen to evaluate LD decay in 537 tetraploid varieties. This study demonstrates that several factors determine LD decay value in a population, which need to be analyzed with great care.

5.1.2 Tuber Yield and Starch Content

Potato tuber yield and starch content are one of the important complex traits for breeding new varieties, influenced by multiple factors which are genetic and environmental. Uncovering allelic variation underlying the traits would allow the identification of diagnostic molecular markers closely associated with the traits. This will facilitate clonal selection at an early stage with precision and speed the breeding process to increase tuber yield and starch content. A total of 448 tetraploid potato clones (varieties and breeding lines) were evaluated for tuber yield, tuber starch content and starch yield. These clones were genotyped by Restriction-site Associated DNA sequencing (RADseq) and the 8.3 K SNP Potato Array (SolCAP). Based on the filtering criteria, RADseq identified 450 to 6,664 genes for different traits, whilst differential SNPs were identified in 275 genes with 8.3 K SolCAP array. GWAS study identified SNP associated with starch content in 117 genes. Physical mapping with both the technologies identified several QTLs for all three traits on all 12 potato chromosomes. This study suggests that thousands of genes control tuber yield and starch content in potato and are distributed unequally and interlinked in the genome (Schönhals et al., 2017).

5.1.3 Plant Maturity and Tuber Flesh Color

Genomic DNA sequence variation and SNP calling in autotetraploid potato is a challenging task. It is important to reach sufficient read depth at genome level, especially to distinguish five different allelic combinations in tetraploid potato. Uitdewilligen et al. (2013) applied Genotype By Sequencing (GBS) in 83 tetraploid potato varieties. The methodology of enriching cultivar-specific DNA sequencing libraries using an in-solution hybridization method (SureSelect) allowed to confine the study to 807 target genes distributed across the genome in 83 cultivars and the reference potato (DM). A total of 1,29,156 sequence variants was detected and allele copy number for each cultivar was obtained. One SNP per 24 bp in exons and 1 SNP per 15 bp in intron regions were obtained in the cultivars, whereas average Minor Allele Frequency (MAF) of a variant was 0.14. In potato cultivars, a large number of rare variants/haplotypes of approximately 61% having MAF < 0.05 and a very high nucleotide diversity and genes under selection were observed. Study also observed marker-trait association for alleles strongly associated with maturity and tuber flesh color (Uitdewilligen et al., 2013).

5.2 Germplasm Diversity and Population Structure

5.2.1 Columbian Germplasm

Information about genetic diversity, structure and Linkage Disequilibrium (LD) and duplicates in a germplasm and breeding collection is very much essential for conservation and maintenance of true-to-type genotypes and breeding strategies in clonally propagated potato crop. Analysis of a total of 144 genotypes showed 57.2% polymorphic SNP markers to establish and differentiate two and three sub-populations of diploid and tetraploid genotypes of potato. They revealed a high level of heterozygosity and diversity in tetraploids, whereas higher LD was observed in diploid sub-populations. The Peruvian tetraploid potatoes had greater diversity and lower LD than those from Columbian genotypes, which had higher degrees of LD. Finally, the study suggests use of SNP genotyping for identification of germplasm on the basis of ploidy, duplicates and diversity analysis for potential use in breeding programs (Berdugo-Cely et al., 2021).

5.2.2 The USA Germplasm

Knowledge about genetic diversity and population structure of germplasm and breeding materials is very essential for any breeding program. Pandey and associates (2021) analyzed 214 advance breeding materials collected over 40 years in the Texas A&M University Potato Breeding Program (USA) and whose genetic makeup and usefulness was unknown so far. A total of 214 tetraploid genotypes were genotyped with the 22 K V3 SNP Potato Array (Infinium Illumina), of which analysis of 10,106 polymorphic SNPs indicated high heterozygosity (0.59) in the genotypes. STRUCURE analysis clearly classified tetraploid clones into three groups—as groups based on market classes and confirmed by principal component analysis.

SNPs were uncovered with the candidate genes controlling tuber flesh and skin color, plant cycle length, tuberization and carbohydrate metabolism in the clones. Core collection of 43 genotypes was developed using Core Hunter software retaining maximum genetic diversity of the whole population and avoiding redundancy for long-term conservation. This analysis suggested that SNP genotyping provides information for management of germplasm collection and breeding lines, selection of diverse parents for breeding programs and genetic fidelity/DNA fingerprinting of clones (Pandey et al., 2021).

Understanding genetic architecture is very important for breeding of any crop determining genomic regions associated with various traits, particularly tuber yield and quality of potato. Zia et al. (2020) performed GWAS analysis, using 12 K SNP potato array (SolCAP) in a panel of 237 tetraploid potato genotypes evaluated for two consecutive years for various morphological and agronomic traits. Population structure and diversity analysis clearly divided these genotypes into four groups, based on origin. Further, marker trait association analysis was carried out, based on Mixed Linear Model using TASSEL software and identified 36 genomic regions in potato. The mean significant LD was 47.5%, reflecting breeding history of the genotypes. These findings provide useful information for strengthening potato breeding programs to enhance production and overcome the challenges.

5.2.3 The International Potato Centre (CIP) Germplasm

Maintenance of genetic purity of clonally propagated potato crops is the most important aspect of germplasm management. A panel of 250 cultivated potato collections consisting of diploid, triploid and tetraploid accessions represented seven cultivated potato taxa. Original mother plants (field-grown) and *in vitro* counterparts of 250 accessions were analyzed using the Infinium 12 K V2 SNP Potato Array. Ellis et al. (2018) fingerprinted these accessions to confirm the genetic fidelity of these accessions and to determine genetic diversity, relatedness and population structure. Mostly, *in vitro* and field-grown mother plants of the same accession were grouped together. However, 11 (4.4%) accessions showed mismatch and in some cases, mixed identity based on SNP was observed in the accessions. SNP genotyping data was used to assess genetic diversity, inter- and intra-specific associations, population structure and hybrid origins. Phylogenetic analysis indicated that the triploids included in this study were genetically similar. The study showed some mismatch in classification of the accessions based on taxonomy and ploidy level. STRUCTURE analysis showed six populations with significant gene flow among the populations, as well as hybrid taxa and accessions. Collectively, SNP genotyping confirmed the genetic identity of the accessions and highlighted diversity in these CIP collections. This study highlights importance of maintenance of genetic identity of potato collection, using SNP genotyping in germplasm genebanks for breeding and other applications.

5.2.4 European Germplasm

Vos et al. (2015) developed the Infinium 20 K SNP array (named SolSTW array) consisting of non-redundant 15,138 SNPs identified previously, and 4,454 SNPs from SolCAP. They used 569 potato genotypes consisting of various segments (marker classes, breeding history, and origin form different countries) and data was analyzed, using the software 'fitTetra' designed for autotetraploids. The study analyzed genetic variations contributed through allelic introgression in breeding the cultivars. SNP markers linked to R genes were identified in the cultivars especially bred after 1945. The study showed that 96% of the genetic variants in ancestral cultivars (pre-1945) were present in the modern cultivars (post-1945). The analysis identified footprints of the breeding history in contemporary breeding material, such as identification of introgression segments, selection and founder signatures. This 20 K SNP array is a useful resource for characterization and identification of trait-specific SNP markers for breeding (Vos et al., 2015).

Conventional breeding methods rely upon phenotypic evaluation and clonal selection, with a few interventions of marker-assisted selection for disease-resistance genes. Advancement in next-generation genome sequencing technologies allows generation of genome sequencing data for a large number of genotypes. This will help to develop diagnostic markers for potato breeders to facilitate precision breeding for agronomic traits. Hamilton et al. (2011) sequenced cDNA from three elite potato cultivars—'Atlantic', 'Premier Russet' and 'Snowden' to generate a large number of SNPs. As a result of data filtering and quality check, a total of 5,75,340 SNPs were detected in these cultivars. Additionally, 2,358 SNPs were identified in other varieties (Bintje, Kennebec and Shepody) from available Sanger sequencing data. Finally, 69,011 SNPs of high confidence level were identified in all the six cultivars after reference genome mapping used for the Infinium genotyping platform. As many as 96 SNPs were used for allelic diversity analysis in 248 potato germplasm lines, of which 82 were found informative for further analysis. This clearly distinguished chip processing potatoes from other classes. Collectively, the SNPs identified in this study will pave pathways for high-throughput genotyping of potato germplasm and breeding lines to enable efficient molecular breeding.

5.3 Population Structure Based on Genome/Transcriptome Sequence Data

The cultivated potato (*Solanum tuberosum*) was domesticated approximately 10,000 years ago in the Andes mountains of southern Peru and include over 100 tuber-bearing *Solanum* species. Hardigan et al. (2017) analyzed diverse Solanum germplasm collection of cultivars, wild and landraces in 67 accessions of diploids and tetraploids representing *S. tuberosum* Groups Phureja, Stenotomum, Angigena, Chilotanum, Tuberosum and wild species to assess genomic diversity and impact on domestication of key loci for various traits. A great genetic diversity was observed in diploid and tetraploid of *S. tuberosum*. They identified 2,622 genes under selection, where 14–16% was shared by North American and Andean cultivars, indicating limited improvement in the early days. Adaptation of upland (*S. tuberosum* group Andigena) and lowland (*S. tuberosum* groups, Chilotanum and Tuberosum) populations targeted distinct loci. This study uncovered genes regulating carbohydrate metabolism, glycoalkaloid biosynthesis, the shikimate pathway, the cell cycle and circadian rhythm. Further, haplotype diversity at plant maturity locus StCDF1 showed introgression of alleles from wild *S. microdontum* in long-day varieties. Thus, this study uncovered a historic role of wild Solanum species in diversification of cultivated potatoes.

Wild species are an important source of genetic resources for potato breeding. Li et al. (2018) investigated genome wide diversity analysis in 201 Solanum accessions representing wild and cultivated types. Genomes of all 201 accessions were sequenced and in turn 64,87,006 high-quality SNPs were identified from 167 accessions (146 wild and 21 cultivated diploids) having broad geographical distribution. As expected, genome-wide sequence variation was observed in wild species, particularly in disease-resistance genes than in cultivated species. QTL analysis identified 609 genes under selection, including genes regulating potato tuberization process and loss of bitterness in tubers. Further phylogenetic analysis indicated genetic relationship of wild *S. brevicaule* and *S. candolleanum* with cultivated potato. Genome analysis of *S. candolleanum*, progenitor of cultivated potato, identified loss of 529 genes in cultivated potato. This study provides valuable resources for SNPs identification along with candidate genes for agronomically important traits.

Increasing potato production to meet market and consumer demands with rising population is an important challenge. The current breeding methodology involves a long process of hybridization, phenotypic evaluation over the years and clonal selection. Little emphasis has been given on marker-assisted selection for disease resistance but less on traits of agronomic importance. Hence, increasing information of genome sequencing technologies and genotyping platforms allows discovery of SNP markers for precision breeding. Caruana et al. (2019) applied high-throughput transcriptome sequencing for SNP discovery for genome-wide selection in potato. A collection of 181 tetraploid cultivars were transcriptome sequenced. Following the standard bioinformatics pipeline, a large number of SNPs were identified. A subset of high quality SNPs with wide genome coverage was identified and demonstrated in genomic prediction for highly heritable traits and further application in genomic selection in potato. This genome-wide discovery of SNP markers will provide significant benefits in rapid breeding efforts.

5.4 Disease Resistance

5.4.1 Potato Cyst Nematode

The potato root cyst nematode *Globodera pallida* is a major limiting factor in potato. The PCR-based diagnostic HC marker, PCR-based, has been used mostly for marker-assisted breeding for nematode resistance in potato breeding. This is tightly linked to the gene conferring high resistance to *G. pallida* pathotype Pa2/3, which has been introgressed into cultivated *S. tuberosum* background from wild species *S. vernei*. Additionally, the major QTL controlling nematode resistance has been mapped on chromosome 5 in a hot spot region of pathogens, including late blight. A set of selected 79 tetraploid varieties and breeding clones with presence (41)/absence (38) of the HC marker was genotyped using 100 SNPs distributed at 10 loci over 38 cM on chromosome 5. Finally, based on LD mapping between SNP markers, six LD groups containing two to 18 SNPs were identified. Thus study indicated existence of multiple alleles and associated SNP markers at a single resistance locus or many loci in the hot spot resistance region, for example, 18 SNPs were found to be associated with the HC marker (Achenbach et al., 2009).

A tetraploid mapping population (F_1) of cross between a susceptible potato cv. Picasso and resistant clone P55/7 was used to map nematode resistance *H2* gene. The gene *H2* confers resistance to the *Pa*1 pathotype of potato cyst nematode *Globodera pallida* and derived from wild species of *S. multidissectum*. Genome reduction approaches, RenSeq and GenSeq, were applied to characterize the population through a bulk segregant analysis. These approaches identified sequence variation close to the *H2* gene on chromosome 5 at the top end side. Further, allele-specific KASP markers located the gene H2 at 4.7 Mb on the distal short arm of chromosome 5, corresponding to 1.4 MB and 6.1 MB of the reference potato genome (Strachan et al., 2019).

5.4.2 Potato Wart

Potato wart is a soil-borne disease caused by *Synchytrium endobioticum* and is managed by use of fungicides, resistant varieties and cultural practices. The use of molecular markers has increased considerably and is linked to *Sen1* gene in breeding varieties against this disease. Recently, Prodhomme et al. (2020) identified haplotype-specific diagnostic SNP markers associated with pathotype 1 (D1) wart disease-resistance breeding in potato. A genome-wide association study was conducted in a panel of 330 common potato varieties and identified haplotype-specific SNP markers on chromosome 11 at the location of *Sen1* locus. The study emphasized the investigation of SNP markers through false positive and false negative analyses and validation in two full-sib populations, suggesting important steps for molecular breeding. The study confirms that *Sen1* gene appears to be the key source of wart (pathotype 1) resistance in these potato varieties. They suggest that the GWAS approach is instrumental in identifying haplotype-specific SNP markers for genomics-assisted breeding in potato.

Potato wart is an important pathogen causing considerable crop damage. Since chemical control is not effective and unfriendly to the environment, therefore it is a quarantine pathogen and also emergence of new pathotypes pose major problems in pathogen management. Therefore, use of resistant varieties is the most effective method, applying diagnostic markers to speed up the breeding cycles. A biparental progeny of 133 F_1 individuals were genotyped through 195 SSR and 8.3 K SNP array. A major resistance locus was identified as multi-allelic gene Sen1/RSe-XIa on potato chromosome 11 through linkage analysis conferring resistance to all four pathotypes 1, 2, 6 and 18. Additionally, seven minor independent loci with small effects were also identified effective in wart resistance. This study suggests that SNP markers close to the resistance loci could be used in molecular breeding to provide wart resistance in potato (Obidiegwu et al., 2015).

5.4.3 Late Blight

Wild species are one of the richest sources of late blight disease resistance. Molecular markers linked to the late blight resistance genes have played key roles in rapid potato breeding. F_1 populations of two wild species. *S. microdontum* and *S. pampasense* were used for mapping studies showing continuous variation of late blight resistance over the years in field testing. Both populations were used to create high density genetic maps using SNP markers. QTLs were mapped in both populations and consistently expressed over multiple years the populations. In *S. microdontum*-derived population,

QTLs were identified on chromosome 5, 6 and 10 to express consistently for three years and explained 21–47% of total phenotypic variation. In another *S. pampasense*-derived population, QTLs were found on chromosomes 11 and 12 for two years. Overall, R gene cluster and associated SNP markers were known in the populations, suggesting marker-assisted breeding strategies for breeding potato against late blight resistance in potato (Meade et al., 2020).

Late blight is one of the most devastating diseases of potato as it is quantitative and polygenic controlled and positively correlated with plant maturity. Although considerable knowledge is available on quantitative resistance to late blight but it is very limited at molecular level resistance with plant maturity. Hence, it is essential to develop diagnostic markers for Maturity Correlated Resistance (MCR) to late blight for efficient selection of improved varieties. A panel of 184 tetraploid varieties were genotyped, using nearly 9,000 SNPs from three different sources (8.3 K SNP potato array, candidate genes from the functions in jasmonate pathway, candidate genes from RNA-seq analysis of contrasting genotypes). GWAS analysis revealed 27 SNPs strongly associated with MCR. A few SNPs were located in the candidate gene functions involved in quantitative resistance, such as lipoxygenase (jasmonate pathway), 3-hydroxy-3-methylglutaryl coenzyme A reductase (mevalonate pathway), P450 protein (terpene biosynthesis), transcription factor and a homolog of *S. venturii*-derived resistance gene. Finally, this study suggests that candidate gene and SNP array-based genotyping methods allowed identification of novel resistance loci and strongly associated SNPs with MCR in potato for genomics-assisted breeding (Mosquera et al., 2016).

5.4.4 Common Scab

Potato common scab caused by the bacterium *Streptomyces scabies* is one of the most important diseases of potato. The use of resistant varieties is the common method to control this disease. Yuan et al. (2020) investigated genome-wide association studies, using 12 K SNP potato array (SolCAP) in a panel of 143 clones consisting of advanced breeding materials and commercial varieties and phenotypes of this disease over multiple years. GWAS identified three resistance QTLs on potato chromosomes 2, 4, and 12 using GWASpoly (R package), explaining 21%, 19% and 26%, respectively phenotypic variations. This study suggests that these findings would be useful in marker-assisted breeding for common scab resistance in potato.

5.5 *Processing Traits*

5.5.1 Tuber Starch Content

Potato starch is an important quality parameter, depending upon crystalline nature, morphology and other physical and chemical properties of starch granules. Starch properties are dependent on traits like starch yield, amylase/amylopectin and phosphorylation, which are controlled by genes. Khlestkin et al. (2020) examined genomic regions in potato associated with starch granule morphology, using SNP markers. A total of 90 potato varieties were investigated by using 22 K SNP array, of which, 15,214 scorable SNPs were used for genome-wide analysis and 53 SNPs were found significantly associated with starch morphology and yield component traits on chromosomes 1, 2, 4, 5, 6, 7, 9, 11 and 12. Eight SNPs on chromosome 1 and 19 SNPs on chromosome 2 and 37 SNPs were located in protein coding regions. The study noticed that starch phosphorylation process is important for granule shape and is regulated by the GWD gene-imparting key roles in starch phosphorylation-dephosphorylation and other processes. It also suggests that starch yield is a polygenic trait and many SNPs are associated in the potato genome (Khlestkin et al., 2020).

In another study, a total of 448 tetraploid potato clones (varieties and breeding lines) were evaluated for tuber yield, tuber starch content and starch yield. These clones were genotyped by restriction site associated DNA sequencing (RADseq) and the 8.3 K SNP Potato Array (SolCAP). Based on the filtering criteria, RADseq identified 450 to 6,664 genes for different traits, whilst differential SNPs were identified in 275 genes with 8.3 K SolCAP array. The GWAS study identified SNPs associated with starch content in 117 genes. Physical mapping with both the technologies identified several QTLs for all three traits on all 12 potato chromosomes. This study suggests that thousands of genes control tuber yield and starch content in potato and are distributed unequally and interlinked in the genome (Schönhals et al., 2017).

5.5.2 Starch Phosphate Content

Natural variation in potato starch phosphate content is available in literature. Phosphorylated starch is more stable at high temperatures and has more water-holding capacity than the raw starch. Therefore, the study focuses at improving phosphate content in starch by breeding methods or transgenic manipulations. Khlestkin et al. (2019) carried out GWAS study in 90 potato varieties, using 22 K SNP potato array to identify genomic regions associated with starch phosphorylation in potato, of which, a total of 15,214 scorable SNPs were identified. In that, 17 significant SNPs were associated with potato starch phosphorus content and 14 were assigned to eight genomic regions on chromosomes 1, 4, 5, 7, 8, 10 and 11 belonging to protein-coding regions. The Principal Component Analysis (PCA) showed that third and eighth components played significant roles in variation of starch phosphorous content in the varieties. The study suggests that validation of SNPs and discovery of more SNPs in surrounding genomic regions would be helpful in developing diagnostic SNPs to accelerate potato breeding for starch phosphorous content (Khlestkin et al., 2019).

5.5.3 Fry Color

Cold-induced sweetening-resistant genotypes are important in the processing industry for chips and fry making. Sprout suppressants are applied to stop sprouting in potato while storage at 10–12°C for table potato, while seed potato are stored at 4°C. For excellent fry color, potatoes are stored at above 8°C. Hence, genomics-assisted breeding is s novel approach to achieve this goal. Byrne et al. (2020) conducted GWAS study using SNP markers in a large collection of genotypes for genomic selection purposes. They identified a major QTL on chromosome 10 for fry color and predicted with moderate accuracy by using genomic markers. This result provides a subset of SNPs for processing traits with moderate predictability for genomic selection in potato.

5.5.4 Protein and Folate Content

Protein content is one of the important quality traits of tuber for value addition and now it is important for innovative breeders. Klaassen et al. (2019) estimated heritability and explored genetic relationship between protein content and tuber under water weight analyzed over multiple years and environments, A panel of 277 potato varieties was genotyped, using 20 K SNP potato array (SolSTW, Illumina) and GWAS analysis identified haplotypes underlying QTLs on chromosomes 3, 5, 7 and 12. Maturity-associated alleles of StCDF1 were co-localized with QTL on chromosome 5. These candidate genes underlying the QTL are relevant for genomics-aided breeding for enhancing protein content in potato.

Potato is the third most consumed human food crop of the world and important for food and nutritional security. It is a fact that micronutrients are essential for human health and therefore increasing the amount of micronutrients, particularly folate content in potato, could alleviate malnutrition, known as 'hidden hunger'. Bali et al. (2018) studied SNP genotyping by using 12 K SNP potato array (SolCAP) for folate content in 94 F_2 progenies of cross between diploid wild species, *S. boliviense* (PI 597736) clone Fol 1.6 and diploid *S. tuberosum* clone USW4self#3. A total of 6,759 high-quality SNPs, having 4,174 (62%) polymorphic SNPs, were used in marker-trait association analysis. 497 significant SNPs were identified as being located on all the chromosomes; of which, 18 SNPs were located within or close to folate metabolism-associated genes. These SNPs have the potential for genomics-enabled breeding for folate content in potato (Bali et al., 2018).

5.5.5 Flower Color

Genotyping by sequencing is a potential technology offering cost-effective SNP discovery and genotyping in tetraploid potato. Bastien et al. (2018) standardized a method of GBS protocols, mainly enzyme combinations (*Ape*KI, and *Pst*I/*Msp*I) in potato, and discovered SNP associated with important traits. Although, *Ape*KI yielded more markers than *Pst*I/*Msp*I but provided lower reads coverage per marker, resulting in more missing data and limiting effective genotyping in the tetraploid crop. Further, accuracy of these SNPs was determined by comparison with SolCAP data and it was found that the match rates between genotype calls was 90.4% and 81.3%, respectively. The quality of GBS data (with *Ape*KI) was assessed through GWAS analysis for flower color in 318 potato clones. The most significant SNP was identified in the *dihydroflavonol 4-reductase* (*DFR*) gene on chromosome 2. This study suggests selection of appropriate enzyme combinations for GBS library and application in genomics-assisted breeding in potato.

6. Conclusion

Potato is supposed to be the ideal crop for application of biotechnologies, like tissue culture and genetic engineering, molecular markers, marker assisted or genome-assisted technologies but these have least been exploited in the crop due to its poly ploidy nature and genome complexity. The increasingly available linked markers to single resistance genes and QTLs for yield as well as other agronomic traits and quality-processing traits may offer potential applications for MAS in

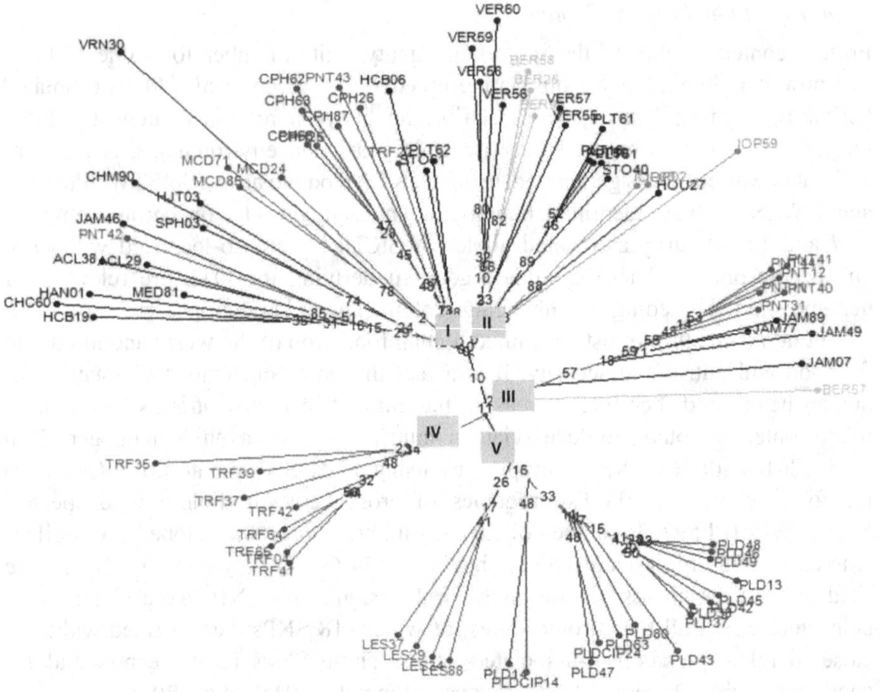

Fig. 1. Molecular diversity in wild potato species (82 accessions) using SSR markers (PGI kit) (*Source*: Tiwari et al., 2019).

developing new potato cultivars. Now genome-level studies offer tremendous opportunities for genome-wide association studies and genomic selection in this crop for targeted speedy breeding. The crop has the potential to be the staple food of the world and its production needs to be improved by applying various advanced molecular techniques to feed the ever-burgeoning population. Even though the cultivated potato is auto tetraploid, much of the genetic research is conducted at the diploid level. Hence, the development of experimental and computational methods for routine and informative high-resolution genetic characterization of polyploids remains an important goal for the realization of many of the potential benefits of the potato genome sequence.

Acknowledgement

I am thankful to the Director, ICAR-Central Potato Research Institute, Shimla, and scientists/technicians/research fellows and other colleagues of the institute for their support under the institute research projects on biotechnology, germplasm, breeding, and seed research on aeroponics. I am also grateful to the funding agencies for support under the externally funded projects (CABin, ICAR-IASRI, New Delhi; ICAR-LBS Young Scientist Award Project, and DBT, Government of India).

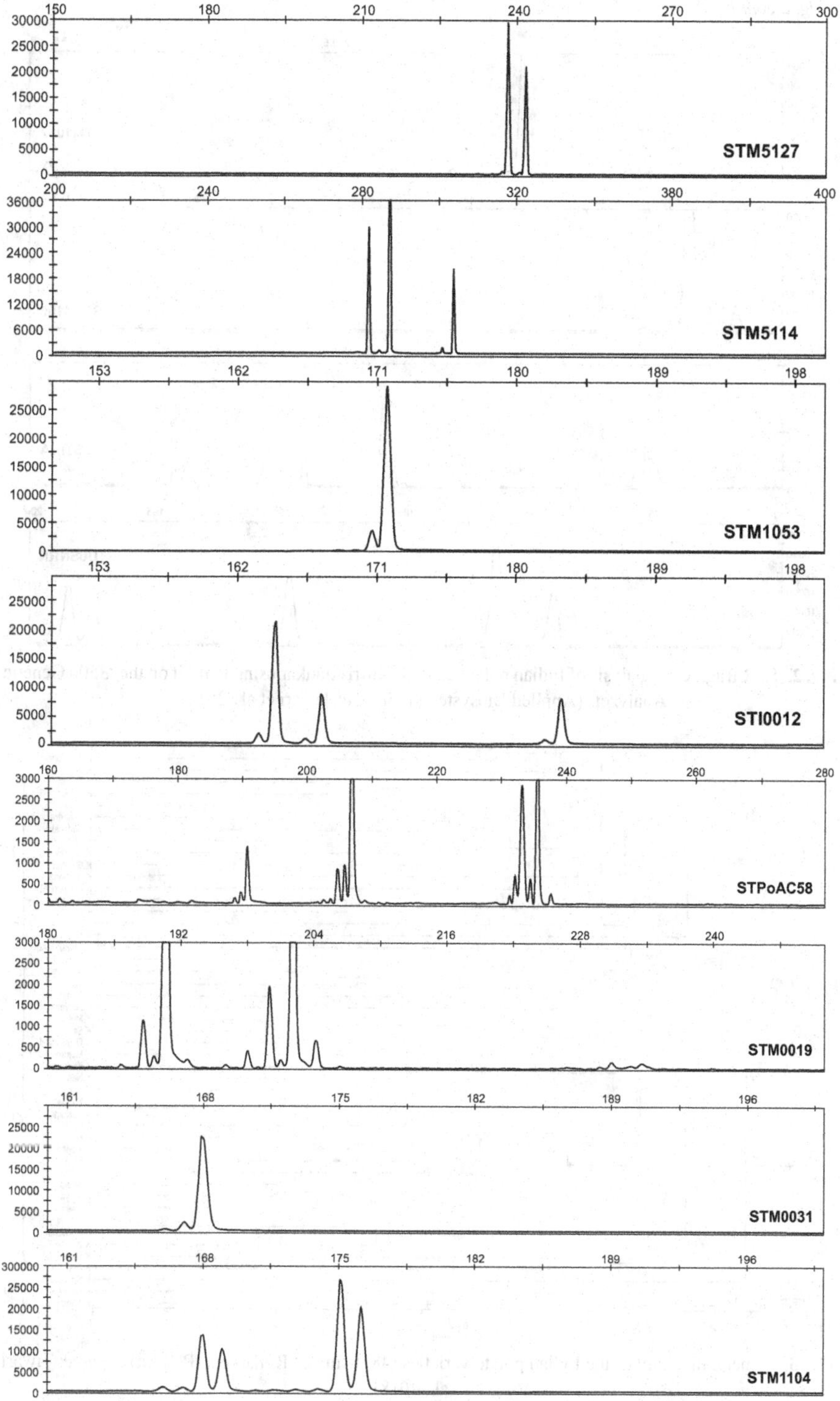
STM5127
STM5114
STM1053
STI0012
STPoAC58
STM0019
STM0031
STM1104

Fig. 2 contd. ...

...Fig. 2 contd.

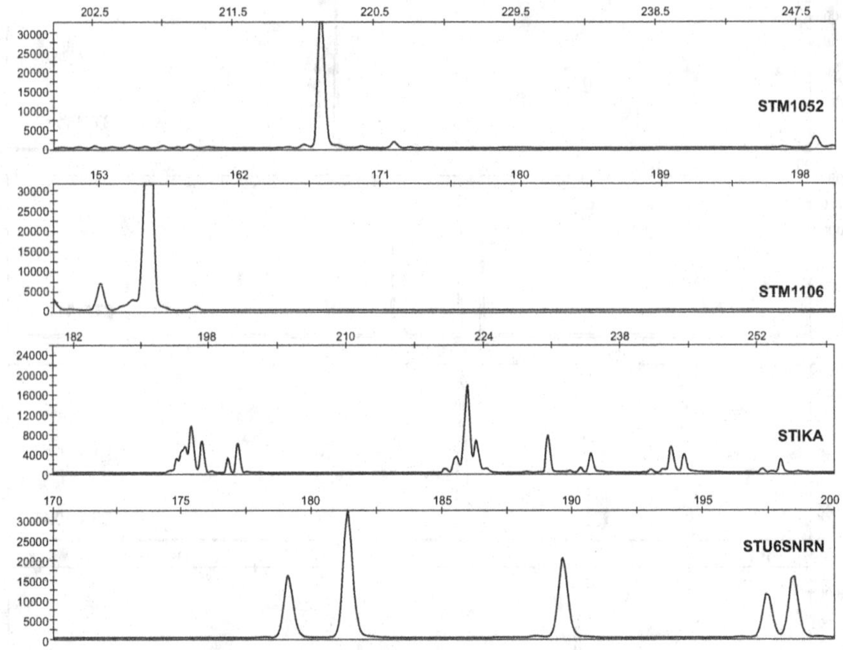

Fig. 2. SSR fragments analysis of Indian potato cultivar Kufri Alankar using PGI kit on the '3500 Genetic Analyzer' (Applied Biosystems) (*Source*: Tiwari et al., 2018).

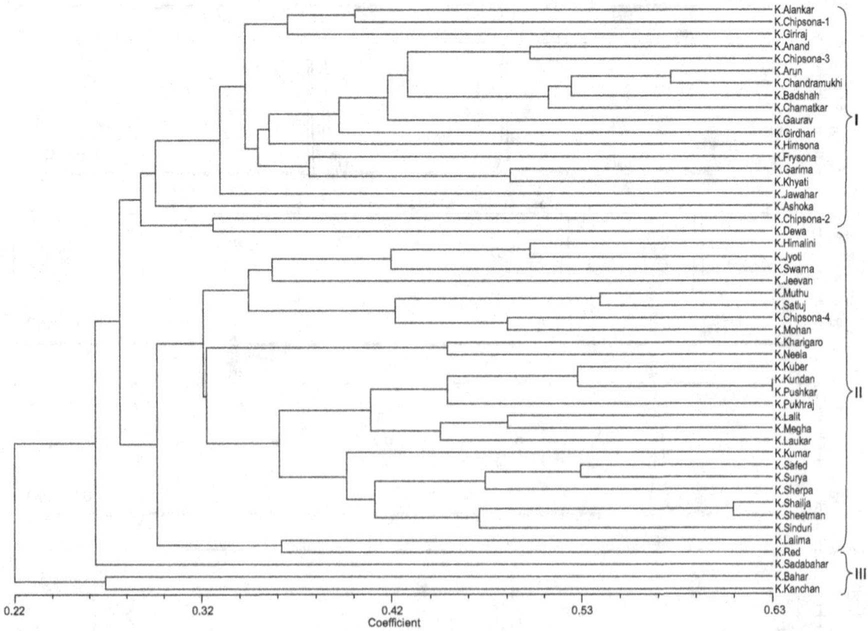

Fig. 3. Genetic diversity of the Indian potato varieties (48) using SSR markers (PGI kit) (*Source*: Tiwari et al., 2018).

References

Achenbach, U., Paulo, J., Ilarionova, E., Lübeck, J., Strahwald, J. et al. (2009). Using SNP markers to dissect linkage disequilibrium at a major quantitative trait locus for resistance to the potato cyst nematode *Globodera pallida* on potato chromosome V. *Theor. Appl. Genet.*, 118(3): 619–629.

Adhikari, S., Saha, S., Biswas, A., Rana, T.S., Bandyopadhyay, T.K. and Ghosh, P. (2017). Application of molecular markers in plant genome analysis: A review. *Nucleus*, Doi: 10.1007/s13237-017-0214-7.

Anisimova, I.N., Alpatieva, N.V., Karabitsina, Y.I. and Gavrilenko, T.A. (2019). Nucleotide sequence polymorphism in the RFL-PPR genes of potato. *J. Genet.*, 98: 87.

Anithakumari, A.M., Dolstra, O., Vosman, B., Visser, R.G.F. and van der Linden, G. (2011). *In vitro* screening and QTL analysis for drought tolerance in diploid potato. *Euphytica*, 181: 357–369.

Anithakumari, A.M., Nataraja, K.N., Visser, R.G. and van der Linden, C.G. (2012). Genetic dissection of drought tolerance and recovery potential by quantitative trait locus mapping of a diploid potato population. *Mol. Breed.*, 30: 1413–1429.

Armstrong, M.R., Vossen, J., Lim, T.Y., Hutten, R., Xu, J. et al. (2019). Tracking disease resistance deployment in potato breeding by enrichment sequencing. *Plant Biotechnol. J.*, 17(2): 540–549.

Bakker, E., Achenbach, U., Bakker, J., van Vliet, J., Peleman, J. et al. (2004). A high-resolution map of the *H1* locus harboring resistance to the potato cyst nematode *Globodera rostochiensis*. *Theor. Appl. Genet.*, 109: 146–152.

Bakker, E., Borm, T., Prins, P., van der Vossen, E., Uenk, G. et al. (2011). A genome-wide genetic map of NB-LRR disease resistance loci in potato. *Theor. Appl. Genet.*, 123: 493–508.

Bali, S., Robinson, B.R., Sathuvalli, V., Bamberg, J. and Goyer, A. (2018). Single Nucleotide Polymorphism (SNP) markers associated with high folate content in wild potato species. *PLoS ONE*, 13(2): e0193415.

Ballvora, A., Hesselbach, J., Niewöhner, J., Leister, D., Salamini, F. and Gebhardt, C. (1995). Marker enrichment and high-resolution map of the segment of potato chromosome VII harbouring the nematode resistance gene Gro1. *Mol. Gen. Genet.*, 249: 82–90.

Ballvora, A., Ercolano, M.R., Weiß, J., Meksem, K., Bormann, C.A. et al. (2002). The *R1* gene for potato resistance to late blight (*Phytophthora infestans*) belongs to the leucine zipper/NBS/LRR class of plant resistance genes. *Plant J.*, 30: 361–371.

Barrell, P.J., Meiyalaghan, S., Jacobs, J.M. and Conner, A.J. (2013). Applications of biotechnology and genomics in potato improvement. *Plant Biotechnol. J.*, 11(8): 907–920.

Bastien, M., Boudhrioua, C., Fortin, G. and Belzile, F. (2018). Exploring the potential and limitations of genotyping-by-sequencing for SNP discovery and genotyping in tetraploid potato. *Genome*, 61(6): 449–456.

Bendahmane, A., Kanyuka, K. and Baulcombe, D.C. (1997). High-resolution genetical and physical mapping of the *Rx* gene for extreme resistance to potato virus X in tetraploid potato. *Theor. Appl. Genet.*, 95: 153–162.

Berdugo-Cely, J.A., Martínez-Moncayo, C. and Lagos-Burbano, T.C. (2021). Genetic analysis of a potato (*Solanum tuberosum* L.) breeding collection for southern Colombia using single nucleotide polymorphism (SNP) markers. *PLoS ONE*, 16(3): e0248787.

Bethke, P.C., Halterman, D.A. and Jansky, S.H. (2019). Potato germplasm enhancement enters the genomics era. *Agronomy*, 9: 575.

Bonierbale, M.W., Plaisted, R.L. and Tanksley, S.D. (1988). RFLP maps based on a common set of clones reveal modes of chromosomal evolution in potato and tomato. *Genetics*, 120: 1095–1103.

Bormann, C.A., Rickert, A.M., Ruiz, R.A.C., Paal, J., Lübeck, J. et al. (2004). Tagging quantitative trait loci for maturity-corrected late blight resistance in tetraploid potato with PCR-based candidate gene markers. *Mol. Plant-Microbe Inter.*, 17: 1126–1138.

Bourke, P.M., Voorrips, R.E., Visser, R.G. and Maliepaard, C. (2015). The double-reduction landscape in tetraploid potato as revealed by a high-density linkage map. *Genetics*, 201(3): 853–863.

Bourke, P.M., Voorrips, R.E., Kranenburg, T., Jansen, J., Visser, R.G. and Maliepaard, C. (2016). Integrating haplotype-specific linkage maps in tetraploid species using SNP markers. *Theor. Appl. Genet.*, 129(11): 2211–2226.

Bradshaw, J.E., Hackett, C.A., Pande, B., Waugh, R. and Bryan, G.J. (2008). QTL mapping of yield, agronomic and quality traits in tetraploid potato (*Solanum tuberosum* subsp. *tuberosum*). *Theor. Appl. Genet.*, 116: 193–211.

Brigneti, G., Garcia-Mas, J. and Baulcombe, D.C. (1997). Molecular mapping of the potato virus Y resistance gene Ry_{sto} in potato. *Theor. Appl. Genet.*, 94: 198–203.

Bryan, G.J., McLean, K., Bradshaw, J.E., De Jong, W.S., Phillips, M. et al. (2002). Mapping QTLs for resistance to the cyst nematode *Globodera pallida* derived from the wild potato species *Solanum vernei*. *Theor. Appl. Genet.*, 105: 68–77.

Bryan, G.J., McLean, K., Pande, B., Purvis, A., Hackett, C.A. et al. (2004). Genetical dissection of H3-mediated polygenic PCN resistance in a heterozygous autotetraploid potato population. *Mol. Breed.*, 14: 105–116.

Byrne, S., Meade, F., Mesiti, F., Griffin, D., Kennedy, C. and Milbourne, D. (2020). Genome-wide association and genomic prediction for fry color in potato. *Agronomy*, 10: 90.

Carley, C.A.S., Coombs, J.J., Douches, D.S., Bethke, P.C., Palta, J.P. et al. (2017). Automated tetraploid genotype calling by hierarchical clustering. *Theor. Appl. Genet.*, 130: 717–726.

Caruana, B.M., Pembleton, L.W., Constable, F., Rodoni, B., Slater, A.T. and Cogan, N. (2019). Validation of genotyping by sequencing using transcriptomics for diversity and application of genomic selection in tetraploid potato. *Front. Plant Sci.*, 10: 670.

Celebi-Toprak, F., Slack, S.A. and Jahn, M.M. (2002). A new gene, Ny_{tbr}, for hypersensitivity to Potato virus Y from *Solanum tuberosum* maps to chromosome IV. *Theor. Appl. Genet.*, 104: 669–674.

Cernák, I., Decsi, K., Nagy, S., Wolf, I., Polgár, Z. et al. (2008). Development of a locus-specific marker and localization of the Ry_{sto} gene based on linkage to a catalase gene on chromosome XII in the tetraploid potato genome. *Breed. Sci.*, 58: 309–314.

Chen, X., Lewandowska, D., Armstrong, M.R., Baker, K., Lim, T.Y. et al. (2018). Identification and rapid mapping of a gene conferring broad-spectrum late blight resistance in the diploid potato species *Solanum verrucosum* through DNA capture technologies. *Theor. Appl. Genet.*, 131(6): 1287–1297.

Colton, L.M., Groza, H.I., Wielgus, S.M. and Jiang, J. (2006). Marker-assisted selection for the broad-spectrum potato late blight resistance conferred by gene *RB* derived from a wild potato species. *Crop Sci.*, 46: 589–594.

D'Hoop, B.B., Paulo, M.J., Mank, R.A., van Eck, H.J. and van Eeuwijk, F.A. (2008). Association mapping of quality traits in potato (*Solanum tuberosum* L.). *Euphytica*, 161: 47–60.

Dahal, K., Li, X.Q., Tai, H., Creelman, A. and Bizimungu, B. (2019). Improving potato stress tolerance and tuber yield under a climate change scenario—A current overview. *Front. Plant Sci.*, 10: 563.

de Boer, J.M., Borm, T.J.A., Jesse, T., Brugman, B., Tang, X. et al. (2011). A hybrid BAC physical map of potato: A framework for sequencing a heterozygous genome. *BMC Genomic*, 12: 594.

De Jong, W., Forsyth, A., Leister, D., Gebhardt, C. and Baulcombe, D.C. (1997). A potato hypersensitive resistance gene against potato virus X maps to a resistance gene cluster on chromosome 5. *Theor. Appl. Genet.*, 95: 246–252.

De Koeyer, D., Chen, H. and Gustafson, V. (2011). Molecular breeding for potato improvement. pp. 41–67. *In*: Bradeen, J.M. and Kole, C. (eds.). *Genetics, Genomics and Breeding of Potato*. Science Publishers, Enfield, New Hampshire.

Dong, F., Song, J., Naess, S.K., Helgeson, J.P., Gebhardt, C. and Jiang, J. (2000). Development and applications of a set of chromosome-specific cytogenetic DNA markers in potato. *Theor. Appl. Genet.*, 101(7): 1001–1007.

El-Kharbotly, A., Leonards-Schippers, C., Huigen, D.J., Jacobsen, E., Pereira, A. et al. (1994). Segregation analysis and RFLP mapping of the *R1* and *R3* alleles conferring race-specific resistance to *Phytophthora infestans* in progeny of dihaploid potato parents. *Mol. Gen. Genet.*, 242: 749–754.

El-Kharbotly, A., Jacobs, J.M., Hekkert, B.T., Stiekema, W.J., Pereira, A. et al. (1996). Localization of Ds-transposon containing T-DNA inserts in the diploid transgenic potato: Linkage to the *R1* resistance gene against *Phytophthora infestans* (Mont.) de Bary. *Genome*, National Research Council, Canada, 39: 249–257.

Ellis, D., Chavez, O., Coombs, J., Soto, J., Gomez, R. et al. (2018). Genetic identity in genebanks: application of the SolCAP 12K SNP array in fingerprinting and diversity analysis in the global in trust potato collection. *Genome*, 61(7): 523–537.

Ewing, E.E., Simko, I., Smart, C.D., Bonierbale, M.W., Mizubuti, E.S.G. et al. (2000). Genetic mapping from field tests of qualitative and quantitative resistance to *Phytophthora infestans* in a population derived from *Solanum tuberosum* and *Solanum berthaultii*. *Mol. Breed.*, 5: 25–36.

Feingold, S., Lloyd, J., Norero, N., Bonierbale, M.W. and Lorenzen, J. (2005). Mapping and characterization of new EST-derived microsatellites for potato (*Solanum tuberosum* L.). *Theor. Appl. Genet.*, 111: 456–466.

Finkers-Tomczak, A., Danan, S., van Dijk, T., Beyene, A., Bouwman, L. et al. (2009). A high-resolution map of the *Grp1* locus on chromosome V of potato harbouring broad-spectrum resistance to the cyst nematode species *Globodera pallida* and *Globodera rostochiensis*. *Theor. Appl. Genet.*, 119: 165–173.

Flint-Garcia, S.A., Thornsberry, J.M. and Buckler, E.S. (2003). Structure of linkage disequilibrium in plants. *Annu. Rev. Plant Biol.*, 54: 357–374.

Flis, B., Hennig, J., Strzelczyk-Zyta, D., Gebhardt, C. and Marczewski, W. (2005). The *Ry-f_{sto}* gene from *Solanum stoloniferum* for extreme resistant to Potato virus Y maps to potato chromosome XII and is diagnosed by PCR marker $GP122_{718}$ in PVY resistant potato cultivars. *Mol. Breed.*, 15: 95–101.

Fulton, T.M., van der Hoeven, R., Eanneta, N.T. and Tanksley, S.D. (2002). Identification, analysis, and utilization of conserved ortholog set markers for comparative genomics in higher plants. *Plant Cell*, 14: 1457–1467.

Galek, R., Rurek, M., De Jong, W.S., Pietkiewicz, G., Augustyniak, H. and Sawicka-Sienkiewicz, E. (2011). Application of DNA markers linked to the potato *H1* gene conferring resistance to pathotype *Ro1* of *Globodera rostochiensis*. *J. Appl. Genet.*, 52: 407–411.

Gebhardt, C., Ritter, E., Debener, T., Schachtchabel, U. and Walkemeier, B. (1989). RFLP analysis and linkage mapping in *Solanum tuberosum*. *Theor. Appl. Genet.*, 78: 65–75.

Gebhardt, C., Ritter, E., Barone, A., Debener, T., Walkemeier, B. et al. (1991). RFLP maps of potato and their alignment with the homoeologous tomato genome. *Theor. Appl. Genet.*, 83: 49–57.

Gebhardt, C., Mugniery, D., Ritter, E., Salamini, F. and Bonnel, E. (1993). Identification of RFLP markers closely linked to the H1 gene conferring resistance to *Globodera rostochiensis* in potato. *Theor. Appl. Genet.*, 85: 541–544.

Gebhardt, C., Ballvora, A., Walkemeier, B., Oberhagemann, P. and Schüler, K. (2004). Assessing genetic potential in germplasm collections of crop plants by marker-trait association: A case study for potatoes with quantitative variation of resistance to late blight and maturity type. *Mol. Breed.*, 13: 93–102.

Gebhardt, C., Bellin, D., Henselewski, H., Lehmann, W., Schwarzfischer, J. and Valkonen, J. (2006). Marker-assisted combination of major genes for pathogen resistance in potato. *Theor. Appl. Genet.*, 112: 1458–1464.

Gebhardt, C. (2011). Population genetics and association mapping. pp. 133–152. *In*: Bradeen, J.M. and Kole, C. (eds.). *Genetics, Genomics and Breeding of Potato*. Science Publishers, Enfield, New Hampshire.

Ghislain, M., Spooner, D.M., Rodríguez, F., Villamón, F., Núñez, J. et al. (2004). Selection of highly informative and user-friendly microsatellites (SSRs) for genotyping of cultivated potato. *Theor. Appl. Genet.*, 108: 881–890.

Hackett, C.A., McLean, K. and Bryan, G.J. (2013). Linkage analysis and QTL mapping using SNP dosage data in a tetraploid potato mapping population. *PLoS ONE*, 8(5): e63939.

Hämäläinen, J.H., Watanabe, K.N., Valkonen, J.P.T., Arihara, A., Plaisted, R.L. et al. (1997). Mapping and marker assisted selection for a gene for extreme resistance to potato virus Y. *Theor. Appl. Genet.*, 94: 192–197.

Hämäläinen, J.H., Sorri, V.A., Watanabe, K.N., Gebhardt, C. and Valkonen, J.P.T. (1998). Molecular examination of a chromosome region that controls resistance to potato Y and A potyviruses in potato. *Theor. Appl. Genet.*, 96: 1036–1043.

Hamilton, J.P., Hansey, C.N., Whitty, B.R., Stoffel, K., Massa, A.N. et al. (2011). Single nucleotide polymorphism discovery in elite North American potato germplasm. *BMC Genomics*, 12: 302.

Hardigan, M.A., Crisovan, E., Hamilton, J.P., Kim, J., Laimbeer, P. et al. (2016). Genome reduction uncovers a large dispensable genome and adaptive role for copy number variation in asexually propagated *Solanum tuberosum*. *The Plant Cell*, 28(2): 388–405.

Hardigan, M.A., Laimbeer, F., Newton, L., Crisovan, E., Hamilton, J.P. et al. (2017). Genome diversity of tuber-bearing Solanum uncovers complex evolutionary history and targets of domestication in the cultivated potato. *Proc. Natl. Acad. Sci., U.S.A.*, 114(46): E9999–E10008.

He, J., Zhao, X., Laroche, A., Lu, Z.-X., Liu, H. and Li, Z. (2014). Genotyping-by-sequencing (GBS), an ultimate marker-assisted selection (MAS) tool to accelerate plant breeding. *Front. Plant Sci.*, 5: 484.

Hosaka, K. and Hanneman Jr., R.E. (1998). Genetics of self-compatibility in a self-incompatible wild diploid potato species *Solanum chacoense*. 2. Localization of an S locus inhibitor (*Sli*) gene on the potato genome using DNA markers. *Euphytica*, 103: 265–271.

Huang, B., Spooner, D.M. and Liang, Q. (2018). Genome diversity of the potato. *Proc. Natl. Acad. Sci., U.S.A.*, 115(28): E6392–E6393.

Jacobs, J.M.E., Eck, H.J., Horsman, K., Arens, P.F.P., Verkerk-Bakker, B. et al. (1996). Mapping of resistance to the potato cyst nematode *Globodera rostochiensis* from the wild potato species *Solanum vernei*. *Mol. Breed.*, 2: 51–60.

Jacobs, J.M., van Eck, H.J., Arens, P., Verkerk-Bakker, B., Hekkert, B.T.L. et al. (1995). A genetic map of potato (*Solanum tuberosum*) integrating molecular markers, including transposons, and classical markers. *Theor. Appl. Genet.*, 9: 289–300.

Jacobs, M.M.J., Vosman, B., Vleeshouwers, V.G.A.A., Visser, R.G.F., Henken, B. and van den Berg, R.G. (2010). A novel approach to locate *Phytophthora infestans* resistance genes on the potato genetic map. *Theor. Appl. Genet.*, 120: 785–796.

Jupe, F., Pritchard, L., Etherington, G.J., Mackenzie, K., Cock, P.J.A. et al. (2012). Identification and localization of the NB-LRR gene family within the potato genome. *BMC Genomics*, 13: 75.

Kasai, K., Morikawa, Y., Sorri, V.A., Valkonen, J.P.T., Gebhardt, C. and Wantabe, K.N. (2000). Development of SCAR markers to the PVY resistance gene Ry_{adg} based on a common feature of plant disease resistance genes. *Genome*, 43: 1–8.

Khan, M.A., Saravia, D., Munive, S., Lozano, F., Farfan, E., Eyzaguirre, R. and Bonierbale, M. (2015). Multiple QTLs linked to agro-morphological and physiological traits related to drought tolerance in potato. *Plant Mol. Biol. Rep.*, 33(5): 1286–1298.

Khlestkin, V.K., Rozanova, I.V., Efimov, V.M. and Khlestkina, E.K. (2019). Starch phosphorylation associated SNPs found by genome-wide association studies in the potato (*Solanum tuberosum* L.). *BMC Genetics*, 20(Suppl 1): 29.

Khlestkin, V.K., Erst, T.V., Rozanova, I.V., Efimov, V.M. and Khlestkina, E.K. (2020). Genetic loci determining potato starch yield and granule morphology revealed by genome-wide association study (GWAS). *Peer J.*, 8: e10286.

Klaassen, M.T., Willemsen, J.H., Vos, P.G., Visser, R.G.F., van Eck, H.J. et al. (2019). Genome-wide association analysis in tetraploid potato reveals four QTLs for protein content. *Mol. Breed.*, 39: 151.

Kloosterman, B., Abelenda, J., Gomez, M.M., Oortwijn, M., de Boer, J.M. et al. (2013). Naturally occurring allele diversity allows potato cultivation in northern latitudes. *Nature*, 495: 246–250.

Kuhl, J.C., Hanneman, R.E. and Havey, M.J. (2001). Characterization and mapping of *Rpi1*, a late-blight resistance locus from diploid (1EBN) Mexican *Solanum pinnatisectum*. *Mol. Gen. Genet.*, 265: 977–985.

Kuhl, J.C. (2011). Mapping and tagging of simply inherited traits. pp. 90–112. *In*: Bradeen, J.M. and Kole, C. (eds.). *Genetics, Genomics and Breeding of Potato*. Science Publishers, Enfield, New Hampshire.

Leister, D., Ballvora, A., Salamini, F. and Gebhardt, C. (1996). A PCR based approach for isolating pathogen resistance genes from potato with potential for wide application in plants. *Nat. Genet.*, 14: 421–429.

Leonards-Schippers, C., Gieffers, W., Salamini, F. and Gebhardt, C. (1992). The *R1* gene conferring race-specific resistance to *Phytophthora infestans* in potato is located on potato chromosome V. *Mol. Gen. Genet.*, 233: 278–283.

Li, X., van Eck, H.J., Rouppe van der Voort, J.N.A.M., Huigen, D.-J., Stam, P. and Jacobsen, E. (1998). Autotetraploids and genetic mapping using common AFLP markers: The R2 allele conferring resistance to *Phytophthora infestans* mapped on potato chromosome 4. *Theor. Appl. Genet.*, 96: 1121–1128.

Li, Y., Colleoni, C., Zhang, J., Liang, Q., Hu, Y. et al. (2018). Genomic analyses yield markers for identifying agronomically important genes in potato. *Mol. Plant*, 11(3): 473–484.

Lindqvist-Kreuze, H., Gastelo, M., Perez, W., Forbes, G.A., De Koeyer, D. and Bonierbale, M. (2014). Phenotypic stability and genome-wide association study of late blight resistance in potato genotypes adapted to the tropical highlands. *Phytopathology*, 104: 624–633.

Lopez-Pardo, R., Barandalla, L., Ritter, E. and de Galarreta, J.I.R. (2013). Validation of molecular markers for pathogen resistance in potato. *Plant Breed.*, 132: 246–251.

Machida-Hirano, R. (2015). Diversity of potato genetic resources. *Breed. Sci.*, 65(1): 26–40.

Malosetti, M., van der Linden, C.G., Vosman, B. and van Eeuwijk, F.A. (2007). A mixed-model approach to association mapping using pedigree information with an illustration of resistance to *Phytophthora infestans* in potato. *Genetics*, 175: 879–889.

Manrique-Carpintero, N.C., Coombs, J.J., Pham, G.M., Laimbeer, F., Braz, G.T. et al. (2018). Genome reduction in tetraploid potato reveals genetic load, haplotype variation, and loci associated with agronomic traits. *Front. Plant Sci.*, 9: 944.

Marczewski, W., Hennig, J. and Gebhardt, C. (2002). The Potato virus S resistance gene Ns maps to potato chromosome VIII. *Theor. Appl. Genet.*, 105: 564–567.

Massa, A.N., Manrique-Carpintero, N.C., Coombs, J., Haynes, K.G., Bethke, P.C. et al. (2018). Linkage analysis and QTL mapping in a tetraploid russet mapping population of potato. *BMC Genetics*, 19(1): 87.

Meade, F., Hutten, R., Wagener, S., Prigge, V., Dalton, E. et al. (2020). Detection of novel QTLs for late blight resistance derived from the wild potato species *Solanum microdontum* and *Solanum pampasense*. *Genes*, 11: 732.

Meksem, K., Leister, D., Peleman, J., Zabeau, M., Salamini, F. and Gebhardt, C. (1995). A high-resolution map of the vicinity of the *R1* locus on chromosome V of potato based on RFLP and AFLP markers. *Mol. Gen. Genet.*, 249: 74–81.

Mihovilovich, E., Aponte, M., Lindqvist-Kreuze, H. and Bonierbale, M. (2014). An RGA-derived SCAR marker linked to PLRV resistance from *Solanum tuberosum* ssp. Andigena. *Plant Mol. Biol. Rep.*, 32: 117–128.

Milbourne, D., Meyer, R.C., Collins, A.J., Ramsay, L.D., Gebhardt, C. and Waugh, R. (1998). Isolation, characterization and mapping of simple sequence repeat loci in potato. *Mol. Gen. Genet.*, 259: 233–245.

Milczarek, D., Flis, B. and Przetakiewicz, A. (2011). Suitability of molecular markers for selection of potatoes resistant to *Globodera* spp. *Am. J. Potato Res.*, 88: 245–255.

Moloney, C., Griffin, D., Jones, P., Bryan, G., McLean, K. et al. (2010). Development of diagnostic markers for use in breeding potatoes resistant to *Globodera pallida* pathotype Pa2/3 using germplasm derived from *Solanum tuberosum* ssp. *andigena* CPC 2802. *Theor. Appl. Genet.*, 120: 679–689.

Mori, K., Sakamoto, Y., Mukojima, N., Tamiya, S., Nakao, T. et al. (2011). Development of a multiplex PCR method for simultaneous detection of diagnostic DNA markers of five disease and pest resistance genes in potato. *Euphytica*, 180: 347–355.

Mosquera, T., Alvarez, M.F., Jiménez-Gómez, J.M., Muktar, M.S., Paulo, M.J. et al. (2016). Targeted and untargeted approaches unravel novel candidate genes and diagnostic SNPs for quantitative resistance of the potato (*Solanum tuberosum* L.) to *Phytophthora infestans* causing the late blight disease. *PLoS ONE*, 11(6): e0156254.

Nachtigall, M., König, J. and Thieme, R. (2018). Mapping of a novel, major late blight resistance locus in the diploid (1EBN) Mexican *Solanum pinnatisectum* Dunal on chromosome VII. *Plant Breed.*, 137: 433–442.

Niewöhner, J., Salamini, F. and Gebhardt, C. (1995). Development of PCR assays diagnostic for RFLP marker alleles closely linked to alleles *Gro1* and *H1*, conferring resistance to the root cyst nematode *Globodera rostochiensis* in potato. *Mol. Breed.*, 1: 65–78.

Oberhagemann, P., Chatot-Balandras, C., Schäfer-Pregl, R., Wegener, D., Palomino, C. et al. (1999). A genetic analysis of quantitative resistance to late blight in potato: towards marker-assisted selection. *Mol. Breed.*, 5: 399–415.

Obidiegwu, J.E., Sanetomo, R., Flath, K., Tacke, E., Hofferbert, H.R. et al. (2015). Genomic architecture of potato resistance to *Synchytrium endobioticum* disentangled using SSR markers and the 8.3 k SolCAP SNP genotyping array. *BMC Genet.*, 16: 38.

Ortega, F. and Lopez-Vizcon, C. (2012). Application of molecular marker assisted selection (MAS) for disease resistance in a practical potato breeding programme. *Potato Res.*, 55: 1–13.

Ottoman, R., Hane, D., Brown, C., Yilma, S., James, S. et al. (2009). Validation and implementation of marker assisted selection (MAS) for PVY resistance in a tetraploid potato breeding program. *Am. J. Potato Res.*, 86: 304–314.

Paal, J., Henselewski, H., Muth, J., Meksem, K., Menéndez, C.M. et al. (2004). Molecular cloning of the potato Gro1-4 gene conferring resistance to pathotype Ro1 of the root cyst nematode *Globodera rostochiensis*, based on a candidate gene approach. *The Plant J.*, 38: 285–297.

Pajerowska-Mukhtar, K., Stich, B., Achenbach, U., Ballvora, A., Lübeck, J. et al. (2009). Single nucleotide polymorphisms in the allene oxide synthase 2 gene are associated with field resistance to late blight in populations of tetraploid potato cultivars. *Genetics*, 181: 1115–1127.

Pandey, J., Scheuring, D.C., Koym, J.W., Coombs, J., Novy, R.G. et al. (2021). Genetic diversity and population structure of advanced clones selected over forty years by a potato breeding program in the USA. *Sci. Rep.*, 11(1): 8344.

Park, T.-H., Vleeshouwers, V.G.A.A., Hutten, R.C.B., van Eck, H.J., van der Vossen, E. et al. (2005). High-resolution mapping and analysis of the resistance locus *Rpi-abpt* against *Phytophthora infestans* in potato. *Mol. Breed.*, 16: 33–43.

Park, T.-H., Foster, S., Brigneti, G. and Jones, J.D.G. (2009). Two distinct potato late blight resistance genes from *Solanum berthaultii* are located on chromosome 10. *Euphytica*, 165: 269–278.

Pineda, O., Bonierbale, M.W., Plaisted, R.L., Brodie, B.B. and Tanksley, S.D. (1993). Identification of RFLP markers linked to the *H1* gene conferring resistance to the potato cyst nematode *Globodera rostochiensis*. *Genome*, 36: 152–156.

Prashar, A., Hornyik, C., Young, V., McLean, K., Sharma, S.K. et al. (2014). Construction of a dense SNP map of a highly heterozygous diploid potato population and QTL analysis of tuber shape and eye depth. *Theor. Appl. Genet.*, 127(10): 2159–2171.

Prevost, A. and Wilkinson, M.J. (1999). A new system of comparing PCR primers applied to ISSR fingerprinting of potato cultivars. *Theor. Appl. Genet.*, 98: 107–112.

Prodhomme, C., Vos, P.G., Paulo, M.J., Tammes, J.E., Visser, R. et al. (2020). Distribution of P1(D1) wart disease resistance in potato germplasm and GWAS identification of haplotype-specific SNP markers. *Theor. Appl. Genet.*, 133(6): 1859–1871.

Ramakrishnan, A.P., Ritland, C.E., Sevillano, R.H.B. and Riseman, A. (2015) Review of potato molecular markers to enhance trait selection. *Am. J. Potato Res.*, 92: 455–472.

Rauscher, G.M., Smart, C.D., Simko, I., Bonierbale, M., Mayton, H. et al. (2006). Characterization and mapping of *RPi-ber*, a novel potato late blight resistance gene from *Solanum berthaultii*. *Theor. Appl. Genet.*, 112: 674–687.

Raymundo, R., Asseng, S., Robertson, R., Petsakos, A., Hoogenboom, G. et al. (2018). Climate change impact on global potato production. *Eur. J. Agron.*, 100: 87–98.

Ritter, E., Debener, T., Barone, A., Salamini, F. and Gebhardt, C. (1991). RFLP mapping on potato chromosomes of two genes controlling extreme resistance to potato virus X (PVX). *Mol. Gen. Genet.*, 227: 81–85.

Rouppe van der Voort, J.N.A.M., Van Eck, H.J., Draaistra, J., van Zandvoort, P.M., Jacobsen, E. and Bakker, J. (1998). An online catalogue of AFLP markers covering the potato genome. *Mol. Breed.*, 4: 73–77.

Rouppe van der Voort, J., Janssen, G., Overmars, H., van Zandvoort, P., van Norel, A. et al. (1999). Development of a PCR-based selection assay for root-knot nematode resistance (Rmc1) by a comparative analysis of the *Solanum bulbocastanum* and *S. tuberosum* genome. *Euphytica*, 106: 187–195.

Sato, M., Nishikawa, K., Komura, K. and Hosaka, K. (2006). Potato Virus Y resistance gene, *Ry-chc*, mapped to the distal end of potato chromosome IX. *Euphytica*, 149: 367–372.

Sattarzadeh, A., Achenbach, U., Lübeck, J., Strahwald, J., Tacke, E. et al. (2006). Single nucleotide polymorphism (SNP) genotyping as basis for developing a PCR based marker highly diagnostic for potato varieties with high resistance to *Globodera pallida* pathotype Pa2/3. *Mol. Breed.*, 18: 301–312.

Schönhals, E.M., Ding, J., Ritter, E., Paulo, M.J., Cara, N. et al. (2017). Physical mapping of QTL for tuber yield, starch content and starch yield in tetraploid potato (*Solanum tuberosum* L.) by means of genome wide genotyping by sequencing and the 8.3 K SolCAP SNP array. *BMC Genomics*, 18(1): 642.

Schumann, M.J., Zeng, Z.B., Clough, M.E. and Yencho, G.C. (2017). Linkage map construction and QTL analysis for internal heat necrosis in autotetraploid potato. *Theor. Appl. Genet.*, 130(10): 2045–2056.

Sharma, S.K., Bolser, D., de Boer, J., Sønderkær, M., Amoros, W. et al. (2013). Construction of reference chromosome-scale pseudomolecules for potato: integrating the potato genome with genetic and physical maps. *G3 (Bethesda)*, 3(11): 2031–2047.

Sharma, S.K., MacKenzie, K., McLean, K., Dale, F., Daniels, S. and Bryan, G.J. (2018). Linkage disequilibrium and evaluation of genome-wide association mapping models in tetraploid potato. *G3 (Bethesda)*, 8: 3185–3202.

Simko, I., Costanzo, S., Haynes, K.G., Christ, B.J. and Jones, R.W. (2004). Linkage disequilibrium mapping of a *Verticillium dahliae* resistance quantitative trait locus in tetraploid potato (*Solanum tuberosum*) through a candidate gene approach. *Theor. Appl. Genet.*, 108: 217–224.

Skupinová, S., Vejl, P., Sedlák, P. and Domkářová, J. (2002). Segregation of DNA markers of potato (*Solanum tuberosum* ssp. *tuberosum* L.) resistance against Ro1 pathotype *Globodera rostochiensis* in selected F1 progeny. *Rostlinná Výroba*, 48(11): 480–485.

Slater, A.T., Cogan, N.O.I. and Forster, J.W. (2013). Cost analysis of the application of marker-assisted selection in potato breeding. *Mol. Breed.*, 32: 299–310.

Śliwka, J., Jakuczun, H., Lebecka, R., Marczewski, W., Gebhardt, C. et al. (2006). The novel, major locus *Rpi-phu1* for late blight resistance maps to potato chromosome IX and is not correlated with long vegetation period. *Theor. Appl. Genet.*, 113: 685–695.

Śliwka, J., Wasilewicz-Flis, I., Jakuczun, H. and Gebhardt, C. (2008). Tagging quantitative trait loci for dormancy, tuber shape, regularity of tuber shape, eye depth and flesh colour in diploid potato originated from six *Solanum* species. *Plant Breed.*, 127: 49–55.

Śliwka, J., Jakuczun, H., Kamiński, P. and Zimnoch-Guzowska, E. (2010). Marker-assisted selection of diploid and tetraploid potatoes carrying *Rpi-phu1*, a major gene for resistance to *Phytophthora infestans*. *J. Appl. Genet.*, 51: 133–140.

Smilde, W.D., Brigneti, G., Jagger, L., Perkins, S. and Jones, J.D. (2005). *Solanum mochiquense* chromosome IX carries a novel late blight resistance gene *Rpi-moc1*. *Theor. Appl. Genet.*, 110: 252–258.

Song, J., Dong, F.f. and Jiang, J. (2000). Construction of a bacterial artificial chromosome (BAC) library for potato molecular cytogenetics research. *Genome*, 43(1): 199–204.

Song, Y.-S., Hepting, L., Schweizer, G., Hartl, L., Wenzel, G. and Schwarzfischer, A. (2005). Mapping of extreme resistance to PVY (Ry(sto)) on chromosome XII using anther-culture-derived primary dihaploid potato lines. *Theor. Appl. Genet.*, 111: 879–887.

Song, Y.-S. and Schwarzfischer, A. (2008). Development of STS markers for selection of extreme resistance (Ry_{sto}) to PVY and maternal pedigree analysis of extremely resistant cultivars. *Am. J. Potato Res.*, 85: 392–393.

Sorri, V.A., Watanabe, K.N. and Valkonen, J.P.T. (1999). Predicted kinase-3a motif of a resistance gene analogue as a unique marker for virus resistance. *Theor Appl Genet.*, 99: 164–170.

Strachan, S.M., Armstrong, M.R., Kaur, A., Wright, K.M., Lim, T.Y. et al. (2019). Mapping the H2 resistance effective against *Globodera pallida* pathotype Pa1 in tetraploid potato. *Theor. Appl. Genet.*, 132(4): 1283–1294.

Szajko, K., Chrzanowska, M., Witek, K., Strzelczyk-Zyta, D., Zagórska, H. et al. (2008). The novel gene *Ny-1* on potato chromosome IX confers hypersensitive resistance to *Potato virus Y* and is an alternative to *Ry* genes in potato breeding for PVY resistance. *Theor. Appl. Genet.*, 116: 297–303.

Tanksley, S.D., Ganal, M.W., Prince, J.P., de Vicente, M.C., Bonierbale, M.W. et al. (1992). High density molecular linkage maps of the tomato and potato. *Genetics*, 132: 1141–1160.

The Potato Genome Sequencing Consortium. (2011). Genome sequence and analysis of the tuber crop potato. *Nature*, 475: 189–195.

Tiwari, J.K., Sundaresha, S., Singh, B.P., Kaushik, S., Chakrabarti, S.K. et al. (2013). Molecular markers for late blight resistance breeding of potato: an update. *Plant Breed.*, 132: 237–245.

Tiwari, J.K., Ali, N., Devi, S., Kumar, V., Zinta, R. and Chakrabarti, S.K. (2018). Development of microsatellite markers set for identification of Indian potato varieties. *Sci. Hortic.*, 231: 22–30.

Tiwari, J.K., Ali, S., Devi, S., Zinta, R., Kumar, V. and Chakrabarti, S.K. (2019). Analysis of allelic variation in wild potato (Solanum) species by simple sequence repeat (SSR) markers. *3 Biotech.*, 9: 262.

Tomczyńska, I., Stefańczyk, E., Chmielarz, M., Karasiewicz, B., Kamiński, P. et al. (2014). A locus conferring effective late blight resistance in potato cultivar Sárpo Mira maps to chromosome XI. *Theor. Appl. Genet.*, 127: 647–657.

Uitdewilligen, J.G., Wolters, A.M., D'hoop, B.B., Borm, T.J., Visser, R.G. and van Eck, H.J. (2013). A next-generation sequencing method for genotyping-by-sequencing of highly heterozygous autotetraploid potato. *PLoS ONE*, 8(5): e62355.

van Eck, H.J., Rouppe van der Voort, J., Draaistra, J., van Zandvoort, P., van Enckevort, E., Segers, B. et al. (1995). The inheritance and chromosomal localization of AFLP markers in a non-inbred potato offspring. *Mol. Breed.*, 1: 397–410.

Van Os, H., Andrezejewski, S., Bakker, E., Barrera, I., Bryan, G.L. et al. (2006). Construction of a 10,000-marker ultradense genetic recombination map of potato: providing a framework for accelerated gene isolation and a genome-wide physical map. *Genetics*, 173: 1075–1087.

Velásquez, A.C., Mihovilovich, E. and Bonierbale, M. (2007). Genetic characterization and mapping of major gene resistance to potato leafroll virus in *Solanum tuberosum* ssp. *andigena*. *Theor. Appl. Genet.*, 114: 1051–1058.

Vos, P.G., Uitdewilligen, J.G., Voorrips, R.E., Visser, R.G. and van Eck, H.J. (2015). Development and analysis of a 20 K SNP array for potato (*Solanum tuberosum*): An insight into the breeding history. *Theor. Appl. Genet.*, 128(12): 2387–2401.

Vos, P.G., Paulo, M.J., Voorrips, R.E., Visser, R.G., van Eck, H.J. and van Eeuwijk, F.A. (2017). Evaluation of LD decay and various LD-decay estimators in simulated and SNP-array data of tetraploid potato. *Theor. Appl. Genet.*, 130(1): 123–135.

Vos, P., Hogers, R., Bleeker, M., Reijans, M., van de Lee, T. et al. (1995). AFLP: A new technique for DNA fingerprinting. *Nucleic Acids Res.*, 23: 4407–4414.

Vossen, J.H., van Arkel, G., Bergervoet, M., Jo, K.R., Jacobsen, E. and Visser, R.G. (2016). The *Solanum demissum R8* late blight resistance gene is an Sw-5 homologue that has been deployed worldwide in late blight resistant varieties. *Theor. Appl. Genet.*, 129(9): 1785–1796.

Watanabe, K. (2015). Potato genetics, genomics, and applications. *Breed. Sci.*, 65: 53–68.

Wenzl, P., Carling, J., Kudrna, D., Jaccoud, D., Huttner, E. et al. (2004). Diversity arrays technology (DArT) for whole-genome profiling of barley. *Proc. Natl. Acad. Sci., U.S.A.*, 101: 9915–9920.

Whitworth, J., Novy, R., Hall, D., Crosslin, J. and Brown, C. (2009). Characterization of broad spectrum potato virus Y resistance in a *Solanum tuberosum* ssp. *andigena*-derived population and select breeding clones using molecular markers, grafting, and field inoculations. *Am. J. Potato Res.*, 86: 286–296.

Witek, K., Strzelczyk-Zyta, D., Hennig, J. and Marczewski, W. (2006). A multiplex PCR approach to simultaneously genotype potato towards the resistance alleles $Ry\text{-}f_{sto}$ and *Ns*. *Mol. Breed.*, 18: 273–275.

Yamanaka, S., Ikeda, S., Imai, A., Luan, Y., Watanabe, J.A. and Watanabe, K.N. (2005). Construction of integrated genetic map between various existing DNA markers and newly developed P450-related PBA markers in diploid potato (*Solanum tuberosum*). *Breed. Sci.*, 55: 223–230.

Yamanaka, Y., Suzuki, E., Tanaka, M., Takeda, Y., Watanabe, J.A. and Watanabe, K.N. (2003). Assessment of cytochorome P450 sequences offers a useful tool for determining genetic diversity in higher plant species. *Theor. Appl. Genet.*, 108: 1–9.

You, Q., Yang, X., Peng, Z., Xu, L. and Wang, J. (2018). Development and applications of a high throughput genotyping tool for polyploid crops: Single nucleotide polymorphism (SNP) array. *Front. Plant Sci.*, 9: 104.

Yuan, J., Bizimungu, B., Koeyer, D.D., Rosyara, U., Wen, Z. and Lagüe, M. (2020). Genome-wide association study of resistance to potato common scab. *Potato Res.*, 63: 253–266.

Zhang, L.H., Mojtahedi, H., Kuang, H., Baker, B. and Brown, C.R. (2007). Marker-assisted selection of Columbia root-knot nematode resistance introgressed from *Solanum bulbocastanum*. *Crop Sci.*, 47: 2021–2026.

Zhou, Q., Tang, D., Huang, W., Yang, Z., Zhang, Y. et al. (2020). Haplotype-resolved genome analyses of a heterozygous diploid potato. *Nature Genet.*, 52(10): 1018–1023.

Zia, M., Demirel, U., Nadeem, M.A. and Çaliskan, M.E. (2020). Genome-wide association study identifies various loci underlying agronomic and morphological traits in diversified potato panel. *Physiol. Mol. Biol. Plants*, 26(5): 1003–1020.

Zimnoch-Guzowska, E., Marczewski, W., Lebecka, R., Flis, B., Schäfer-Pregl, R. et al. (2000). QTL analysis of new sources of resistance to *Erwinia carotovora* ssp. *atroseptica* in potato done by AFLP, RFLP, and resistance-gene-like markers. *Crop Sci.*, 40: 1156.

Chapter 5

Conventional to Genomics-assisted Breeding

1. Conventional Breeding

The cultivated potato (*Solanum tuberosum* L. Group Tuberosum) is autotetraploid ($2n = 4x = 48$) and contains four copies (homologues) of 12 basic ($x = 12$) chromosomes. Potato is highly heterozygous and suffers from acute inbreeding depression. The heterozygosity in commercial cultivars is preserved through clonal propagation, using tubers. Breeding potato is a difficult task as more than 50 desirable traits are to be combined in an ideal modern potato variety. These traits include morphological features, yielding ability, tuber characters, ability to withstand biotic and abiotic stresses, wider adaptability, quality parameters, consumer and industrial acceptability. It is perhaps an impossible task to combine all the traits to obtain an ideal variety because of the complex heterozygous nature of potato. The new variety thus, should be superior to existing one in at least one important characteristic, without being significantly inferior to it in any other important traits. The genetic base of the released potato varieties grown on large areas is relatively narrow as compared to accessible genepool for classical potato breeding. Only a fraction of useful genes from wild species has been successfully introgressed into modern potato varieties (nearly 4,000 varieties worldwide). Modern varieties are products of extensive breeding efforts between different cultivar groups, cultivated and wild potato species. History of classical potato breeding reveals that many of the important varieties took nearly 30 or more years from hybridization and clonal selection for their release. The long breeding cycle in potato is due to quantitative nature of most of the desirable traits, rapid inbreeding depression and low intensity of selection in early generation. The low multiplication rate and amenability of potato clones to pathogens lead to degeneration of stock and reduces the quality of the tubers. Conventional potato breeding scheme involves selection of trait specific parents, hybridization, phenotypic recurrent selection in seedling and clonal generations at targeted locations for a wide range of desirable characters, which nearly take 12–15 years (Luthra et al., 2020). Thus the classical breeding cycle takes a very long time, huge investment, more manpower and delays the accessibility of the targeted variety to stakeholder due to less targeted traits-specific approach. The major objectives in potato breeding are as mentioned below:

- Develop varieties with high tuber yield, early tuber bulking (< 80 days), short to medium duration (80–100 days for sub-tropical plains and plateau regions) and long duration (> 100 days for hilly regions). Depending upon the potato growing regions and consumer's preference and processing demand, target traits may vary in different countries.
- Resistance to various biotic stresses, like diseases and insect-pests, such as late blight, viruses, bacterial wilt, wart, storage rots, potato cyst nematodes, wart, aphid, white fly and tuber moth, etc., abiotic stress tolerance (heat, drought, salinity and cold), improved input use efficient, particularly nitrogen fertilizer and climate resilient varieties.
- White/yellow flesh potato with attractive, medium sized and shallow eye-depth tubers and good keeping quality and low glycoalkaloids are more preferred worldwide. Recently, nutritionally rich (anthocyanin, carotenoids, micronutrients like Fe and Zn, etc.) potatoes are desirable. The native potatoes with variable tuber shapes and multi-colors are popular in South American countries. Additionally, red-skin potato is preferred in eastern regions of India, Nepal, Bangladesh, Bhutan, Pakistan and Philippines.

2. Considerations in Breeding

2.1 Parent Selection

Selection of parents depends on the breeding objectives. Breeders prefer to work with parents having traits of interest, good agronomic attributes and adaptation. Generally, monogenic and dominant controlled traits are easy to transfer as they are highly heritable. The most commonly used parents belong to the cultivated (*S. tuberosum* groups Tuberosum and Andigena) background, while others could be semi-cultivated/wild potato species. Most complex traits like yield and others are controlled by both additive and non-additive genes in potato. Due to highly heterozygous and autotetraploid crop, selection of parents is a challenging task. Since phenotypic selection of parents could be erroneous and ineffective when the trait is non-additive gene controlled, therefore, genetic merit or breeding value of parents is assessed, based on combining ability parameters, particularly in Indian breeding program. Studies show use of good general combining parents in breeding and resulted in superior progeny and later released as new variety (Kumar and Gopal, 2003, 2006; Luthra et al., 2006).

2.2 Progeny Test

Due to non-additive gene control of many agronomic important traits, identification of superior clones is carried out, based on the progeny test. This has been used extensively in potato breeding the world over for a number of traits, like disease-pest resistance and yield-contributing traits (Gopal, 1997; Gopal et al., 2000; Pande et al., 2005). Performing of progeny test is recommended with 80 seedlings

(20 × 4 replications) per progeny or even 20 or 40 genotypes when the population size is larger (Bradshaw and Mackey, 1994). In this way, a superior progeny or cross is identified from which individual elite clone is selected for variety development in successive generations.

2.3 Flowering and Hybridization

Availability of profuse flowering at synchronizing time with fertile male and female flowers is an indispensable requirement for successful hybridization. Important factors that determine flowering behavior are genotype, day-length and temperature. The ideal conditions for flowering and berry setting are photoperiod (14–18 h light) and temperature (15–20°C). Natural flowering occurs under long-day conditions in hills (first week of July onwards in Shimla, Himachal Pradesh, India) and artificial illuminated long-day length in sub-tropical plain (last week of November onwards in Modipuram, Meerut, Uttar Pradesh, India) conditions. Pollens are collected from fresh opening buds or just newly opened flowers after the male fertility test in acetocarmine (2%). In the female flower, generally four to five large-size flower buds per bunch are carefully emasculated for pollination. The freshly collected pollen grains are preferably pollinated with the stigma of the female parent or may be stored for a few days in a refrigerator (6–8°C) or under natural conditions in the hills.

2.4 Berry Harvesting and TPS Extraction

Initiation of berry can be seen within a week's time and berries are harvested after about six to seven weeks of pollination. The harvested berries are allowed to ripen at room temperature for a few weeks. Then the berries are macerated by hand or crushing machine, washed in running water and True Potato Seeds (TPS) are extracted and dried up to 5–6% moisture. TPS are packed in aluminium-foil cover bags, sealed and stored in desiccators containing calcium chloride in the refrigerator (6–8°C).

2.5 Seedling Raising and Clonal Selection

Freshly harvested TPS is treated with gibberellic acid (1500 ppm for 24 h) and then dried, or one-year old TPS is used for raising of seedling nursery. In order to select the best clone for varietal development, about 50,000 seedlings of 25 cross or progenies (2,000 seedlings per cross) are used. Due to vegetative propagation, genetic constitution of the parents fixed in the progenies is maintained in the first clonal generation at the seedling stage itself. Hence, if a clone is obtained with desirable attributes, it can be multiplied for commercial cultivation. Further, clonal selection is followed by successive generations up to 12–15 years to select desirable clones with target traits to develop a new variety.

3. Breeding Strategies

Potato is a vegetatively propagated crop through tubers. Hence, heterosis vigor from parents gets fixed in the first generation itself and later, clonal selection is practiced to select elite genotype or clone with desirable agronomic traits. Conventional breeding strategy broadly involves selection of better parents with traits of interest, hybridization to generate segregating population (F_1), rejection of undesirable genotype in subsequence generations with increase in clonal selection pressure, and finally selection of elite clone with yield and other desirable traits. Finally, yield trials of advance hybrids along with standard check on research farm and multi-locations over the years are conducted to select a superior genotype and release a new variety. Figure 1 depicts the breeding strategies in conventional potato breeding under Indian conditions for table or fresh potato. Tables 1 and 2 outline a few selection criteria and traits to be considered while developing new potato varieties.

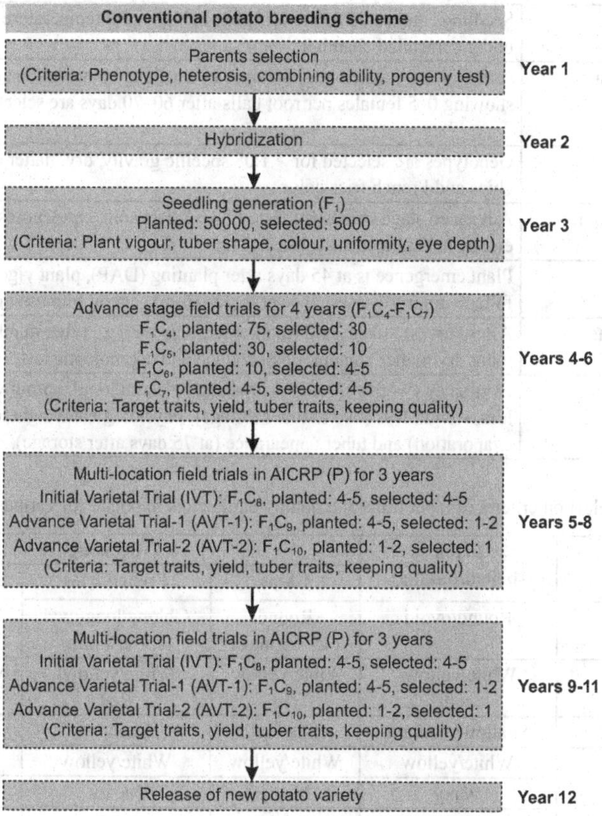

Fig. 1. Conventional breeding scheme in potato.

4. Speed Breeding

Conventional breeding cycle requires over a decade to develop a new variety. The new scheme proposed here, 'Speed Breeding', leads to rapid multiplication of mini-tubers in aeroponics or apical rooted cuttings technologies and enhance clonal selection procedures in less duration to develop new varieties. In brief, it includes trait-specific selection of parents, targeted hybridization, population screening up to F_1C_3 (F_1 hybrid advanced up to third clonal stage), rapid tuber multiplication and

Table 1. Selection criteria at different clonal stages in potato breeding.

SN	Trait	Selection criteria
1.	Seedling stage	Plant vigor, desirable tuber traits (tuber number, color, shape, eye depth, no cracking) and tuber yield is not the criterion here.
2.	Late blight resistance	Seedlings are tested for late blight resistance in controlled chamber by challenge inoculation, or hot spot conditions in hills. Disease score is calculated, based on the area under disease progress curve (AUDPC) value: HR (< 50), R (50–100), MR (101–300) and S (> 300).
3.	Virus resistance	Virus resistance, mainly PVX, PVY and PLRV, are the most damaging in potato crop worldwide and PALCV in India. Resistance is tested by challenge inoculation and serological and molecular methods.
4.	Heat stress tolerance	Seedlings are tested for tuberization at high night temperature (> 20–24°C) under controlled chamber.
5.	Potato cyst nematode resistance	Genotypes are tested in glasshouse by challenge inoculation and genotypes showing 0–5 females per root balls after 60–70 days are selected.
6.	Processing attributes	Genotypes are selected for > 1.08 specific gravity, dry matter content and chips and French fries color.
7.	Nutritional traits	Advanced stage clones are tested for anthocyanin, carotenoids, iron, zinc, etc. by using biochemical assay.
8.	Phenotype	Plant emergence is at 45 days after planting (DAP), plant vigor at 60 DAP, foliage maturity/senescence at 90 DAP and general impression.
9.	Tuber traits	After harvest, tuber yield (marketable size > 20 g), tuber number per plant, tuber dry matter content, tuber rottage and organoleptic test.
10.	Storage weight loss	Dormancy (> 80% sprouts), sprouting (> 2 mm long), sprout weight loss, weight loss (due to tuber rottage), physiological weight loss (due to evaporation) and tuber appearance (at 75 days after storage).

Table 2. Selection criteria of tuber traits and quality parameters in potato for various purposes.

Tuber traits	Fresh (boiled/baking)	Processing		
		Chips	French fries	Flakes
Tuber shape	Round/ovoid	Round (45–85 mm)	Oblong/long-oval (75–110 mm)	Oval/round (30–85 mm)
Skin color	White/yellow/ red	White/yellow/ red	White/yellow/ Red	White/yellow/ red
Eye depth	Shallow/medium	Shallow	Shallow	Shallow
Flesh color	White/yellow	White/yellow	White/yellow	White-cream
Texture*	Waxy	Mealy	Mealy	Mealy
Tuber uniformity	High	High	High	High
Dry matter (%)*	18–20	> 20	> 20	> 20
Reducing sugars (mg/100 g FW)	NA	< 100	< 150	< 150
Glycoalkaloids (mg/100 g FW)	< 15	< 15	< 15	< 15
Phenols	High	Less	Less	Less
Defects	Minimum	Minimum	Minimum	Minimum
Keeping quality	Good	Good	Good	Good
Bruising resistance	High	High	High	High

* For baking purposes, dry matter: > 20%, and Texture: Mealy

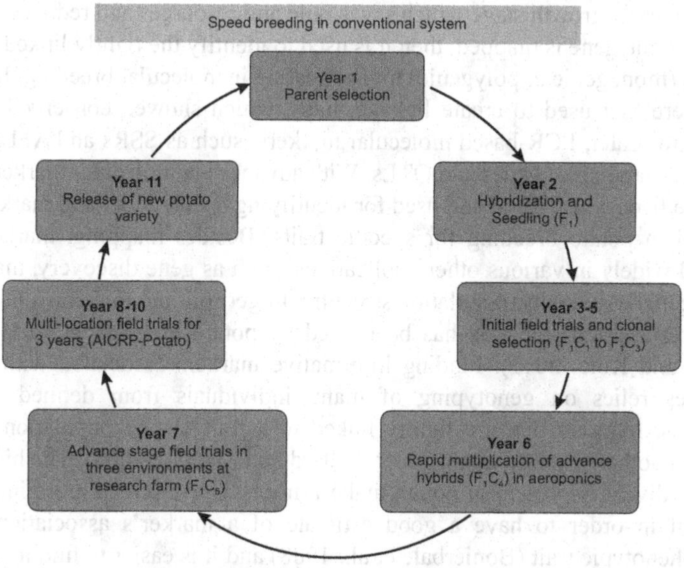

Fig. 2. Speed breeding scheme in potato.

fast multi-location evaluation. This may result in delivery of end product in a shorter period (Fig. 2). Some key points are:

- The first requirement for speed breeding is multiplication of F_1C_4 advanced hybrids in environment-controlled aeroponics during offseason and seed multiplication of the selected hybrids in main crop season.
- Multi-location evaluation of the selected advance stage hybrids (F_1C_5) at research farms in three environments for desirable agronomic and target traits. Advancement of identified promising F_1C_5 advanced stage hybrids for desirable traits.
- Multi-location testing of advanced stage hybrids (F_1C_6) across the country and then release as a new variety.
- The process is expected to shorten the breeding cycle in potato at least by one to two years.

5. Marker-assisted Selection

The genus *Solanum* is one of the richest sources of biotic and abiotic stress resistance/tolerance genes in potato. Conventional breeding has been applied the world over to breed varieties to overcome these problems. A large collection of invaluable materials has been found in wild, semi-cultivated and cultivated potatoes, and further beneficial traits have been introgressed into cultivated background. However, due to long and time-consuming conventional breeding method, results are less satisfactory. For the past 40 years, focus has been on mapping and identification of resistance genes/QTL from wild/cultivated species using various molecular markers and later deploying these tightly linked molecular markers (< 10 cM) with a target gene for Marker Assisted Selection (MAS) for targeted traits in potato. This allows indirect selection at early growth stage which saves time and resources and reduces breeding cycle. Once the gene is mapped, then it is used to identify the tightly linked markers with traits (monogenic or polygenic) for application in molecular breeding. In potato, RFLPs were first used to create linkage maps, which showed conserved markers with tomato. Later, PCR-based molecular markers, such as SSRs and AFLPs, were applied for mapping of genes and QTLs. With advances in molecular markers, many maps have been constructed and used for identifying specific loci and markers have been used in potato breeding for specific traits. Besides mapping, markers have been used widely in various other applications, such as gene discovery, map-based cloning, genetic diversity, population structure, fingerprinting, to name a few.

Marker-assisted selection has been used by potato breeders in various ways (Bradeen and Kole, 2011). Finding informative markers associated with specific phenotypes relies on genotyping of many individuals from defined breeding populations. Markers that are tightly linked to a trait in one population may be useless in another (Collard et al., 2005; Collard and Mackill, 2008). The high levels of genetic diversity present in potato make it necessary to screen more individuals than usual in order to have a good estimate of a marker's association with a specific phenotypic trait (Bonierbale et al., 1988) and it is easier to find informative markers in diploid than in tetraploid varieties. In addition, the age and type of the polyploidy-inducing event affects the researchers' abilities to identify valuable markers (Carputo and Frusciante, 2011). Older, natural ploidy events have more mutations and more heterozygosity than artificial polyploid cultivars. Molecular markers linked to traits can be identified, using databases of cultivar qualities, such as tuber shape and flesh color (D'hoop et al., 2008), but for novel trait associations, the phenotyping must be accomplished *de novo*. To date, many markers have been mapped in potato (http://solgenomics.net/breeders/index.pl), but relatively few have successfully been used in MAS programs, possibly due to the highly heterogeneous genome of potato. As the cost for screening markers has decreased, new methods are being developed. Genomic Selection (GS) is a novel method that calculates genomic estimated breeding values based on molecular genetic information from the whole genome and shows promise for breeding complex traits (Heslot et al., 2012), but it has yet to be applied in potato. As the costs for sequencing diminish, new methods for screening potatoes will be developed.

MAS is the selection of genotypes on the basis of linked molecular marker with the target trait at an early stage, irrespective of season/environment. Markers tightly linked to the gene of interest can be used in such cases to identify progenies with minimum introgression of wild genome, thereby decreasing the number of breeding generations as well as recurrent selection cycles. By and large, MAS has been found more effective in the case of introgression of major dominant gene, preferably qualitative controlled. In usual practice, introgression of traits from wild or cultivated donor parents to a commercial cultivar is achieved by recurrent back crossing and selection cycles. MAS also provides a uniform and objective scoring of offspring and avoids cumbersome phenotypic scoring until the end of the breeding program. Tightly linked markers for many qualitative and quantitative traits of potato have been made available for MAS and a wide range of molecular markers have been deployed in potato breeding for resistance to traits, like late blight, viruses and potato cyst nematodes.

Molecular markers offer the ability to identify genotypes with traits of interest at early stage of selection before reaching the field level; hence, reduced breeding cycle. In the current scenario, consumer-driven and market-specific varieties are required in most countries. Hence, under these conditions, early development of varieties is important to save money, time and resources. To illustrate, Slater et al. (2013) compared cost between MAS and conventional breeding, indicating a saving of 10% cost for Potato Cyst Nematode (PCN) resistance breeding through MAS, depending on the marker-trait association. Most of the markers reported so far in potato are associated with mainly late blight, viruses and PCN resistance (R) genes. Identification of diagnostic markers depends on the screening of a large number of population and marker-trait associations as it is easy in diploid population than tetraploid. Markers tightly linked with the target trait can be deployed across the population from one to another. Overall, MAS allows a significant decrease in field exposure cycles by early stage selection and thus greatly accelerates the speed of breeding for new cultivars (Ramakrishnan et al., 2015). A time line of application of markers in potato is outlined below:

- 1980's: The first genetic map of potato was completed in the late 1980's using RFLP markers based on classical recombination frequencies and linkage groups (Bonierbale et al., 1988; Gebhardt et al., 1989; Ritter et al., 1991; Tanksley et al., 1992).
- 1990's: Beginning of use of PCR-based markers for trait selection in potato (Meksem et al., 1995).
- 2000's: Wider application of MAS in potato breeding using PCR-based markers, like RAPD, SCAR and SSR (Barone, 2004; Bradeen and Kole, 2011).
- 2009: Ghislain et al. (2009) discovered a set of 24 SSR markers (two SSR per chromosome) for application in molecular characterization in potato.
- 2010's: Discovery of novel SNP markers (8.3 K SolCAP SNP Potato array) to facilitate rapid breeding and characterization of potato and its wider application (Hamilton et al., 2011).
- 2011: The potato genome sequence was deciphered and accelerated the process of marker development.
- 2014: Genome-wide association mapping for late blight resistance genes (Lindqvist-Kreuze et al., 2014). Till now, several markers have been mapped in potato and are being applied widely.

6. MAS for Biotic Stress Resistance

6.1 Late Blight Resistance

Late blight, caused by *Phytophthora infestans*, is the most devastating disease of potato. The pathogen quickly develops resistance to defence mechanisms of potatoes. So, for more than a century, breeders are continually making efforts at searching new resistance genes. Eleven R-genes introgressed into *S. tuberosum* from *S. demissum* have been characterized for resistance to late blight (Black et al., 1953) and other sources as well—*S. bulbocastanum* (Sokolova et al., 2011). Plenty of markers have been identified and tested for linkage to disease phenotypes. Disease resistance genes often produce proteins that have regions with Nucleotide Binding Sites (NBS) and Leucine Rich Repeats (LRR). Hundreds of NBS loci have been identified and they could be used in future screening studies, but these markers require conversion to PCR-based markers and testing on breeding populations segregated for resistance before being useful on a broad range of varieties (Jupe et al., 2013). There are five RFLP, three AFLP and numerous SSR markers linked to late blight (Kuhl, 2011). For example, markers linked to late blight resistant genes—*R1, R3, R6, R7* from *S. demissum* mapped on to chromosome 11 and *R2* gene on chrmosome 4 have been used for their introgression into cultivated potato. Similarly, *R* gene from *S. berthaultti* (chromosome 10) and *RB* gene from *S. bulbocastanum* (chromosome 8) have also been brought into the tetraploid cultivated potato by using marker-assisted introgression technique. Reports of identifying numerous *R* genes and linked markers conferring resistance to late blight in potato have surfaced but are yet to be used for introgression into cultivated (Table 3; review by Tiwari et al., 2013).

6.2 Virus Resistance

There are over three dozens of potato viruses that infect potato, but only a few are economically important, the most notable being potato virus Y (PVY), potato virus X (PVX), potato apical leaf curl virus (PALCV) and potato leaf roll virus (PLRV). Resistance to these viruses has been bred into commercial potatoes from several sources. Extreme resistance to PVY, characterized by symptomless plants, is frequently bred from either *S. tuberosum* ssp. *andigena* (Ry_{adg} gene; Hämäläinen et al., 1998) or from *S. stoloniferum* (Ry_{sto} gene; Song et al., 2005). Similarly, extreme resistance to PVX is conferred by several *Rx* genes, including *Rx1* from *S. tuberosum* ssp. *andigena* and *Rx2* from *S. acaule* (Mori et al., 2011). These genes confer complete, entire plant resistance to attack by PVY and PVX. Such extreme resistance has yet to be found for PLRV in *S. tuberosum*, but some resistance has been bred from *S. tuberosum* ssp. *andigena.* Other wild plants have resistance genes but they have yet to be incorporated into the breeding stock (Novy et al., 2007). Two genes that confer hypersensitive resistance to PVY, Ny_{tbr} and *Ny-1*, were mapped in *S. tuberosum*. Genes responsible for hypersensitive response to PVX were mapped earlier: the gene *Nb* was found to be in a resistance hotspot that also contains *R1* (vs. late blight), *Gpa* (vs. nematode) and *Rx2* (extreme resistance to PVX) (De Jong et al., 1997). Hypersensitive response to PVX was mapped to the *Ns* gene in *S. tuberosum* ssp. *andigena* (Marczewski et al., 2002). The potential for more valuable genes being found increases if intensive genetic association methods are used, such as high-density association mapping studies, EST analyses and/or SNP association mapping studies. The majority of markers for virus resistance presented are PCR-based with only one AFLP marker and five RFLP markers (Table 4). While some success has been achieved by using these virus resistance markers for marker-assisted selection, more studies identifying linkage of markers to disease

Table 3. Molecular markers for late blight resistance breeding in potato.

Gene/ QTL	Marker	Marker type	Primer sequence (5' → 3')	References
R1	R1-1205	SCAR	F: CACTCGTGACATATCCTCACTA R: GTAGTACCTATCTTATTTCTGCAAGAAT	Sokolova et al. (2011)
	BA47f2	SCAR	F: TAACCAACATTATCTTCTTTGCC R: GAATTTGGAGAGGGGTTTGCTG	Gebhardt et al. 2004)
	CosA	SCAR	F: CTCATTCAAAATCAGTTTTGATC R: GAATGTTGAATCTTTTTGTGAAGG	Gebhardt et al. 2004)
	R1F/R (76-2sf2/76-2SR)	AS	F: CACTCGTGACATATCCTCACTA R: CAACCCTGGCATGCCACG	Ballvora et al. 2002)
	GP76	SCAR	F: ATGAAGCAACACTGATGCAA R: TTCTCCAATGAACGCAAACT	Oberhagemann et al. (1999)
	SPUD237 (*Alu*I)	CAPS	F: TTCCTGCTGATACTGACTAGAAAACC R: AGCCAAGGAAAAGCTAGCATCCAAG	De Jong et al. (1997)
	GP21 (*Alu*I)	CAPS	F: AGTGAGCCAGCATAGCATTACTTG R: GGTTGGTGGCCTATTAGCCATGC	De Jong et al. (1997)
	GP179	SCAR	F: GGTTTTAGTGATTGTGCTGC R: AATTTCAGACGAGTAGGCACT	Meksem et al. (1995)
R3 (*R3a* & *R3b*)	R3-1380	SCAR	F: TCCGACATGTATTGATCTCCCTG R: AGCCACTTCAGCTTCTTACAGTAGG	Sokolova et al. (2011)
	SHa-F/ SHa-R	AS	F: ATCGTTGTCATGCTATGAGATTGTT R: CTTCAAGGTAGTGGGCAGTATGCTT	Huang et al. (2005)
	*R3b*F4/ *R3b*R5	AS	F: GTCGATGAATGCTATGTTTCTCGAGA R: ACCAGTTTCTTGCAATTCCAGATTG	Rietman (2011)
RB/ Rpi-blb1	RB-629/638	SCAR	F: AATCAAATTATCCACCCCAACTTTTAAAT R: CAAGTATTGGGAGGACTGAAAGGT	Sokolova et al. (2011)
	RB-1223	SCAR	F: ATGGCTGAAGCTTTCATTCAAGTTCTG R: CAAGTATTGGGAGGACTGAAAGGT	Pankin et al. (2011)
	CT88 (Primer 1/ primer 1')	SCAR	F: CACGAGTGCCCTTTTCTGAC R: ACAATTGAATTTTTAGACTT	Colton et al. (2006)
Rpi-abpt	*R2*-F1/*R2*-R3	AS	F: GCTCCTGATACGATCCATG R: ACGGCTTCTTGAATGAA	Kim et al. (2012)
	Th2	CAPS	F: AGGATTTCAGTATGTCTCG R: TCCATTGTTGATTGCCCCT	Park et al. (2005b)
Rpi-ber1	CT214 (*Dde*I)	CAPS	F: GAACGCGAAAGAGTGCTGATAG R: CCCGCTGCCTATGGAGAGT	Tan et al. (2010)
	TG63 (*Bme*1390I)	CAPS	F: TCCAATTGCCAGACGAA R: TAGAGAAGGCCCTTGTAAGTTT	Tan et al. (2010)
	Q133	SCAR	F: ATCATCTCCTCAAAGAATCAAG R: ATCTCCCCATTGACAACCAA	Tan et al. (2010)
Rpi-mcd1	TG339 (*Mnl*I)	CAPS	F: GCTGAACGCTATGAGGAGATG R: TGAGGTTATCACGCAGAAGTTG	Tan et al. (2010)
Rpi-phu1	GP94 (OPB07+TG/ GT)	RAPD	F: GAAACGGGTG + TG/GT	Śliwka et al. (2006)

Table 3 contd. ...

...Table 3 contd.

Gene/ QTL	Marker	Marker type	Primer sequence (5' → 3')	References
QTL_ *phu-stn*	OPA17	RAPD	GACCGCTTGT	Wickramasinghe et al. (2009)
	OPA03	RAPD	AGTCAGCCAC	Wickramasinghe et al. (2009)
	GP198F/ R	SCAR	F: GTAATTTGCGAGGAAGGAGAAG R: TCACTTTGGTGCTTCTGTCG	Wickramasinghe et al. (2009)
	GP198F-1/R	AS	F: TTTGCTTACTCTTGTTGTATG R: TCACTTTGGTGCTTCTGTCG	Wickramasinghe et al. (2009)
Rpi-sto1	Ssto-448	SCAR	F: GTGGAACGCCGTCCATCCTTAG R: TGCATAGGTGGTTAGATGTATGTTTGATTA	Sokolova et al. (2011)
Rpi-avl1	N2527	AS	F: GAAACACAGGGGAATATTCACC R: CCATRTCTTGWATTAAGTCATGC	Verzaux (2010)
Rpi-cap1	CP58 (*Msp*I)	CAPS	F: ATGTATGGTTCGGGATCTGG R: TTAGCACCAACAGCTCCTCT	Jacobs et al. (2010)
Rpi-dlc1	GP101 (*Alu*I)	CAPS	F: GGCATTTCTATGGTATCAGAG R: GCTTAACATGCAAAGGTTAAA	Golas et al. (2010)
	S1d5-a	AS	F: CGCCTCTTTCTCTGAATTTC R: GATCTGGGATGGTCCATTC	Golas et al. (2010)
Rpi-mcq1	TG328 (*Alu*I)	CAPS	F: AATTAAATGGAGGGGGTATC R: GTAGTATTCTAGTTAAACTACC	Smilde et al. (2005)
Rpi-snk1.1 and *Rpi-snk1.2*	Th21 (*Mbo*I)	CAPS	F: ATTCAAAATTCTAGTTCCGCC R: AACGGCAAAAAAGCACCAC	Jacobs et al. (2010)
Rpi-ver1	CD67 (*Hpy*CH4IV, *Ssi*I)	CAPS	F: CCCCTGCAAATCCGTACATA R: CCATACGAGTTGAGGGATCG	Jacobs et al. (2010)
Rpi-vnt1.1, and *Rpi-vnt1.3*	TG35(*Hha*I/ *Xap*I)	CAPS	F: CACGGAGACTAAGATTCAGG R: TAAAGGTGATGCTGATGGGG	Pel et al. (2009)
	NBS3B	AS	F: CCTTCCTCATCCTCACATTTAG R: GCATGCCAACTATTGAAACAAC	Pel et al. (2009)

Adapted and modified from Tiwari et al. (2013)

phenotypes need to be undertaken to find makers that will directly benefit breeding programs.

6.3 Potato Cyst Nematodes Resistance

Potato cyst (*Globodera* sps.) and root knot (*Meloidogyne* sps.) nematode are the two important pests that infect potato and cause severe losses. Nematodes cause significant crop losses due to reduction in tuber size and other damages. Resistance to nematode attack is one of the most desirable traits in potato breeding. Approximately, 17 genes have been mapped that are associated with resistance to potato cyst nematodes *G. pallida* and *G. rostochiensis* (Bakker et al., 2004). Four genes in particular confer

Table 4. Molecular markers for virus resistance breeding in potato.

Gene	Marker name	Marker type	Primer sequence (5' → 3')	References
Potato virus Y (PVY)				
Ry_{adg}	$ADG1_{356}$	RGL	F: CACACTCTCGTATCAGTTTGA	Hämäläinen et al. (1998)
			R: ATTTAATAGCGTGACAGTCAAC	
	$ADG2_{354}$	RGL	F: ATACACTCATCTAAATTTGATGG	Hämäläinen et al. (1998)
			R: ACTTAACTGCATCATGTTCAAG	
	$ADG2_{310}$ (*Bbv*I)	CAPS	F: ATACTCTCATCTAAATTTGATGG	Sorii et al. (1999)
			R: ACTGAACAGCATCATGTTCAAG	
	$RYSC3_{321}$	SCAR	F: ATACACTCATCTAAATTTGATGG	Kasai et al. (2000)
			R:AGGATATACGGCATCATTTTTCCGA	
Ry_{sto}	M5	AS	F: GACTGCGTACATGCAGTG	Brigneti et al. (1997)
			R: GATGAGTCCTGAGTAACAA	
	M45	AS	F: GACTGCGTACATGCAGCT	Brigneti et al. (1997)
			R: GATGAGTCCTGAGTAAGGA	
	M17	AS	F: GACTGCGTACATGCAGTG	Brigneti et al. (1997)
			R: GATGAGTCCTGAGTAACAT	
	M6	AS	F: GACTGCGTACATGCAGCT	Brigneti et al. (1997)
			R: GATGAGTCCTGAGTAAGAA	
Ry_{sto}	$GP122_{406}$ (*Eco*RV)	CAPS	F: CAATTGGCTCCCGACTATCTACAG	Heldák et al. (2007)
			R: ACAATTGCACCACCTTCTCTTCAG	
	$STM0003_{111}$	SSR	F: GGAGAATCATAACAACCAG	Song et al. (2005)
			R: AATTGTAACTCTGTGTGTGTG	
			R: GATGAGTCCTGAGTAACGA	
	YES3-$3A_{341}$	STS	F: TAACTCAAGCGGAATAACCC	Song et al. (2005)
			R:AATTCACCTGTTTACATGCTTCTTGTG	
	YES3-$3B_{284}$	STS	F: TAACTCAAGCGGAATAACCC	Song et al. (2005)
			R: CATGAGATTGCCTTTGGTTA	
Ry-f_{sto}	$UBC857_{980}$	ISSR	ACACACACACACACAC(CT)G	Flis et al. (2005)
	$GP122_{718}$ (*Eco*RV)	CAPS	F: TATTTTAGGGGTACTTCTTTCTTA	
			R: GATACTTCCAACCGCTTCAC	
	$GP204_{800}$ (*Taq*I)	CAPS	F: CATAGATGGCTCAAACAACTC	Flis et al. (2005)

Table 4 contd. ...

...Table 4 contd.

Gene	Marker name	Marker type	Primer sequence (5' → 3')	References
			R: GTGGAAACATGGCTTACC	
	$GP269_{650}$ (*Dde*I)	CAPS	F: TCGCAATGAAAGATAAGC	Flis et al. (2005)
			R: TGTGATAAAGAGTGTAGCAGTC	
	$GP81_{400}$	STS	F: GCAGCGTTTCCTACAAT	Flis et al. (2005)
			R: AGAGACTAATGCTGAAAAT	
	$GP122_{564}$ (*Eco*RV)	CAPS	F: TATTTTAGGGGTACTTCTTTCTTATGTT	Witek et al. (2006)
			R:CTGTCAAAAAAATTCACTTGCATAACTAC	
Ry_{chc}	38-530 (OPC-01)	RAPD	TTCGAGCCAG	Hosaka et al. (2001)
Ny-1	$SC895_{1139}$	SCAR	F: GGTAGCTCTTGATCTCGTCTT R: GTAGCTCTTGATCACCCATTT	Szajko et al. (2008)
	$GP41_{443}$	SCAR	F: GTTGGTACCAGGCTTGTT	Szajko et al. (2008)
			R: CATTCGGTGCTTTAGGAT	
	C2_ $At3g16840_{1100}$ (*Taq*I)	COSII	F: TCCAGTGTCCAAAGAAAGAAAA R: ATGCTCATGTCCCCGAAACC	Szajko et al. (2008)
Potato leaf roll virus (PLRV)				
Plrv.1 (QTL)	$Nl27_{1164}$	SCAR	F: TAGAGAGCATTAAGAAGCTGC	Marczewski et al. (2001a)
			R: TTTTGCCTACTCCCGGCATG	
Plrv.4 (QTL)	$UBC864_{600}$	ISSR	ATGATGATGATGATGATG	Marczewski et al. (2004)
	$UBC864AC_{600}$	SCAR	ATGATGATGATGATGATGAC	Marczewski et al. (2004)
Rl_{adg}	$E35M48_{192}$	AFLP	F: GACTGCGTACCAATTCACA	Velásquez et al. (2007)
			R: GATGAGTCCTGAGTAACAC	
Rlr_{etb}	C2_ $At1g42990_{1100}$ (*Alu*I)	COSII	F: ATGACCCCGTCGATAAGAAGCG R: ACCTCACAGCTGCATCTCTATTCCTC	Kelley et al. (2009)
Potato virus X				
Rx	IPM3 (*DdeI*)	CAPS	F: AGTAGTTTCAGGCTAGTG	Bendahmane et al. (1997)
			R: CAACATCACTTGATCAGAC	
	IPM4 (T*aq*I)	CAPS	F: GTACTGGAGAGCTAGTAGTGATCA	Bendahmane et al. (1997)
			R: ACCACTGGCAAATGGCCATACGA	
	IPM5 (*DdeI*)	CAPS	F: AGCTCCATTCGTGACGAT	Bendahmane et al. (1997)
			R: AGCTTCGATAATTCTAAATTTG	

Table 4 contd. ...

...Table 4 contd.

Gene	Marker name	Marker type	Primer sequence (5' → 3')	References
Nb	GM339$_{330}$	Allele Sp.	F: GGTAGTTGGACGAGCATAT	Marano et al. (2002)
			R: CTCACTTTTAGACCAGATTT	
	GM637$_{220}$	Allele Sp.	F: GCAGAAGATCGGATAGCAAAC	Marano et al. (2002)
			R: GTAACGAGTTGAAGTTACTGA	
	GP21 (*Alu*I)	CAPS	F: GGTTGGTGGCCTATTAGCCATGC	De Jong et al. (1997)
			R: AGTGAGCCAGCATAGCATTACTTG	
	SPUD237 (*Alu*I)	CAPS	F: TTCCTGCTGATACTGACTAGAAAACC	De Jong et al. (1997)
			R: AGCCAAGGAAAAGCTAGCATCCAAG	
Potato virus S				
Ns	OPE15$_{550}$	RAPD	ACGCACAACC	Marczewski et al. (1998)
	OPJ13$_{500}$	RAPD	CCACACTACC	Marczewski et al. (1998)
	OPG17$_{450}$	RAPD	ACGACCGACA	Marczewski et al. (1998)
	OPH19$_{900}$	RAPD	GTCAGGGCAA	Marczewski et al. (1998)
	SCG17$_{321}$	SCAR	F:ACGACCGACACTCAAATTTGTACAAGAAA	Marczewski et al. (2001b)
			R: GATGCCCCGACAGAGGAAG	
	SCG17$_{448}$ (*Mun*I)	CAPS	F: ACGACCGACACTCAAATTTGTACA	Marczewski et al. (2001b)
			R: ACGACCGACAAGAGGACCAAGGGAATAAC	
	UBC811$_{660}$ UBC811$_{950}$	ISSR ISSR	GAGAGAGAGAGAGAGAC GAGAGAGAGAGAGAGAC	Marczewski (2001)
	SC811$_{260}$ (*Mbo*I)	CAPS	F: CGAACAAAATACGTAATGCATTGAATAA	Witek et al. (2006)
			R: GACCTATATCAGTCCCTTCTAATCCACTAT	
Potato virus M				
Rm	GP283$_{320}$ (*Dde*I)	CAPS	F: CCCTCCCCATGAAAAAGGTA	Marczewski et al. (2006)
			R: GCAACTTCCTGTCCGAATGT	
	GP250$_{510}$ (*Xap*I)	CAPS	F: AGTTCAACACCAGTAGGAC	Marczewski et al. (2006)
			R: GACATCAAGTTACCTATGAC	
Gm	SC878$_{885}$	SCAR	F: GGATGGATGGATGAGGAGGAAACT	Marczewski et al. (2006)
			R: CCGACTAGCGATTTGGATGC	

Adapted and modified from Tiwari et al. (2012)

Table 5. Molecular markers for potato cyst nematode breeding in potato.

Resistance gene	Nematode spp., pathotype	DNA marker	Marker type	Primer sequence (5' → 3')	References
H1	*G. rostochiensis*, Ro1,6	TG689	SCAR	F: TAAAACTCTTGGTTATAGCCTAT R: CAATAGAATGTGTTGTTTCACCAA	Milczarek et al. (2011), Lopez-Pardo et al. (2013)
H1	*G. rostochiensis*, Ro1,5	239E4left	CAPS/ *Alu*I	F: GGCCCCACAAACAAGAAAAC R: AGGTACCTCCATCTCCATTTTGTAAG	Bakker et al. (2004) Pajerowska-Mukhtar et al. (2009) Milczarek et al. (2011)
H1	*G. rostochiensis*, Ro1,4	N146	SCAR	F: AAGCTCTTGCCTAGTGCTC R: AGGCGGAACATGCCATG	Mori et al. (2011)
H1	*G. rostochiensis*, Ro1,4	N195	SCAR	F: TGGAAATGGCACCCACTA R: CATCATGGTTTCACTTGTCAC	Mori et al. (2011)
H1	*G. rostochiensis*, Ro1,4	CP113	SCAR	F1: GCGTTACAGTCGCCGTAT R1: GTTGAAGAAATATGGAATCAAA F2: GCCTTACAGTCGCCGTAT R2: GTTGAAGAAATATGGAATCAAA	Niewöhner et al. (1995) Skupinová et al. (2002) Milczarek et al. (2011)
H1	*G. rostochiensis*, Ro1,4	TG689, TG689indel12	SCAR	F1: GGCGTTACAGTCGCCGTAT R1: GTTGAAGAAATATGGAATCAAA F2: TAAAACTCTTGGTTATAGCCTAT R2: CAATAGAATGTGTTGTTTCAC	Galek et al. (2011)
Gro1-4	*G. rostochiensis*, Ro1	Gro1-4	SCAR	F: TCTTTGGAGATACTGATTCTCA R: CGACCTAAAATGAAAAGCATCT	Gebhardt et al. (2006), Paal et al. (2004) Milczarek et al. (2011)
GroVI (allelic to *H1*)	*G. rostochiensis*, Ro1	U14	SCAR	F: GGGCTTGTATAAGACCTCCGAGAGG R: CCCTTCCTTGGGTAGTTTGAGCG	Jacobs et al. (1996), Milczarek et al. (2011)
GroVI (allelic to *H1*)	*G. rostochiensis*, Ro1	X02	SCAR	F: CCACCAAACCCATAAAGCTGC R: TGTGAATTGGTATGAATCTGCAACC	Jacobs et al. (1996) Milczarek et al. (2011)

GpaIV$_{adg}$	*G. pallida*, Pa2/3	STM3016-122/177	SSR	F: TCAGAACACCGAATGGAAAAC R: GCTCCAACTTACTGGTCAAATCC	Bryan et al. (2004), Milbourne et al. (1998), Moloney et al. (2010)
GpaIV$_{adg}$	*G. pallida*, Pa2/3	C237	CAPS/ *Taq*I	F: GCAGTCCTAATTGCACGTAACA R: CTTACTTGGGCAACCCAGAAT	Moloney et al. (2010)
RGp5-vrnHC, allelic to *Grp*1	*G. pallida*, Pa2/3	HC (snp212-T/ snp444-G)	SNP	F: ACACCACCTGTTTGATAAAAAACT R: GCCTTACTTCCCTGCTGAAG	Sattarzadeh et al. (2006), Lopez-Pardo et al. (2013)
Gpa2	*G. pallida*, pa2	77R	CAPS/ *Hae*III	F: CTCGAGGGATTGAATCCAAATTAT R: GGAAGCAGAATACTCCTGACTACT	Rouppe van der Voort et al. (1997) Milczarek et al. (2011)
Gpa2	*G. pallida*, pa2	GP34	SCAR	F: GTTGCTAGGTAAGCATGAAGAAG R: GTTATCGTTGATTTCTCGTTCCG	Rouppe van der Voort et al. (1997)
Gpa2	*G. pallida*, pa2	GP34	CAPS/ *Taq*I	F: CGTTGCTAGGTAAGCATGAAGAAG R: GTTATCGTTGATTTCTCGTTCCG	Bendahmane et al. (1997)
Gpa2	*G. pallida*, pa2	Gpa2-1	STS	F: TTTAGCACGGAATGTGGGGA R: GTTTCCCCATCAAAACTCAC	Asano et al. (2012)
Gpa2	*G. pallida*, pa2	Gpa2-2	STS	F: GCACTTAGAGACTCATTCCA R: ACAGATTGTTGGCAGCGAAA	Asano et al. (2012)
Gpa5	*G. pallida*, pa5	HC-I	SCAR	F: GTAGTACATCAACATACATTTTGCGG R: GCCTTACTTCCCTGCTGAAG	Asano et al. (2021)
Gpa5	*G. pallida*, pa5	HC-I2	SCAR	F: GTAGTACATCAACATACATTTTGCGG R: GGTGACTAAGATGGAAATCAGAG	Asano et al. (2021)
Grp1	*G. pallida*, *G. rostochiensis*, Ro5, Pa2,3	TG432	CAPS/ *Rsa*I	F: GGACAGTCATCAGATTGTGG R: GTACTCCTGCTTGAGCCATT	Finkers-Tomczak et al. (2009), Milczarek et al. (2011)
-	*G. pallida*	SPUD1636	SCAR	F: GTGCGCACAGGGTAAAACC R: ACCTTAGCGGATGAAAGCC	Bryan et al. (2002), Milczarek et al. (2011)

strong resistance to *Globodera* species: *H1*, *GroV1*, *Gro1* (*Gro1.4* on chromosme 3 and *Gro1.2* on chromosme 10) and *Gpa2* (Bakker et al., 2004). Two sets of markers have been mapped for the emerging threat by root knot nematodes—*M. chitwoodi*, *M. fallax*, and *M. hapla* (Milczarek et al., 2011). Twenty-six markers are associated with nine genes with the *H1* gene accounting for 11 of those markers. The *H1* gene was introduced into *S. tuberosum* ssp. *tuberosum* commercial varieties from *S. tuberosum* ssp. *andigena* in the 1950's, and is still quite effective (Gebhardt et al., 1993). A hint for the use of these markers is to use them first and if more markers are desired, to choose markers that map closest to the *H1* gene, namely, CMI and CP113. The marker SPUD1636 is linked to a *S. vernei* resistance QTL and may be useful if *S. vernei*-related germplasm is screened (Milczarek et al., 2011). The *Rmc1* gene was identified in *S. bulbocastanum* (Brown et al., 1996), conferring resistance to the Columbia root-knot nematode. Two markers, N146 and N195, are part of a useful multiplex protocol designed to identify markers linked to five genes (*H1, Rx1, R1, R2*, R_{ych}). As they are multiplexed, they would be ideal to use for testing the presence of genes during breeding trials (Mori et al., 2011). Molecular markers applied in potato cyst nematode resistance breeding through MAS are summarized in Table 5.

7. MAS for Abiotic Stress Tolerance and Quality Traits

7.1 Drought Stress

Drought is an important environmental stress for potato crop. However, MAS for abiotic stress tolerance traits lags far behind than for disease resistance genes because tolerance to these abiotic stressors is often mediated by many interacting genes and factors (Watanabe et al., 2011). Gene expression studies have attempted to quantify genetic characteristics that make certain strains, varieties and species more tolerant to drought stress (Mane et al., 2008). Potato responses to drought are mediated by a host of genes, including transcription factors and receptor kinases (Ambrosone et al., 2011). Some drought-resistance genes are related to root-shoot ratios, which are controlled by relatively few QTLs and may be potential targets for marker development (Kumari et al., 2011). Several metabolites are associated with drought tolerance and perhaps could be used to screen breeding clones (Evers et al., 2010). Responses to drought are highly varied however and results from these studies are highly dependent on the experimental conditions used. With the availability of complete potato genome sequence, potential development of markers will be facilitated.

7.2 Cold/Low Temperature Stress

Like above, cold stress is another problem in potato and tolerance to cold varies significantly among different varieties and species of potato. *S. tuberosum* is notoriously sensitive to freezing (below 0°C), though it is tolerant to temperatures of 2–12°C. Many of the same genes are often expressed in response to cold stress as in response to drought or salt stress (Shinozaki and Yamaguchi-Shinozaki, 2007). Genes induced by cold include regulation of carbohydrates, photosynthesis and detoxification, among many others (Oufir et al., 2008). Changes in lipid unsaturation in combination with expression of protective proteins can also increase cold tolerance (De Palma et al., 2008). *S. commersonii* and *S. acaule* are more cold tolerant than *S. tuberosum*, while somatic hybrids with *S. tuberosum* display some freezing tolerance (Carputo et al., 2007). In addition, several studies have introduced genes from yeast, Arabidopsis, or cyanobacteria to increase cold or freezing tolerance (Demin et al., 2008) and no markers have yet been identified to use for MAS.

7.3 Tuber Quality Traits

Potato tuber quality traits are sometimes under simple genetic control but often they are multigenic, complicating MAS (Slater et al., 2014a). Some quality traits are controlled by only a few genes, such as eye depth and tuber shape while other traits, such as yield, chip color and specific gravity, are controlled by many different genes on different chromosomes (Li et al., 2013). Due to the potential savings in time and field evaluation, researchers are trying to develop protocols for MAS for some of these multigenic traits. For example, it might be possible to use polymorphisms in the sucrose synthase gene to screen for light chip frying color, but the technique has only been tested on 70 cultivars and not on breeding populations (Kawchuk et al., 2008). Markers for selecting starch yield and chip quality have been discovered that could be useful in breeding programs (Fischer et al., 2013; Li et al., 2013). Combinations of markers can improve results for multigenic traits, but the costs associated with genotyping many QTLs can negate any potential benefits from MAS in such situations (Moreau et al., 2000; Slater et al., 2013); however, next-generation technologies and genetic screening arrays are reducing the costs of genotyping and may enable breeders to regularly screen genetically complex traits (Slater et al., 2014a,b).

8. Genomic Selection

In conventional breeding, breeders have to deal with more than 40 traits (biotic, abiotic, quality, yield and other agronomic traits), which takes 12–15 years to develop a new variety. However, marker-assisted selection has been the powerful tool in plant breeding and has been applied in potato breeding for a wide range of traits, like late blight, potato viruses and potato cyst nematode resistance. But MAS has limited application, particularly for complex inheritance traits, like yield and abiotic stresses. Besides, bi-parental linkage mapping has also been applied for various agronomic important and disease-pest resistance traits in potato. With the progress in advanced sequencing technologies, computational tools and sequencing of more plant genomes, genomics-aided crop improvement through next-generation breeding technology is gaining interest (Zia et al., 2017; Ortiz, 2020). Genomic Selection (GS) or genome-wide selection or genomic-assisted breeding is a novel breeding method to resolve the issues associated with MAS and to accelerate genetic gain in a short time span. Briefly, GS allows integration of technologies, like high-throughput genotyping (HTG) using high-density molecular markers, like SNP chips or genotyping by sequencing (GBS)-led and discovered new SNP combined with phenotyping (preferably high-throughput phenotyping, HTP) to enhance genetic gain in less time (Slater et al., 2017; Li et al., 2018).

GS is the strategy to predict breeding model at whole-genome level for rapid breeding. Unlike linkage mapping, GS works on the principle of Linkage Disequilibrium (LD) with a minimum of one marker per trait loci in a breeding population. GS accelerates the breeding cycle with increase in genetic gain per unit time and reduces cost as well. GS is more established in animal breeding than in crop plants. The success of GS method depends on the extent of genetic similarity between a Training Population (TP) and a Breeding Population (BP) within the LD, between marker and trait loci. First, we determine TP, which includes a set of genotypes or a large collection of germplasm/varieties/breeding materials phenotyped in targeted environments and used to investigate future segregating generation to develop new varieties. Initially, TP is both genotyped and phenotyped for target environment and then GS model is predicted, based on the HTG and HTP data of TP for marker genotypes and trait phenotypes. After that, a new BP or segregating population is generated, like conventional breeding scheme, from which selection has to be made and BP is genotyped (not phenotyped). Later, the predicted GS model is applied to the BP genotype data and Genomic Estimated Breeding Value (GEBV) is estimated and prediction accuracy is determined. Collectively, GS estimates all genetic variances in every individual based on the GEBV and once GEBV is predicted, it can be considered for selection of new lines (Fig. 3). Thus, based on the GEBV, future breeding line is selected for the target environment and then cross-validated for phenotype and other agronomic traits before release of new variety and reach to farmers.

GS determines genetic association and diversity in various landraces/varieties/breeding lines and wild species with variations in topography and ecology. With the identification of genome rearrangements and SNP discovery at whole genome level, GS can be applied in near future. GS has been successfully applied in animals and reported to some extent in plants, like maize, wheat, sugar beet, etc. Specific SNPs or haplotypes can be used for GS, as predicted frequency of SNP in potato is 1 in 24 base pair in the exons. As of now, application of GS in potato so far is limited, which might be due to unavailability of SNP markers distributed throughout the genome, trait association, SNP calling rate and software uses in this tetraploid and heterozygous nature of the crop. Effect of SNP and haplotypes is searched throughout the genome to GEBV in a TP and is then used to predict the performance of other breeding lines by avoiding phenotypic selection at every generation. TP should be genotyped through as many SNP markers as possible.

8.1 Advantages of GS

- GS can resolve the problems associated with complex traits and not specific to a particular trait, as in MAS. It reduces the length of breeding cycles to enhance genetic gain and selection efficiency with limited resources. It reduces cost of phenotype screening and increases gain per unit time.
- GS handles single gene or multiple genes governing traits with complex nature, unlike MAS which is applicable to a limited number of QTLs/genes. It investigates whole genome level-based variances and avoids complications associated with GWAS and QTLs related markers. It adds advantage where no reference genome sequence is available to genotype a large breeding population.

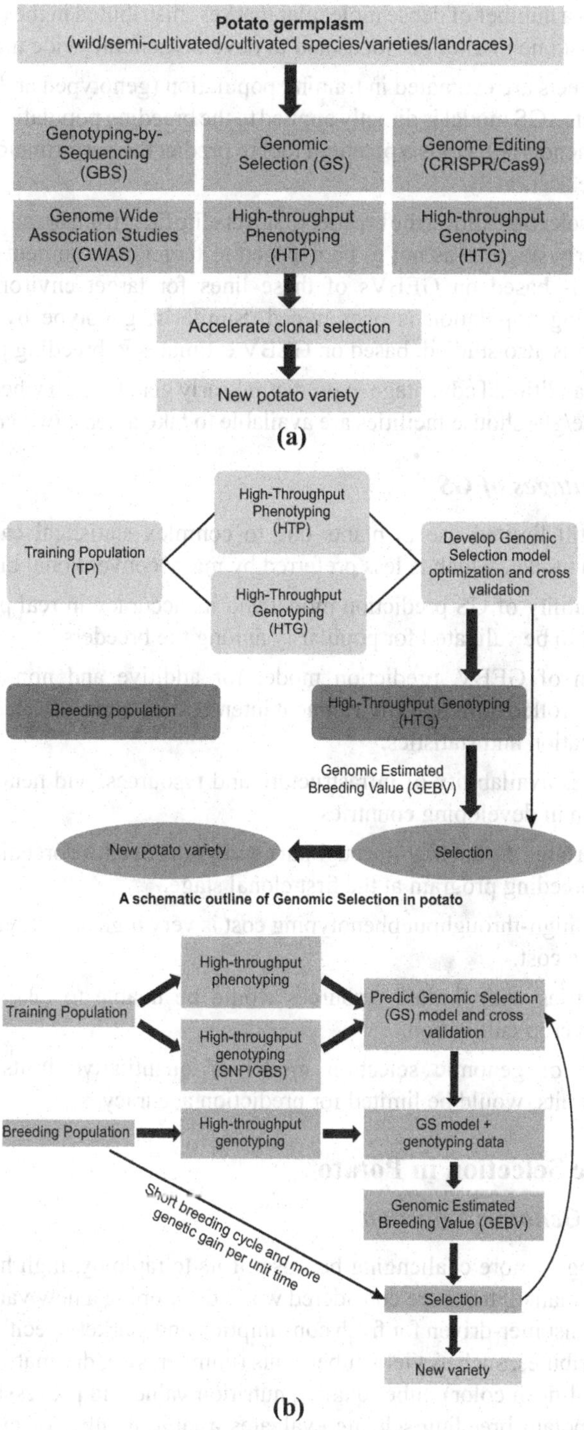

Fig. 3. An outline of modern genomics tools (a) and genomic selection (b).

- GS applies a number of dense molecular markers distributed in the entire genome with at least one marker per locus and to develop genome-wide marker maps.
- Marker effects are estimated in training population (genotyped and phenotyped) and predicted GS model is directly applied to the breeding population (genotyped, without phenotype). GS is a potential tool to predict the performance of breeding lines more accurately.
- Genomic selection allows the breeder to select elite lines from breeding population at very early stage so as not to be exposed to target environment/phenotyping. Selection is based on GEBVs of these lines for target environment, where only training population is phenotyped. Similarly, genotype by environment interaction is also studied, based on GEBV estimates in breeding population.
- GS offers additional advantage in predicting early genetic gain where off-season/ poly-house/glasshouse facilities are available to take at least two crops in a year.

8.2 Disadvantages of GS

- GS has still limited use in plants due to complex statistical calculation and prediction model, which is less preferred by many conventional breeders.
- Practical utility of GS prediction model and its accuracy in real plant breeding still needs to be validated for popularity among the breeders.
- Estimation of GEBV, prediction model for additive and non-additive gene action-controlled traits and environment interaction require advanced knowledge of computation and statistics.
- GS requires availability of infrastructure and resources, and hence has limited application in developing countries.
- Requires a huge cost of high-throughput genotyping of entire breeding population in every breeding program at the first clonal stage.
- Currently, high-throughput phenotyping cost is very high and maybe in future, it would save cost.
- The countries with limited resources would be unable to take advantage of off-season crop cultivation.
- Prediction of genomic selection model of quantitative traits, particularly complex traits, would be limited for prediction accuracy.

9. Genomic Selection in Potato

9.1 Need of Genomic Selection

Potato breeding is more challenging because of its tetraploidy, high heterozygosity and also more than 40 traits are considered while developing a new variety. Many of the traits are consumer-driven for fresh consumption and market-specific, particularly processing attributes, such as yield, tuber traits (number, size, dry matter content, eye depth, skin-and-flesh color), tuber quality, nutrition value and processing traits. The conventional potato breeding scheme evaluates a large number of clones over the successive generations and a smaller number at multiple locations to select an elite clone with desirable traits, taking over a decade in developing a new variety. In other word, large numbers of progenies are evaluated phenotypic with concurrent increase in number of plants or clones of a superior progeny and reducing the total population size through increasing selection pressure.

Prior to 1980s, intense phenotypic selection method was mostly used, which was later discouraged due to rejection of superior as well as inferior clones, particularly the low-heritability traits which are not correlated with tuber yield. Later, this phenotypic selection method was discouraged. With the genetics principle, Bradshaw practiced the progeny test to select the better parents for hybridization. Progeny test estimates the value of the individuals for the traits under investigation, without any prior knowledge about genes. This method was practiced the world over by potato breeders for a long time to increase the yield and other desirable traits. With the advancement in statistical genetics, the Best Linear Unbiased Prediction (BLUP) method allows determination of accurate genotypic or breeding value or genetic merit of pedigree information for all types of family (full-sib, half-sib and others) via additive genetic variance. BLUP has showed a clear advantage in potato over the prior method, like progeny test and phenotypic selection including low-heritability traits also (Slater et al., 2014a). Moreover, MAS has been used intensively in potato breeding for qualitative traits, like disease resistance, that are controlled by major genes or QTLs. MAS combined with estimated breeding values at second field generation could be useful for complex traits also. This will drastically reduce breeding cycle from more than 10 years to four years (Slater et al., 2014b). Genomic selection has the ability to reduce the breeding cycle and accelerate genetic gain in potato.

Genomic selection has been successfully applied in animals and major crops to achieve the expected genetic gain. Recent research has focused on re-sequencing of more potatoes after the potato genome sequence. The potato genome sequence has led to the discovery of over 39,000 genes in potato governing multiple traits involved in biotic and abiotic stress resistance/tolerance. Further, it could help in identifying a large number of markers for GS. Given that improvement for complex traits, like the yield, is difficult, a considerable research work on genomic prediction models has been demonstrated in potato for disease resistance, tuber quality, chipping and yield components. Nevertheless, with the increasing information on thousands of SNPs resources, distributed across the entire potato genome, estimation of breeding value is now possible for genomic prediction models in potato. The factors affecting genomic prediction accuracy in achieving genetic gains are tetraploid potato are ploidy, high heterozygosity, LD decay between marker and trait loci, dense marker number, TP size and trait heritability. Over the time, significant research progress has been made in increasing knowledge on improving genetic gain in potato. Genome re-sequencing of more potatoes, like wild and cultivated species, next-generation potato improvement through Genomic Selection (GS) is important to accelerate the genetic gain and fasten the breeding cycle in less time. The rapid advancement in genotyping techniques (SNP and haplotypes), HTP and trait association tools would lead to the reality of GS in potato in future. Figure 4 depicts genomic selection in potato breeding.

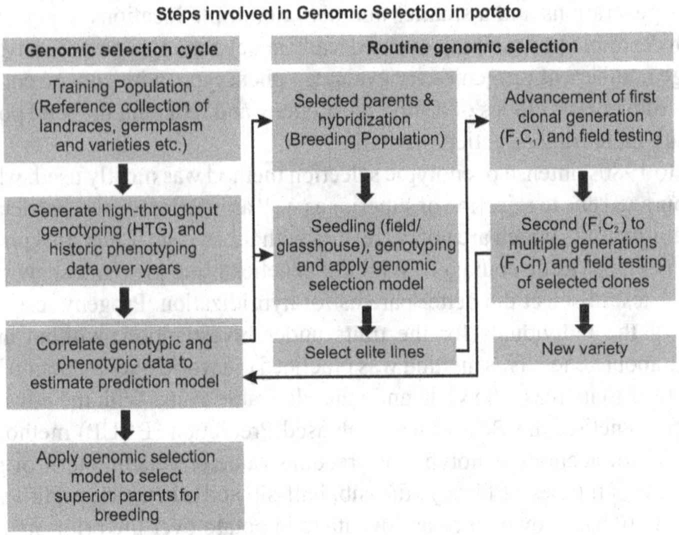

Fig. 4. A schematic outline of genomic selection in potato breeding.

9.2 Considerations in Genomic Prediction

Meuwissen et al. (2001) first proposed the genomic selection method. They estimated markers effects in terms of breeding values in a reference population (called 'training population') which was completely genotyped, using genome-wide molecular markers and phenotyped as well. Then genetic merit was used to predict future phenotype in an uncharacterized or new breeding population, based on the breeding values. Genomic selection analyses all markers together to capture the most genetic variance, unlike limited number of genes/QTLs in MAS. Genomic selection is a more preferred system than GWAS and QTL analysis which have problems of overestimation of markers effects (Tables 6 and 7). To establish genomic selection in potato breeding, the following points should be considered:

- Genomic selection requires a huge number of molecular markers, particularly SNP distributed across the potato genome because genome-wide SNP covers most alleles in the genome considering QTL in linkage disequilibrium (LD). With successive progress in molecular marker systems, a considerable number of molecular marker maps have been developed in potato (Bonierbale et al., 1988; Gebhardt et al., 1989; Milbourne et al., 1998; Tanksley et al., 1992). Further, progress led to the construction of a dense linkage map of potato, using 10,000 ALFP markers (Van Os et al., 2006), which was applied to construct the physical map for the potato genome sequence assembly.
- In recent years, a few SNP array chips (8.3 K, 13 K and 21 K) covering whole genome of potato have been developed for high-throughput genotyping (Uitdewilligen et al., 2013). Genome-Wide Association Study (GWAS) determines the genes having large effects on applying SNP chips, which may miss some relevant information as they do not capture the entire allele frequency. Therefore, a low-cost genotyping by sequencing technology allows to discover

Table 6. Genomic prediction model and accuracy in potato.

Trait	**Narrow sense heritability (h^2)**	**Pedigree best linear unbiased prediction (BLUP) accuracy**	**Reference population size (*Np*)**			
			500	1000	2000	5000
Breeder's visual preference	0.23	0.33	0.19	0.27	0.37	0.53
Yield	0.56	0.19	0.29	0.40	0.52	0.70
Bioling color	0.73	0.44	0.33	0.44	0.57	0.74
Maturity	0.86	0.78	0.36	0.47	0.61	0.77

Note: Adapted from Slater et al. (2016)

Table 7. Comparison of genomic prediction model with selection intensity in potato.

Generation	**Details**	**Selection intensity**			**Genomic selection model (seedling population size)**	
		Intense	Moderate	Mild	GS 5000	GS 2000
G0 (seedling generation)	No. of Seedlings	1,00,000	20,000	6,667	5,000	2,000
	Selection rate (%)	0	0	0	50	50
G1 (first clonal generation)	No. of Seedlings	1,00,000	20,000	6,667	2,500	1,000
	Selection rate (%)	2	10	30	80	80
G2 (second clonal generation)	No. of genotypes	2000	2,000	2,000	2,000	800
Total cost*		6.33	2.57	1.94	1.38	0.55

* Cost in lakh Australian dollars (adapted from Slater et al., 2016)

genome-wide SNP in any organism (Elshire et al., 2011; The Potato Genome Sequencing Consortium, 2011). This has been applied in potato to identify SNP for many traits.

- The possible allele constitutions of genotypes in autotetraploid potato are AAAA, AAAB, AABB, ABBB, and BBBB. A method of quantifying allele doses in autotetraploid has been illustrated by Gidskehaug et al. (2011), based on SNP genotyping, whereas Uitdewilligen et al. (2013) and Ashraf et al. (2014) described them, based on GBS. Although, alleles calling in tetraploid potato crop is a highly challenging task. Advancement in computation tools and genome sequencing of more potatoes would be feasible in future to determine more dense molecular markers for genomic selection.
- Training population should include widely cultivated varieties and germplasm collection, including semi-cultivated species and wild species that are well genotyped and phenotyped for target and relevant traits. Historical cultivars and potential parents should be integrated in the training population to get a high prediction value.
- Training or reference population should be well characterized for as many traits under different climatic conditions (tropical, temperate and sub-tropical) in multiple locations. Also, develop historic phenotypic data to develop algorithms to correlate genotypic and phenotypic data with greater prediction accuracy.

- A substantial number of genome-wide distributed SNP markers should be available in either cultivars or landraces or germplasm to be used as training population.
- More diverse population requires a larger set of training or reference population having variability for those traits. Trait heritability is a major factor to determine the genetic factor for any unknown breeding population. Traits with lower heritability would require a much larger training population size.
- Algorithms should be determined for parents to be used in a future breeding program. The genomic data from at least 10% of superior seedling generation should be estimated for future as parents. This would predict future phenotype rather than tedious phenotyping of the whole population.
- In later stage, routine phenotyping of new germplasm to maintain accuracy of genomic selection.
- Hybridization, seedling raising, clonal selection and application of genomic selection model prediction accuracy are applied in breeding population to select elite clones in less time. Genotypic data is used to predict equation for future phenotype as conventional phenotyping normally takes long time. Best parents give maximum genetic gain and accuracy in genomic selection strategy.
- It is likely that pot-grown experiments under glasshouse conditions could further hasten the breeding cycle by raising two crops per year. So, early prediction of desirable traits and parental combinations is possible.
- Clonally propagated crop like potato shows a high degree of heterozygosity and most genomic prediction models include additive (dominance) effect which has been found effective and could play potential roles. Inclusion of non-additive effects (dominance and epistasis) can also be possible in potato varieties which are widely grown, but still has limited application in crop plants.
- *Addition of new germplasm*: Establishment of genomic selection materials should be completed before addition of new germplasm into the genomic selection strategy. Wild species are always sources of new traits, mainly disease-pests resistance, abiotic stress tolerance and tuber nutritive value. Hence, native potatoes, progenitors, landraces, wild relatives, commercial cultivars, advance breeding lines, reference collection are the potential sources of desirable traits. Cultivated background potatoes would be easy to include in training population than the wild germplasm. Genotyping and phenotyping of new materials would be necessary to predict new traits or germplasm.
- A number of traits can be included in this method via high-throughput phenotyping (phenomics), post-harvest traits, such as nutritive, quality and processing traits (reducing sugar, sucrose, starch, acrylamide content, etc.), biotic (disease pest resistance) and abiotic (heat, drought, salinity, nutrient use efficiency) stress resistance/tolerance traits; also integration of new traits based on modern phenomics tools for a large number of traits, metabolites, proteomes, transcriptomes and ionomes, etc. Then develop prediction equations for the trait of interest and apply the prediction equations to the larger set of population to develop a selection index for a number of traits, and allow integration of new traits into the genomic selection.

9.3 Application of Genomic Prediction in Potato

The first genomic selection study was successfully demonstrated recently in potato for late blight and common scab resistance, using the Infinium 8303 SNP array (Enciso-Rodriguez et al., 2018). A set of 381 genotypes was analyzed with field phenotyping data for late blight (273 genotypes with 1,763 disease records) and common scab (370 genotypes with 3,885 disease records) collected for seven and nine years, respectively. They reported moderately high heritability estimates (0.46 ± 0.04, for late blight; and 0.45 ± 0.017, common scab), and genotype × year interaction effect was high and low for late blight and common scab, respectively. More than 90% of variances could be captured, using an additive model. In particular, the highest prediction accuracy was obtained in common scab, using an additive model with small but statistically significant prediction accuracy for late blight using a model with both additive and dominance effects. Further, they identified SNPs in the hot spot region for late blight resistance and a new locus in the WRKY transcription factor for common scab, using whole-genome regression model. Genomic prediction and accuracy may be beneficial when many traits are selected using same genotyping data. Overall, studies suggest that accurate genomic prediction is possible in tetraploid potato crop but it necessitates a very large reference population of potato to capture the entire allelic diversity in whole germplasm collection (Table 8).

10. Conclusion

Conventional breeding has played a significant role in improvement of this clonally-propagated potato crop over the past many decades in breeding varieties for food and nutritional security. However, this method takes 10–12 years to release a variety with desirable agronomic traits through clonal selection over the generations. Since, potato suffers from various biotic and abiotic stresses while breeding varieties all these problems along with agronomic and tuber traits have to be taken into consideration. Selection of contrasting parents with diverse genetic backgrounds is essential in breeding new cultivars. The advances in genomics, particularly genomic selection, has shown promising results for a few traits so far in fast-breeding technology but currently it is at very initial stage in this tetraploid potato. The applications of sequencing technologies, SNP genotyping and genome-wide association studies for accurate genomic prediction are important in genomic selection, using a set of diverse germplasm to accelerate potato breeding for both simple and complex traits.

Acknowledgement

I am thankful to the Director, ICAR-Central Potato Research Institute, Shimla, and scientists/technicians/research fellows and other colleagues of the institute for their support under the institute research projects on biotechnology, germplasm, breeding, and seed research on aeroponics. I am also grateful to the funding agencies for support under the externally funded projects (CABin, ICAR-IASRI, New Delhi; ICAR-LBS Young Scientist Award Project, and DBT, Government of India).

Table 8. Examples of successful work on genomic selection studies in potato.

Target trait	Reference genotypes	High-throughput genotyping system	Heritability, genomic model prediction and accuracy	References
Disease resistance				
Late blight resistance	184 tetraploid potato clones and 96 advanced breeding clones (three years' data)	SolCAP 8.3 k SNP array	Estimated broad sense heritability of 0.77 and prediction accuracy of 0.66–0.68 with various prediction models (additive/dominance/ epistasis effects) and methods.	Stich and van Inghelandt (2018)
Late blight resistance	273 genotypes with 1763 disease records of seven years (early generation and advanced tetraploid breeding lines)	Infinium 8303 Potato Array	Explained moderately high genomic heritability of about 0.46 ± 0.04 using 4110 codominant SNPs. Additive model (A) captured most of the genetic variance (0.94). Cross-validation (CV) resulted in estimated prediction correlations of 0.31 (CV1) and 0.41–0.74 (CV2).	Enciso-Rodriguez (2018)
Common scab resistance	370 genotypes with 3885 disease records of nine years (early generation and advanced tetraploid breeding lines)	Infinium 8303 Potato Array	Explained moderately high genomic heritability of about 0.45 ± 0.02 using 4110 codominant SNPs. Additive model (A) captured most of the genetic variance (0.98). Cross-validation (CV) resulted an estimated prediction correlations of 0.22–0.27 (CV1) and 0.46–0.76 (CV2).	Enciso-Rodriguez (2018)
Maturity corrected resistance	184 tetraploid potato clones and 96 advanced breeding clones (three years' data)	SolCAP 8.3 k SNP array	Estimated broad sense heritability 0.68 and prediction accuracy of 0.75–0.77 with various prediction models (additive/ dominance/ epistasis effects) and methods.	Stich and van Inghelandt (2018)
Plant phenotype				
Plant maturity	184 tetraploid potato clones and 96 advanced breeding clones (three years' data)	SolCAP 8.3 k SNP array	Estimated broad sense heritability 0.92 and prediction accuracy of 0.63–0.71 with various prediction models (additive/dominance/ epistasis effects) and methods.	Stich and van Inghelandt (2018)
Maturity	Reference population size (N_p): 500, 1000, 2000 and 5000	-	Genomic prediction calculated narrow-sense heritability of 0.86 and prediction accuracy of 0.78 (between 0.36 to 0.77 depending up on the population size).	Slater et al. (2016)
Breeders' visual preference	Reference population size (N_p): 500, 1000, 2000 and 5000	-	Genomic prediction calculated narrow-sense heritability of 0.23 and prediction accuracy of 0.33 (between 0.19 to 0.53 depending up on the population size).	Slater et al. (2016)
Number of stems per plant	190 potato varieties	78,111 SilicoDArT markers	Cross-validation accuracy across models was 0.05 (0.01–0.13).	Habyarimana et al. (2017)

Tuber traits				
Tuber starch content	184 tetraploid potato clones and 96 advanced breeding clones (three years' data)	SolCAP 8.3k SNP array	Estimated broad sense heritability 0.92–0.95 and prediction accuracy of 0.39–0.87 with various prediction models (additive/dominance/ epistasis effects) and methods.	Stich and van Inghelandt (2018)
Tuber starch yield	184 tetraploid potato clones and 96 advanced breeding clones (three years' data)	SolCAP 8.3k SNP array	Estimated broad sense heritability 0.78 and prediction accuracy of 0.45–0.50 with various prediction models (additive/dominance/ epistasis effects) and methods.	Stich and van Inghelandt (2018)
Starch content	762 offspring (biparental crosses of 18 tetraploid parents) and 74 clones of a test panel	Genotyping-by-sequencing (1,71,859 SNPs)	Prediction accuracy was 0.56 in training population and 0.30–0.31 in test panel.	Sverrisdóttir et al. (2017)
Dry matter content	762 tetraploid mapping population clones (crosses of 18 potato varieties) and 292 breeding clones (test panel)	Genotyping-by-sequencing (1,67,637 SNPs)	Cross-validated prediction ranged between 0.75–0.83 in combined three populations but prediction across the population was lower (0.37–0.71).	Sverrisdóttir et al. (2018)
Tuber grades as diameter	190 potato varieties	78,111 SilicoDArT markers	Cross-validation accuracy across models was 0.32 (0.15–0.41).	Habyarimana et al. (2017)
Number of tubers per plant	190 potato varieties	78,111 SilicoDArT markers	Cross-validation accuracy across models was 0.17 (0.13–0.23).	Habyarimana et al. (2017)
Tuber flesh colour	190 potato varieties	78,111 SilicoDArT markers	Cross-validation accuracy across models was above or equal to 0.7 except when using Bayes C (0.59).	Habyarimana et al. (2017)
Tuber dry matter content	190 potato varieties	78,111 SilicoDArT markers	Cross-validation accuracy across models was 0.65 (0.54–0.68).	Habyarimana et al. (2017)
Yield				
Yield	Reference population size (N_p): 500, 1000, 2000 and 5000	-	Genomic prediction calculated narrow-sense heritability of 0.56 and prediction accuracy of 0.19 (between 0.29–0.70, depending up on the population size).	Slater et al. (2016)
Tuber yield	184 tetraploid potato clones and 96 advanced breeding clones (three years' data)	SolCAP 8.3 k SNP array	Estimated broad sense heritability 0.77, and prediction accuracy of 0.47–0.53 with various prediction models (additive/dominance/ epistasis effects) and methods.	Stich and van Inghelandt (2018)

Table 8 contd. ...

...Table 8 contd.

Target trait	Reference genotypes	High-throughput genotyping system	Heritability, genomic model prediction and accuracy	References
Total yield	190 potato varieties	78,111 SilicoDArT markers	Cross-validation accuracy across models was 0.37 (0.22–0.41).	Habyarimana et al. (2017)
Total yield	571 round and white clones	SolCAP potato SNP array (v1 containing 8303 SNP markers, or new v2)	Of total genetic variance, SNPs estimated 45% additive genetic variance with prediction accuracy between 0.06–0.63.	Endelman et al. (2018)
Processing traits				
Bioling colour	Reference population size (N_p): 500, 1000, 2000 and 5000	-	Genomic prediction calculated narrow-sense heritability of 0.73 and prediction accuracy of 0.44 (between 0.33–0.74 depending up on the population size).	Slater et al. (2016)
Chipping quality	762 offspring (biparental crosses of 18 tetraploid parents) and 74 clones of a test panel	Genotyping-by-sequencing (1,71,859 SNPs)	Prediction accuracy was 0.73 in training population, and 0.42–0.43 in test panel.	Sverrisdóttir et al. (2017)
Chipping quality	762 tetraploid mapping population clones (crosses of 18 potato varieties) and 292 breeding clones (test panel)	Genotyping-by-sequencing (1,67,637 SNPs)	Cross-validated prediction ranged between 0.39–0.79 in combined three populations but prediction across the population was lower (0.28–0.48).	Sverrisdóttir et al. (2018)
Specific gravity	571 round and white clones	SolCAP potato SNP array (v1 containing 8303 SNP markers, or new v2)	Of total genetic variance, SNPs estimated 20% additive genetic variance with prediction accuracy between 0.25–0.63.	Endelman et al. (2018)
Chip fry color	571 round and white clones	SolCAP potato SNP array (v1 containing 8303 SNP markers, or new v2)	Of total genetic variance, SNPs estimated 45% additive genetic variance with prediction accuracy between 0.40–0.45.	Endelman et al. (2018)
Fry colour	499 breeding lines (TP) and 56 testing panel	Genotyping-by-sequencing (46,406 SNPs)	Average prediction ability ranged between 0.11–0.77 for fry color 'off-the-field' and 0.24–0.66 for fry color after long-term storage at 4.5°C using four models *viz.*, rrBLUP (ridge regression best linear unbiased Predictor), Bayes A, Bayessian Lasso and Random Forest.	Byrne et al. (2020)

References

Ambrosone, A., Costa, A., Martinelli, R., Massarelli, I., Simone, V. et al. (2011). Differential gene regulation in potato cells and plants upon abrupt or gradual exposure to water stress. *Acta Physiol. Plant.*, 33: 1157–1171.

Asano, K., Kobayashi, A., Tsuda, S, Nishinaka, M. and Tamiya, S. (2012). DNA marker-assisted evaluation of potato genotypes for potential resistance to potato cyst nematode pathotypes not yet invading into Japan. *Breeding Science*, 62: 142–150.

Asano, K., Shimosaka, E., Yamashita, Y., Narabu, T., Aiba, S. et al. (2021). Improvement of diagnostic markers for resistance to *Globodera pallida* and application for selection of resistant germplasms in potato breeding. *Breed. Sci.*, 71: 354–364.

Ashraf, B.H., Jensen, J., Asp, T. and Janss, L.L. (2014). Association studies using family pools of outcrossing crops based on allele-frequency estimates from DNA sequencing. *Theor. Appl. Genet.*, 127: 1331–1341.

Bakker, E., Achenbach, U., Bakker, J., van Vliet, J., Peleman, J. et al. (2004). A high-resolution map of the H1 locus harbouring resistance to the potato cyst nematode *Globodera rostochiensis*. *Theor. Appl. Genet.*, 109: 146–152.

Ballvora, A., Ercolano, M.R., Weiß, J., Meksem, K., Bormann, C.A. et al. (2002). The *R1* gene for potato resistance to late blight (*Phytophthora infestans*) belongs to the leucine zipper/NBS/LRR class of plant resistance genes. *Plant J.*, 30: 361–371.

Barone, A. (2004). Molecular marker-assisted selection for potato breeding. *Am. J. Potato Res.*, 81: 111–117.

Bendahmane, A., Kanyuka, K. and Baulcombe, D.C. (1997). High-resolution genetical and physical mapping of the *Rx* gene for extreme resistance to potato virus X in tetraploid potato. *Theor. Appl. Genet.*, 95: 153–162.

Black, W., Mastenbroek, C., Mills, W.R. and Peterson, L.C. (1953). A proposal for an international nomenclature of races of *Phytophthora infestans* and of genes controlling immunity in *Solanum demissum* derivatives. *Euphytica*, 2: 173–240.

Bonierbale, M.W., Plaisted, R.L. and Tanksley, S.D. (1988). RFLP maps based on a common set of clones reveal modes of chromosomal evolution in potato and tomato. *Genetics*, 120: 1095–1103.

Bradeen, J.M. and Kole, C. (2011). *Genetics, Genomics and Breeding of Potato*. Science Publishers, Enfield, New Hampshire, USA.

Bradshaw, J.E. and Mackay, G.R. (1994). Breeding strategies for clonally propagated potatoes. pp. 467–497. *In*: Bradshaw, J.E. and Mackay, G.R. (eds.). *Potato Genetics*. CAB International, Wallingford, UK.

Brigneti, G., Garcia-Mas, J. and Baulcombe, D.C. (1997). Molecular mapping of the potato virus Y resistance gene Ry_{sto} in potato. *Theor. Appl. Genet.*, 94: 198–203.

Brown, C.R., Yang, C.-P., Mojtahedi, H., Santo, G.S. and Masuelli, R. (1996). RFLP analysis of resistance to Columbia root-knot nematode derived from *Solanum bulbocastanum* in a BC_2 population. *Theor. Appl. Genet.*, 92: 572–576.

Bryan, G.J., McLean, K., Bradshaw, J.E., De Jong, W.S., Phillips, M. et al. (2002). Mapping QTLs for resistance to the cyst nematode *Globodera pallida* derived from the wild potato species *Solanum vernei*. *Theor. Appl. Genet.*, 105: 68–77.

Bryan, G.J., McLean, K., Pande, B., Purvis, A., Hackett, C.A. et al. (2004). Genetical dissection of H3-mediated polygenic PCN resistance in a heterozygous autotetraploid potato population. *Mol. Breed.*, 14: 105–116.

Byrne, S., Meade, F., Mesiti, F., Griffin, D., Kennedy, C. and Milbourne, D. (2020). Genome-wide association and genomic prediction for fry color in potato. *Agronomy*, 10: 90.

Carputo, D., Castaldi, L., Caruso, I., Aversano, R., Monti, L. and Frusciante, L. (2007). Resistance to frost and tuber soft rot in near pentaploid *Solanum tuberosum-S. commersonii* hybrids. *Breed. Sci.*, 57: 145–151.

Carputo, D. and Frusciante, L. (2011). Classical genetics and traditional breeding. pp 20–40. *In*: Bradeen, J.M. and Kole, C. (eds.). *Genetics, Genomics and Breeding of Potato*. Science Publishers, Enfield, New Hampshire, USA.

Collard, B.C.Y., Jahufer, M.Z.Z., Brouwer, J.B. and Pang, E.C.K. (2005). An introduction to markers, quantitative trait loci (QTL) mapping and marker-assisted selection for crop improvement: The basic concepts. *Euphytica*, 142: 169–196.

Collard, B.C.Y. and Mackill, D.J. (2008). Marker-assisted selection: An approach for precision plant breeding in the twenty-first century. *Philosoph., Trans. Royal Soc. London, Series B: Biol. Sci.*, 363: 557–572.

Colton, L.M., Groza, H.I., Wielgus, S.M. and Jiang, J. (2006). Marker-assisted selection for the broad-spectrum potato late blight resistance conferred by gene *RB* derived from a wild potato species. *Crop Sci.*, 46: 589–594.

D'hoop, B.B., Paulo, M.J., Mank, R.A., van Eck, H.J. and van Eeuwijk, F.A. (2008). Association mapping of quality traits in potato (*Solanum tuberosum* L.). *Euphytica*, 161: 47–60.

De Jong, W., Forsyth, A., Leister, D., Gebhardt, C. and Baulcombe, D.C. (1997). A potato hypersensitive resistance gene against potato virus X maps to a resistance gene cluster on chromosome 5. *Theor. Appl. Genet.*, 95: 246–252.

De Palma, M., Grillo, S., Massarelli, I., Costa, A., Balogh, G. et al. (2008). Regulation of desaturase gene expression, changes in membrane lipid composition and freezing tolerance in potato plants. *Mol. Breed.*, 21: 15–26.

Demin, I.N., Deryabin, A.N., Sinkevich, M.S. and Trunova, T.I. (2008). Insertion of cyanobacterial desA gene coding for Δ12-acyl-lipid desaturase increases potato plant resistance to oxidative stress induced by hypothermia. *Russian J. Plant Physiol.*, 55: 639–648.

Elshire, R.J., Glaubitz, J.C., Sun, Q., Poland, J.A., Kawamoto, K. et al. (2011). A robust, simple genotyping-bysequencing (GBS) approach for high diversity species. *PLoS ONE*, 6: e19379.

Enciso-Rodriguez, F., Douches, D., Lopez-Cruz, M., Coombs, J. and de Los Campos, G. (2018). Genomic selection for late blight and common scab resistance in tetraploid potato (*Solanum tuberosum*). *G3 (Bethesda)*, 8(7): 2471–2481.

Endelman, J.B., Carley, C., Bethke, P.C., Coombs, J.J., Clough, M.E. et al. (2018). Genetic variance partitioning and genome-wide prediction with allele dosage information in autotetraploid potato. *Genetics*, 209(1): 77–87.

Evers, D., Lefèvre, I., Legay, S., Lamoureux, D., Hausman, J.F., Rosales, R.O.G. et al. (2010). Identification of drought-responsive compounds in potato through a combined transcriptomic and targeted metabolite approach. *J. Exp. Bot.*, 61: 2327–2343.

Finkers-Tomczak, A., Danan, S., van Dijk, T., Beyene, A., Bouwman, L. et al. (2009). A high-resolution map of the *Grp1* locus on chromosome V of potato harbouring broad-spectrum resistance to the cyst nematode species *Globodera pallida* and *Globodera rostochiensis*. *Theor. Appl. Genet.*, 119: 165–173.

Fischer, M., Schreiber, L., Colby, T., Kuckenberg, M., Tacke, E. et al. (2013). Novel candidate genes influencing natural variation in potato tuber cold sweetening identified by comparative proteomics and association mapping. *BMC Plant Biol.*, 13: 113–128.

Flis, B., Hennig, J., Strzelczyk-Zyta, D., Gebhardt, C. and Marczewski, W. (2005). The $Ry\text{-}f_{sto}$ gene from *Solanum stoloniferum* for extreme resistant to Potato virus Y maps to potato chromosome XII and is diagnosed by PCR marker $GP122_{718}$ in PVY resistant potato cultivars. *Mol. Breed.*, 15: 95–101.

Galek, R., Rurek, M., De Jong, W.S., Pietkiewicz, G., Augustyniak, H. and Sawicka-Sienkiewicz, E. (2011). Application of DNA markers linked to the potato *H1* gene conferring resistance to pathotype Ro1 of *Globodera rostochiensis*. *J. Appl. Genet.*, 52: 407–411.

Gebhardt, C., Ritter, E., Debener, T., Schachtschabel, U., Walkemeier, B. et al. (1989). RFLP analysis and linkage mapping in *Solanum tuberosum*. *Theor. Appl. Genet.*, 78: 65–75.

Gebhardt, C., Mugniery, D., Ritter, E., Salamini, F. and Bonnel, E. (1993). Identification of RFLP markers closely linked to the H1 gene conferring resistance to *Globodera rostochiensis* in potato. *Theor. Appl. Genet.*, 85: 541–544.

Gebhardt, C., Ballvora, A., Walkemeier, B., Oberhagemann, P. and Schüler, K. (2004). Assessing genetic potential in germplasm collections of crop plants by marker-trait association: a case study for potatoes with quantitative variation of resistance to late blight and maturity type. *Mol. Breed.*, 13: 93–102.

Gebhardt, C., Bellin, D., Henselewski, H., Lehmann, W., Schwarzfischer, J. and Valkonen, J.P.T. (2006). Marker-assisted combination of major genes for pathogen resistance in potato. *Theor. Appl. Genet.*, 112: 1458–1464.

Ghislain, M., Núñez, J., Rosario Herrera, M., Pignataro, J., Guzman, F. et al. (2009). Robust and highly informative microsatellite-based genetic identity kit for potato. *Mol. Breed.*, 23: 377–388.

Gidskehaug, L., Kent, M., Hayes, B.J. and Lien, S. (2011). Genotype calling and mapping of multisite variants using an Atlantic salmon iSelect SNP array. *Bioinformatics*, 27: 303–310.

Golas, T.M., Sikkema, A., Gros, J., Feron, R.M.C., van den Berg, R.G. et al. (2010). Identification Hander of a resistance gene *Rpi-dlc1* to *Phytophthora infestans* in European accessions of *Solanum dulcamara*. *Theor. Appl. Genet.*, 120: 797–808.

Gopal, J. (1997). Progeny selection for agronomic characters in early generation of a potato breeding program. *Theor. Appl. Genet.*, 95: 307–311.

Gopal, J., Chahal, G.S. and Minocha, J.L. (2000). Progeny mean, heterosis and heterobeltiosis in *Solanum tuberosum* × *tuberosum* and *S. tuberosum* × *andigena* families under a short day sub-tropic environment. *Potato Res.*, 43: 61–70.

Habyarimana, E., Parisi, B. and Mandolino, G. (2017). Genomic prediction for yields, processing and nutritional quality traits in cultivated potato (*Solanum tuberosum* L.). *Plant Breed.*, 136: 245–252.

Hämäläinen, J.H., Sorri, V.A., Watanabe, K.N., Gebhardt, C. and Valkonen, J.P.T. (1998). Molecular examination of a chromosome region that controls resistance to potato Y and A potyviruses in potato. *Theor. Appl. Genet.*, 96: 1036–1043.

Hamilton, J., Hansey, C., Whitty, B., Stoffel, K., Massa, A. et al. (2011). Single nucleotide polymorphism discovery in elite North American potato germplasm. *BMC Genomics*, 12: 302–313.

Heldák, J., Bežo, M., Štefúnová, V. and Gallikovà, A. (2007). Selection of DNA markers for detection of extreme resistance to potato virus Y in tetraploid potato (*Solanum tuberosum* L.) F_1 progenies. *Czech J. Genet. Plant Breed.*, 43: 125–134.

Heslot, N., Yang, H., Sorrells, M.E. and Jannink, J. (2012). Genomic selection in plant breeding: a comparison of models. *Crop Sci.*, 52: 146–160.

Hosaka, K., Hosaka, Y., Mori, M., Maida, T. and Matsunaga, H. (2001). Detection of a simplex RAPD marker linked to resistance to potato virus Y in a tetraploid potato. *Am. J. Potato Res.*, 78: 191–196.

Huang, S., van der Vossen, E.A.G., Kuang, H., Vleeshouwers, V.G.A.A., Zhang, N. et al. (2005). Comparative genomics enabled the cloning of the *R3a* late blight resistance gene in potato. *The Plant J.*, 42: 251–261.

Jacobs, J.M.E., Eck, H.J., Horsman, K., Arens, P.F.P., Verkerk-Bakker, B. et al. (1996). Mapping of resistance to the potato cyst nematode *Globodera rostochiensis* from the wild potato species *Solanum vernei*. *Mol. Breed.*, 2: 51–60.

Jacobs, M.M.J., Vosman, B., Vleeshouwers, V.G.A.A., Visser, R.G.F., Henken, B. and van den Berg, R.G. (2010). A novel approach to locate *Phytophthora infestans* resistance genes on the potato genetic map. *Theor. Appl. Genet.*, 120: 785–796.

Jupe, F., Witek, K., Verweij, W., Śliwka, J., Pritchard, L. et al. (2013). Resistance gene enrichment sequencing (RenSeq) enables reannotation of the NB-LRR gene family from sequenced plant genomes and rapid mapping of resistance loci in segregating populations. *The Plant J.*, 76: 530–544.

Kasai, K., Morikawa, Y., Sorri, V.A., Valkonen, J.P.T., Gebhardt, C. and Wantabe, K.N. (2000). Development of SCAR markers to the PVY resistance gene Ry_{adg} based on a common feature of plant disease resistance genes. *Genome*, 43: 1–8.

Kawchuk, L.M., Lynch, D.R. and Yada, R.Y. (2008). Marker assisted selection of potato clones that process with light chip color. *Am. J. Potato Res.*, 85: 227–231.

Kelley, K.B., Whitworth, J.L. and Novy, R.G. (2009). Mapping of the potato leafroll virus resistance gene, Rlr_{etb}, from *Solanum etuberosum* identifies interchromosomal translocations among its E-genome chromosomes 4 and 9 relative to the A-genome of *Solanum* L. sect. Petota. *Mol. Breed.*, 23: 489–500.

Kim, H.-J., Lee, H.-R., Jo, K.-R., Mortazavian, S.M.M., Huigen, D.J. et al. (2012). Broad spectrum late blight resistance in potato differential set plants Ma*R8* and Ma*R9* is conferred by multiple stacked *R* genes. *Theor. Appl. Genet.*, 124: 923–935.

Kuhl, J.C. (2011). Mapping and tagging of simply inherited traits. pp. 90–112. *In*: Bradeen, J.M. and Kole, C. (eds.). *Genetics, Genomics and Breeding of Potato*. Science Publishers, Enfield, New Hampshire, USA.

Kumar, R. and Gopal, J. (2003). Combining ability of andigena accessions for yield components and tuber dry matter in third clonal generation. *J. Indian Potato Assoc.*, 30: 3–4.

Kumar, R. and Gopal, J. (2006). Repeatability of progeny mean, combining ability, heterosis and heterobeltiosis in early generations of a potato breeding program. *Potato Res.*, 49(2): 131–141.

Li, L., Tacke, E., Hofferbert, H.-R., Lübeck, J., Strahwald, J. et al. (2013). Validation of candidate gene markers for marker-assisted selection of potato cultivars with improved tuber quality. *Theor. Appl. Genet.*, 126: 1039–1052.

Li, X., Xu, J., Duan, S., Bian, C., Hu, J. et al. (2018). Pedigree-based deciphering of genome-wide conserved patterns in an elite potato parental line. *Front. Plant Sci.*, 9: 690.

Lindqvist-Kreuze, H., Gastelo, M., Perez, W., Forbes, G.A., De Koeyer, D. and Bonierbale, M. (2014). Phenotypic stability and genome-wide association study of late blight resistance in potato genotypes adapted to the tropical highlands. *Phytopathology*, 104: 624–633.

Lopez-Pardo, R., Barandalla, L., Ritter, E. and de Galarreta, J.I.R. (2013). Validation of molecular markers for pathogen resistance in potato. *Plant Breed.*, 132: 246–251.

Luthra, S.K., Sharma, P.C. and Gopal, J. (2006). Identification of superior parents and crosses in potato (*S. tuberosum* L.) by combining ability analysis. *Indian J. Agric. Sci.*, 76(3): 205–208.

Luthra, S.K., Gupta, V.K., Tiwari, J.K., Kumar, V., Bhardwaj, V. et al. (2020). Potato Breeding in India, *CPRI Technical Bulletin No 74* (revised), ICAR-Central Potato Research Institute, Shimla, Himachal Pradesh, India. P. 214.

Mane, S.P., Robinet, C.V., Ulanov, A., Schafleitner, R., Tincopa, L. et al. (2008). Molecular and physiological adaptation to prolonged drought stress in the leaves of two Andean potato genotypes. *Funct. Plant Biol.*, 35: 669–688.

Marano, M.R., Malcuit, I., De Jong, W. and Baulcombe, D.C. (2002). High-resolution genetic map of *Nb*, a gene that confers hypersensitive resistance to potato virus X in *Solanum tuberosum*. *Theor. Appl. Genet.*, 105: 192–200.

Marczewski, W., Ostrowska, K. and Zimnoch-Guzowska, E. (1998). Identification of RAPD markers linked to the *Ns* locus in potato. *Plant Breed.*, 117: 88–90.

Marczewski, W. (2001). Inter-simple sequence repeat (ISSR) markers for the *Ns* resistance gene in potato (*Solanum tuberosum* L.). *J. Appl. Genet.*, 42: 139–144.

Marczewski, W., Flis, B., Syller, J., Schäfer-Preg, R. and Gebhardt, C. (2001a). A major quantitative trait locus for resistance to *Potato leafroll virus* is located in a resistance hotspot on potato chromosome XI and is tightly linked to *N*-Gene-like markers. *Mol. Plant Microbe Interact.*, 14: 1420–1425.

Marczewski, W., Talarczyk, A. and Hennig, J. (2001b). Development of SCAR markers linked to the *Ns* locus in potato. *Plant Breed.*, 120: 88–90.

Marczewski, W., Hennig, J. and Gebhardt, C. (2002). The Potato virus S resistance gene Ns maps to potato chromosome VIII. *Theor. Appl. Genet.*, 105: 564–567.

Marczewski, W., Flis, B., Syller, J., Strzelczyk-Żyta, D., Hennig, J. and Gebhardt, C. (2004). Two allelic or tightly linked genetic factors at the *PLRV.4* locus on potato chromosome XI control resistance to potato leafroll virus accumulation. *Theor. Appl. Genet.*, 109: 1604–1609.

Marczewski, W., Strzelczyk-Zyta, D., Hennig, J., Witek, K. and Gebhardt, C. (2006). Potato chromosomes IX and XI carry genes for resistance to potato virus M. *Theor. Appl. Genet.*, 112: 1232–1238.

Meksem, K., Leister, D., Peleman, J., Zabeau, M., Salamini, F. and Gebhardt, C. (1995). A high-resolution map of the vicinity of the *R1* locus on chromosome V of potato based on RFLP and AFLP markers. *Mol. Gen. Genet.*, 249: 74–81.

Meuwissen, T.H., Hayes, B.J. and Goddard, M.E. (2001). Prediction of total genetic value using genome-wide dense marker maps. *Genetics*, 157: 1819–1829.

Milbourne, D., Meyer, R.C., Collins, A.J., Ramsay, L.D., Gebhardt, C. and Waugh, R. (1998). Isolation, characterization and mapping of simple sequence repeat loci in potato. *Mol. Gen. Genet.*, 259: 233–245.

Milczarek, D., Flis, B. and Przetakiewicz, A. (2011). Suitability of molecular markers for selection of potatoes resistant to *Globodera* spp. *Am. J. Potato Res.*, 88: 245–255.

Moloney, C., Griffin, D., Jones, P., Bryan, G., McLean, K. et al. (2010). Development of diagnostic markers for use in breeding potatoes resistant to *Globodera pallida* pathotype Pa2/3 using germplasm derived from *Solanum tuberosum* ssp. *andigena* CPC 2802. *Theor. Appl. Genet.*, 120: 679–689.

Moreau, L., Lemarié, S., Charcosset, A. and Gallais, A. (2000). Economic efficiency of one cycle of marker-assisted selection. *Crop Sci.*, 40: 329.

Mori, K., Sakamoto, Y., Mukojima, N., Tamiya, S., Nakao, T. et al. (2011). Development of a multiplex PCR method for simultaneous detection of diagnostic DNA markers of five disease and pest resistance genes in potato. *Euphytica*, 180: 347–355.

Niewöhner, J., Salamini, F. and Gebhardt, C. (1995). Development of PCR assays diagnostic for RFLP marker alleles closely linked to alleles *Gro1* and *H1*, conferring resistance to the root cyst nematode *Globodera rostochiensis* in potato. *Mol. Breed.*, 1: 65–78.

Novy, R.G., Gillen, A.M. and Whitworth, J.L. (2007). Characterization of the expression and inheritance of potato leafroll virus (PLRV) and potato virus Y (PVY) resistance in three generations of germplasm derived from *Solanum etuberosum*. *Theor. Appl. Genet.*, 114: 1161–1172.

Oberhagemann, P., Chatot-Balandras, C., Schäfer-Pregl, R., Wegener, D., Palomino, C. et al. (1999). A genetic analysis of quantitative resistance to late blight in potato: towards marker-assisted selection. *Mol. Breed.*, 5: 399–415.

Ortiz, R. (2020). Genomic-led potato breeding for increasing genetic gains: Achievements and outlook. *Crop Breed. Genet. Genom.*, 2(2): e200010.

Oufir, M., Legay, S., Nicot, N., Moer, K., Hoffmann, L. et al. (2008). Gene expression in potato during cold exposure: Changes in carbohydrate and polyamine metabolisms. *Plant Sci.*, 175: 839–852.

Paal, J., Henselewski, H., Muth, J., Meksem, K., Menéndez, C.M. et al. (2004). Molecular cloning of the potato *Gro1-4* gene conferring resistance to pathotype Ro1 of the root cyst nematode *Globodera rostochiensis*, based on a candidate gene approach. *The Plant J.*, 38: 285–297.

Pajerowska-Mukhtar, K., Stich, B., Achenbach, U., Ballvora, A., Lübeck, J. et al. (2009). Single nucleotide polymorphisms in the allene oxide synthase 2 gene are associated with field resistance to late blight in populations of tetraploid potato cultivars. *Genetics*, 181: 1115–1127.

Pande, P.C., Luthra, S.K., Singh, B.P. and Pandey, S.K. (2005). Selection of superior crosses on the basis of progeny mean in potato. *Potato J.*, 32: 37–42.

Pankin, A., Sokolova, E., Rogozina, E., Kuznetsova, M., Deahl, K. et al. (2011). Allele mining in the gene pool of wild *Solanum* species for homologues of late blight resistance gene *RB/Rpi-blb1*. *Plant Genet. Res. Character. Utiliz.*, 9: 305–308.

Park, T.-H., Vleeshouwers, V.G.A.A., Hutten, R.C.B., van Eck, H.J., van der Vossen, E. et al. (2005b). High-resolution mapping and analysis of the resistance locus *Rpi-abpt* against *Phytophthora infestans* in potato. *Mol. Breed.*, 16: 33–43.

Pel, M.A., Foster, S.J., Park, T.-H., Rietman, H., van Arkel, G. et al. (2009). Mapping and cloning of late blight resistance genes from *Solanum venturii* using an interspecific candidate gene approach. *Mol. Plant Microbe Interact.*, 22: 601–615.

Ramakrishnan, A., Ritland, C.E., Sevillano, R.H.B. and Riseman, A. (2015). Review of potato molecular markers to enhance trait selection. *Am. J. Potato Res.*, 92: 455–472.

Rietman, H. (2011). Putting the *Phytophthora infestans* genome sequence at work; identification of many new *R* and *Avr* genes in *Solanum*. PhD Thesis, Wageningen University.

Ritter, E., Debener, T., Barone, A., Salamini, F. and Gebhardt, C. (1991). RFLP mapping on potato chromosomes of two genes controlling extreme resistance to potato virus X (PVX). *Mol. Gen. Genet.*, 227: 81–85.

Rouppe Van Der Voort, J., Wolters, P., Folkertsma, R., Hutten, R., van Zandvoort, P. et al. (1997). Mapping of the cyst nematode resistance locus *Gpa2* in potato using a strategy based on comigrating AFLP markers. *Theor. Appl. Genet.*, 95: 874–880.

Sattarzadeh, A., Achenbach, U., Lübeck, J., Strahwald, J., Tacke, E. et al. (2006). Single nucleotide polymorphism (SNP) genotyping as basis for developing a PCR based marker highly diagnostic for potato varieties with high resistance to *Globodera pallida* pathotype Pa2/3. *Mol. Breed.*, 18: 301–312.

Shinozaki, K. and Yamaguchi-Shinozaki, K. (2007). Gene networks involved in drought stress response and tolerance. *J. Exp. Bot.*, 58: 221–227.

Skupinová, S., Vejl, P., Sedlák, P. and Domkářová, J. (2002). Segregation of DNA markers of potato (*Solanum tuberosum* ssp. *tuberosum* L.) resistance against Ro1 pathotype *Globodera rostochiensis* in selected F_1 progeny. *Potato Res.*, 2002: 480–485.

Slater, A.T., Cogan, N.O.I. and Forster, J.W. (2013). Cost analysis of the application of marker-assisted selection in potato breeding. *Mol. Breed.*, 32: 299–310.

Slater, A.T., Wilson, G.M., Cogan, N.O.I., Forster, J.W. and Hayes, B.J. (2014a). Improving the analysis of low heritability complex traits for enhanced genetic gain in potato. *Theor. Appl. Genet.*, 127: 809–820.

Slater, A.T., Cogan, N.O.I., Hayes, B.J., Schultz, L. and Finlay, M. (2014b). Improving breeding efficiency in potato using molecular and quantitative genetics. *Theor. Appl. Genet.*, 127: 2279–2292.

Slater, A.T., Cogan, N.O., Forster, J.W., Hayes, B.J. and Daetwyler, H.D. (2016). Improving genetic gain with genomic selection in autotetraploid potato. *The Plant Genome*, 9(3): 1–15. Doi: 10.3835/plantgenome2016.02.0021.

Slater, A.T., Cogan, N.O.I., Rodoni, B.C., Daetwyler, H.D., Hayes, B.J. et al. (2017). Breeding differently-the digital revolution: High-throughput phenotyping and genotyping. *Potato Res.*, 60: 337–352.

Śliwka, J., Jakuczun, H., Lebecka, R., Marczewski, W., Gebhardt, C. et al. (2006). The novel, major locus *Rpi-phu1* for late blight resistance maps to potato chromosome IX and is not correlated with long vegetation period. *Theor. Appl. Genet.*, 113: 685–695.

Smilde, W.D., Brigneti, G., Jagger, L., Perkins, S. and Jones, J.D. (2005). *Solanum mochiquense* chromosome IX carries a novel late blight resistance gene *Rpi-moc1*. *Theor. Appl. Genet.*, 110: 252–258.

Sokolova, E., Pankin, A., Beketova, M., Kuznetsova, M., Spiglazova, S. et al. (2011). SCAR markers of the R-genes and germplasm of wild Solanum species for breeding late blight resistant potato cultivars. *Plant Genet. Resour.*, 9: 309–312.

Song, Y.-S., Hepting, L., Schweizer, G., Hartl, L., Wenzel, G. and Schwarzfischer, A. (2005). Mapping of extreme resistance to PVY (Ry(sto)) on chromosome XII using anther-culture-derived primary dihaploid potato lines. *Theor. Appl. Genet.*, 111: 879–887.

Sorri, V.A., Watanabe, K.N. and Valkonen, J.P.T. (1999). Predicted kinase-3a motif of a resistance gene analogue as a unique marker for virus resistance. *Theor. Appl. Genet.*, 99: 164–170.

Stich, B. and van Inghelandt, D. (2018). Prospects and potential uses of genomic prediction of key performance traits in tetraploid potato. *Front. Plant Sci.*, 9: 159.

Sverrisdóttir, E., Byrne, S., Sundmark, E., Johnsen, H.Ø., Kirk, H.G. et al. (2017). Genomic prediction of starch content and chipping quality in tetraploid potato using genotyping-by-sequencing. *Theor. Appl. Genet.*, 130(10): 2091–2108.

Sverrisdóttir, E., Sundmark, E., Johnsen, H.Ø., Kirk, H.G., Asp, T. et al. (2018). The value of expanding the training population to improve genomic selection models in tetraploid potato. *Front. Plant Sci.*, 9: 1118.

Szajko, K., Chrzanowska, M., Witek, K., Strzelczyk-Zyta, D., Zagórska, H. et al. (2008). The novel gene *Ny-1* on potato chromosome IX confers hypersensitive resistance to *Potato virus Y* and is an alternative to *Ry* genes in potato breeding for PVY resistance. *Theor. Appl. Genet.*, 116: 297–303.

Tan, M.Y.A., Hutten, R.C.B., Visser, R.G.F. and van Eck, H.J. (2010). The effect of pyramiding *Phytophthora infestans* resistance genes Rpi-mcd1 and Rpi-ber in potato. *Theor. Appl. Genet.*, 121: 117–125.

Tanksley, S.D., Ganal, M.W., Prince, J.P., de Vicente, M.C., Bonierbale, M.W. et al. (1992). High density molecular linkage maps of the tomato and potato genomes. *Genetics*, 132: 1141–1160.

The Potato Genome Sequencing Consortium. (2011). Genome sequence and analysis of the tuber crop potato. *Nature*, 475: 189–195.

Tiwari, J.K., Gopal, J. and Singh, B.P. (2012). Marker-assisted selection for virus resistance in potato: options and challenges. *Potato J.*, 39: 101–117.

Tiwari, J.K., Sundaresha, S., Singh, B.P., Kaushik, S., Chakrabarti, S.K. et al. (2013). Molecular markers for late blight resistance breeding of potato: an update. *Plant Breed.*, 132: 237–245.

Uitdewilligen, J.G.A.M.L., Wolters, A.A., D'hoop, B.B., Borm, T.J.A., Visser, R.G.F. and van Eck, H.J. (2013). A next-generation sequencing method for genotyping-by-sequencing of highly heterozygous autotetraploid potato. *PLoS ONE*, 8: e62355.

Van Os, H., Andrzejewski, S., Bakker, E., Barrena, I., Bryan, G.J. et al. (2006). Construction of a 10,000-marker ultradense genetic recombination map of potato: Providing a framework for accelerated gene isolation and a genomewide physical map. *Genetics*, 173: 1075–1087.

Velásquez, A.C., Mihovilovich, E. and Bonierbale, M. (2007). Genetic characterization and mapping of major gene resistance to potato leafroll virus in *Solanum tuberosum* ssp. *andigena*. *Theor. Appl. Genet.*, 114: 1051–1058.

Verzaux, E. (2010). *Resistance and Susceptibility to Late Blight in Solanum: Gene Mapping, Cloning and Stacking*. PhD Thesis, Wageningen University, Wageningen.

Watanabe, K., Kikuchi, A., Shimazaki, T. and Asahina, M. (2011). Salt and drought stress tolerances in transgenic potatoes and wild species. *Potato Res.*, 54: 319–324.

Wickramasinghe, W.M.D.K., Qu, X.S., Costanzo, S., Haynes, K.G. and Christ, B.J. (2009). Development of PCR-based markers linked to quantitative resistance to late blight in a diploid hybrid potato population of *Solanum phureja* × *S. stenotomum*. *Am. J. Potato Res.*, 86: 188–195.

Witek, K., Strzelczyk-Zyta, D., Hennig, J. and Marczewski, W. (2006). A multiplex PCR approach to simultaneously genotype potato towards the resistance alleles *Ry-f*$_{sto}$ and *Ns*. *Mol. Breed.*, 18: 273–275.

Zia, M.A.B., Naeem, M., Demirel, U. and Caliskan, M.E. (2017). Next generation breeding in potato. *Ekin J.*, 3(2): 1–33.

Chapter 6

Omics Approaches in Potato

1. Introduction

Potato is an important crop for food and nutritional security of the world's increasing population. Despite the rich *Solanum* genepool, potato has a narrow genetic base and hence it requires use of diverse genetic materials for improvement by applying advanced technologies. Moreover, the potato crop also suffers from various insect-pests (aphids, mites, thrips, hoppers, potato cyst nematodes, etc.), diseases (late blight, viruses, bacterial wilt, storage rots, etc.), environmental problems of abiotic stresses, such as heat, drought, cold and salinity and processing quality as well as nutritional traits. Chemical sprays are adopted to manage insect-pests and diseases but in many cases, they are ineffective or unavailable, such as for viruses. As an example, Irish famine in 1845 saw massive losses of potato crop in Ireland due to the fungus *Phytophthora infestans* causing late blight disease. This resulted in death of millions of Irish people due to starvation and unavailability of food. Therefore, molecular characterizations of genetic materials are required for alterative applications of biotechnology and breeding methods. A number of omics approaches, mainly genomics, transcriptomics, proteomics, metabolomics, ionomics and phenomics are now available for biotechnological application in potato to meet the food demand and sustainable agriculture. Among these, genomics has been discussed earlier and phenomics will be elaborated separately. Here the focus is on aplicaiton of transcriptomics, proteomics, metabolomics and ionomics approaches in potato research.

The term 'functional genomics' referes to the development and application of genome-wide experimental approaches to assess gene function by making use of information and reagents provided by structural genomics. It involves the use of high-throughput methods to study entire gene sets. Indirect information on cellular or developmental function can be obtained from spatial and temporal expression patterns, for example, the presence of mRNA and/or protein in different cell types, during development, during pathogen infection, or in different environments. The subcellular localization and post-translational modification of proteins and metabolites can also be informative. The techniques used for functional genomics worldwide include microarray which has now been replaced with RNA sequencing and functional validation of genes by real-time PCR, transgenics regeneration (gene overexpression or RNAi: RNA interference) and VIGS (Virus Induced Gene Silencing) techniques. The introduction of real-time PCR technology has significantly improved and simplified the quantification of nucleic acids, with this technology becoming an invaluable tool for many scientists working in different disciplines. RT-PCR has gained importance especially in the field of molecular diagnostics, gene expression and copy number detection assays.

With the completion of genome sequencing of potato, focus has now shifted from sequencing to delineating the biological functions of genes, proteins and metabolites. In recent years, significant advances have been made in genome sequencing technology and with increased application of omics approaches to deal cellular and molecular components in plants. The potato genome sequence provides unprecedented molecular information about potato, which is most important in plant biology and provides genes related to various desirable traits for crop improvement (The Potato Genome Sequencing Consortium, 2011). Further, the biological information, such as DNA, RNA, proteins and metabolites is linked to an array of metabolic pathways leading to the target traits. To uncover the complete pathways, omics approaches are greatly required to detect the traits and regulate the mechanism at the cellular, tissue and organism levels. A combination of omics tools is required to deliver reliable information which would lead to breeding new crop cultivars. The several omics data are very huge, consisting of different samples/time points/ environments and so cannot be handled manually. The omics data are therefore handled by using a pipeline of bioinformatics approaches. All prior information of genome sequence data is very useful in predicting future crop phenotype at the very early generation itself, without additional phenotypes. In other words, omics approaches allow selection of future genotypes based on high-throughput genotyping of potato accessions without additional phenotyping (phenomics) over multiple years/seasons and locations. Thus integration of omics data is very essential to accelerate potato breeding (Fig. 1).

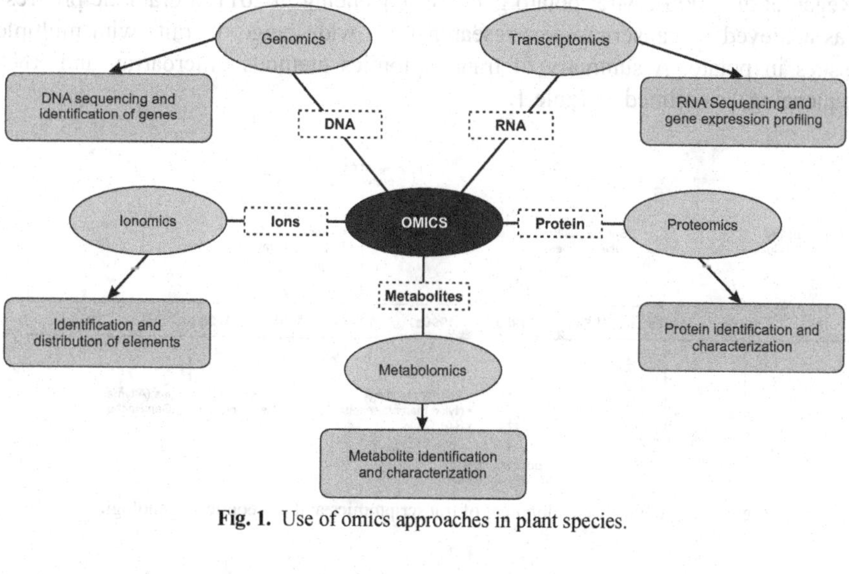

Fig. 1. Use of omics approaches in plant species.

2. Transcriptomics

Transcriptomics is the study of transcripts and it includes coding RNA (messenger RNA) and non-coding RNA molecules of any organism. A number of transcriptomics studies have been carried out in potato for multiple traits of importance, such as biotic and abiotic stresses, processing and quality traits. This has resulted in genes discovery and their characterization for numerous traits and opening up of new avenues for gene manipulation and molecular breeding in potato. Micrarrray and RNA sequencing (RNA-seq) methods have been used the world over for transcriptomics studies. A timeline is shown in Fig. 2. At the begninjng, in late 1970s, libraries of silkmoth mRNAs were prepared and converted into cDNA for storage, using reverse transcriptoase. In the 1980s, Expressed Sequence Tags (ESTs) sequencing based on the Sanger sequencing method was followed. This was predominant till the discovery of Solexa/Illumina sequencing by synthesis technologies. During the 1990s, the word 'transcriptome' was used for the first time and quantitation using nother blotting, nylon membrane arrays and reverse transcriptase quantitative PCR (RT-PCR) became popular methods. In 1995, development of ESTs methodology using Sanger sequencing-based transcriptomics method, called 'Serial Analysis of Gene Expression (SAGE)' was developed. By 2000s, discovery of microarray (2006 by NimbleGen Roche) and RNA-sequencing (RNA-seq) in 2008 by Illumina technologies became popular the world over for transcriptomcs research. Later, RNAs-seq became the most popular transcriptome sequencing technology in the world.

Before the invention of potato genome sequencing, a transcriptomics study identified 20,756 Expressed Sequence Tags (ESTs) from cDNA library of mRNA in leaf and root tissues of potato for abiotic stresses (heat, drought, cold and salt) (Rensink et al., 2005). Later, the Potato Oligo Consortium (POCI) was constituted, using 44,000 probes representing 42,034 unigenes in potato (Kloosterman et al., 2008). The POCI array was integrated into the Canadian potato breeding program, using functional genomics to improve disease resistance and tuber quality traits (Regan et al., 2006). After potato genome sequencing in 2011, a dramatic progress was achieved in transcriptomics research for a wide range of traits with multiple tissues in potato. A summary of transcriptomics methods (microarray and RNA sequencing) is outlined in Table 1.

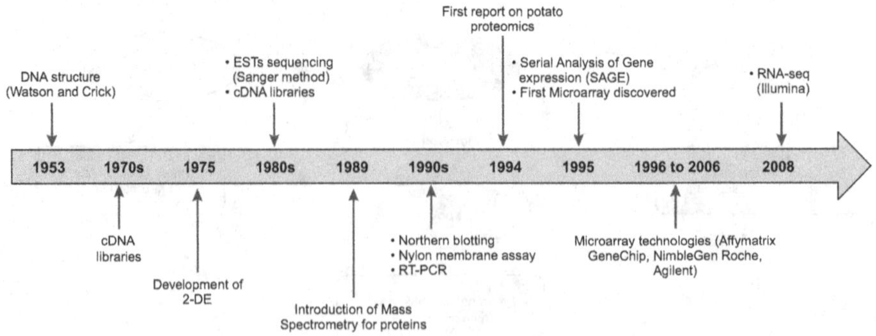

Fig. 2. A timeline of development of transcriptomics and proteomics technologies.

Table 1. Comparison of transcriptomics methods.

Parameters	**Microarray**	**RNA sequencing**
Data throughput	Higher	High
Input template	High ~ 1 µg mRNA	Low (~ 1 ng total RNA)
Sample preparation and data analysis	Low labour intensive	High labour intensive
Prior knowledge on genome	Reference transcripts required for probes preparation	Not required (but genome sequences useful)
Quantitation accuracy	> 90% (depending upon fluorescence detection accuracy)	~ 90% (depending upon sequences coverage)
Seqeunce resolution	Dedicated arrays can detect splice variants (limited by probe design and cross-hybridisation)	Able to detect SNPs and variants
Sensitivity	10^{-3} (depending upon fluorescence detection)	10^{-6} (depending upon sequence coverage)
Dynamic range	103–104 (depending upon fluorescence saturation)	> 10^5 (depending upon sequence coverage)
Reproducibility	> 99%	> 99%
Sequencing platform	• Affymetrix GeneChip array • NimbleGen array • Agilent array	• Roche 454 (Switzerland) (2005) • Illumina (USA) (2006) • SOLiD (Thermo Fisher Scientific, USA) (2008) • Ion Torrent (Thermo Fisher Scientific, USA) (2010) • PacBio (Pacbio, USA) (2011)

Source: Modified and adapted from Lowe et al. (2017)

2.1 Microarray Technology

Microarray consists of short oligonucletides called ‘probes’, which are arrayed in a solid substrate (glass). Transcript abundance is determined by hybridization of fluorescently labeled transcripts to these probes. Microarrays technology allows the assay of thousands of transcripts at lesser cost and time, uisng spotted oligonucletide arrays and Affimatrix high-density arrays. Further, this was advanced by the use of advancement in fluorescence detection techniques for sensitivity and accuracy of measurement. Two types of microarrays have been used: low-density spotted arrays and high-density probe arrays. High-density arrays were popularized by the Affymetrix GeneChip array in which each transcript is quantified by numerous 25-mer probes. NimbleGen arrays are also high-density arrays which permit flexible manufacture of arrays in small or large numbers. These arrays have hundreds of thousands of 45–85 mer probes and are hybridized with a one-color labeled sample for expression analysis (Fig. 3).

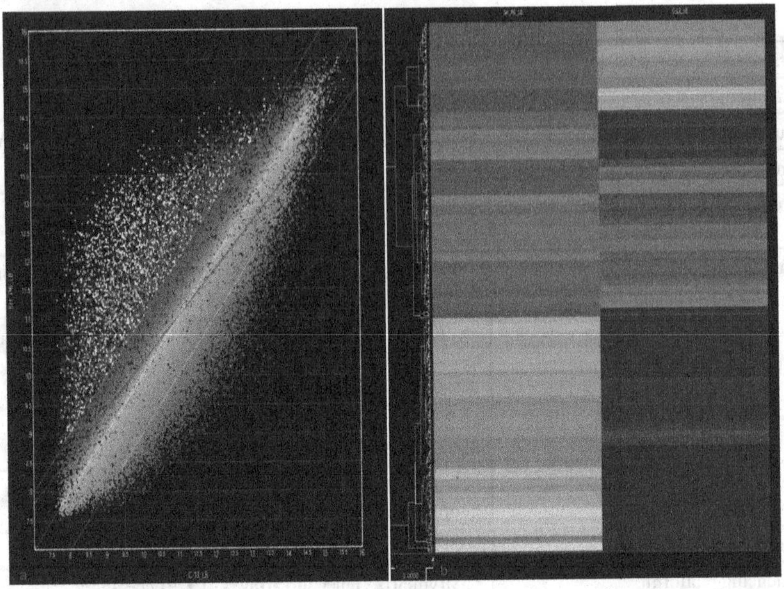

Fig. 3. Scatter plot (a) and Heat map (b) of differentially expressed genes in potato using microarray technology.

2.2 Transcriptome Sequencing (RNA-sequencing)

RNA-seq refers to the sequencing of transcript cDNAs synthesized from total RNA of plant tissues. It is the combination of a high-throughput sequencing methodology with computational methods to capture and quantify transcripts present in an RNA extract. The nucleotide sequences generated are typically around 100 bp in length, but can range from 30 bp to over 10,000 bp, depending on the sequencing method used. RNA-seq leverages deep sampling of the transcriptome with many short fragments from a transcriptome to allow computational reconstruction of the original RNA transcript by aligning reads to a reference genome or de novo assembly. RNA-seq may be used to identify genes within a genome or identify which genes are active at a particular point in time, while read counts can be used to accurately model the relative gene expression level. RNA-seq methodology has constantly improved, primarily through the development of DNA sequencing technologies to increase throughput, accuracy and read length. Since the first descriptions in 2006 and 2008, RNA-seq has been rapidly adopted and overtook microarrays as the dominant transcriptomics technique in 2015. RNA-seq has been highly influenced by the development of high-throughput sequencing technologies. The first RNA-seq work was published in 2006 using the 454 Roche technology. Subsequently this progressed and by 2008, the advent of Solexa/Illumina technollgies revolutionized the RNA-seq research in plant sciences.

3. Applications of Transcriptomics

3.1 Biotic Stress Resistance

Transcriptome sequencing (RNA-seq) is widely applied for transcripts analysis as well as novel gene discovery in potato (Massa et al., 2011). A summary of transcriptomics research is outlined in Table 2. In a study on PVY infection, RNA-seq identified differentially expressed and significant upregulated genes (407) for resistance to PVY^{O} in potato cv. Premier Russet (Goyer et al., 2015). The identified genes were associated with ABC transporter, a MYC2 transcription factor, a VQ-motif containing protein and others. In another study, 265 differential genes were identified in potato against PVY, PVA and PLRV infection (Osmani et al., 2019). For broad resistance to late blight, 3,354 DEGs (differentially expressed genes) were identified in potato by applying RNA-seq analysis, which encoded transcription factors, NBS-LRR genes and protein kinases (Yang et al., 2018). Likewise, genes were also identified for resistance to nematode (*Meloidogyne chitwoodi*) into the cultivated potato introgressed from wild potato species (*S. bulbocastanum*). Of the total, 61 genes were observed common in different time points of 48 h, seven days and 21 days after nematode inoculation (Bali et al., 2019).

A typical microarray experiment involves the hybridization of an mRNA molecule to the DNA template from which it originated. Microarray analysis for late blight resistant Indian potato cv. Kufri Girdhari revealed up-regulation of 2,344 genes post-inoculation as compared to pre-inoculation stage. Molecular chaperones played a critical role in controlling resistance in Kufri Girdhari (Sundaresha et al., 2014). Besides, late blight resistance genes in somatic hybrid [C-13 (+) *S. pinnatisectum*] were also identified by using microarray. A total of 5,810 statistically significant genes ($p \leq 0.05$) were identified, of which 2,101 genes ($\geq$ 2-fold) were up-regulated and 3,709 genes were down-regulated. It was observed that defence responsive genes played a key role in late blight resistance mechanism in potato somatic hybrid clone P-7 (Singh et al., 2016a). Apical leaf curl disease, caused by tomato leaf curl New Delhi virus-[potato] (ToLCNDV-[potato]), is one of the most important viral diseases of potato in India. Microarray analysis showed that a total of 1,111 genes and 2,588 genes were differentially regulated in response to virus infection in Kufri Bahar (resistant) and Kufri Pukhraj (susceptible), respectively. These altered transcripts were involved in stress responses, signal transduction pathways, protein binding, cellular transport and metabolic process (Jeevalatha et al., 2017).

3.2 Abiotic Stress Tolerance

Global gene expression (transcriptome) profile has been analyzed for abiotic stress tolerance, mainly drought, heat, cold and salinity stresses in potato, like the model plant species (Table 2). In potato, transcriptome profiling reveals genes involved under drought stress and drought followed by water stimulus conditions (Gong et al., 2015; Barra et al., 2019). After analyzing the RNA sequencing data, these researchers identified thousands of differentially expressed genes homologous to the known genes in Arabidopsis, such as abscisic acid 8'-hydroxylases, auxin responsive

Table 2. Examples of some recent transcriptomics, proteomics, metabolomics and ionomics, and integrated omics studies in potato.

Trait	Genotype/stress	Platform	Key findings	References
Transcriptomics				
Biotic stress				
Late blight (*Phytophthora infestans*)	cv. SD20 and sampling at 0, 24, 48, and 72 h post-inoculation	Illumina HiSeqX10	Identified 3354 differentially expressed genes (DEGs) for late blight ressitance associated with transcription factors, protein kinases, NBS-LRR genes, multi-signaling pathways of salicylic acid, jasmonic acid and ethylene signaling pathways such as WRKY, ERF and MAPK.	Yang et al. (2018)
Potato virus Y (PVY^{O} and PVY^{NTN})	cv. Premier Russet and Russet Burbank, and sampling at 4 and 10 h post-inoculation	Illumina HiSeq2000	Identified 139-489 DEGs, of which a few important genes were predicted to encode for a putative ABC transporter, a MYC2 transcription factor, a VQ-motif containing protein, a non-specific lipid-transfer protein and a xyloglucan endotransglucosylase-hydroxylase.	Goyer et al. (2015)
Bacterial wilt (*Ralstonia solanacearum* strain UY031)	Wild spcies (*S. commersonii* lines F118 and F97) sampling at 3–4 days after inoculation	Illumina-Solexa Genome Analyzer II	Identified 2978 novel transcripts and a high proportion of all genes were altered only in F118, while phythormone-related genes were highly induced in F97. Hormone-related genes indicated that both ET and JA were induced in the susceptible accession F97, but not in the resistant accession F118, while SA-related genes were downregulated in both accessions after pathogen infection.	Zuluaga et al. (2015)
Late blight (*P. infestans*), bacterial wilt (*R. solanacearum*), and *Potato virus Y*	cv. Helan 15 and sampling at 2 days after inoculation	Illumina HiSeqTM2500	Identified 75,500 unigenes, including 6,945 resistance genes and 11,878 transcription factors, and pathways associated genes and reported hub unigenes to regulate plant immune responses, such as *FLAGELLIN-SENSITIVE 2* and chitinases.	Cao et al. (2020)
Common scab (*Streptomyces scabies*)	cv Hindenburg (resistatnt) and Green Mountain (susceptible) growth under 8/16 h photoperiod at 22°C	Illumina HiSeq2500	Out of 25,548 annotated expressed genes, 1,064 DEGs differentiated resistant and susceptible cultivars, and identified a set of 273 co-regulated differentially-expressed genes in 34 pathways that are more likely reflect the genetic differences of the cultivars and metabolic mechanisms involved in the scab pathogenesis and resistance.	Fofana et al. (2020)
Colorado Potato Beetle (*Leptinotarsa decemlineata*)	cv. Shepody and sampling after 24 h of 0.5 to 40 µg spinosad treatment	Ion Torrent	Identified 13,281 DEGs, including venom carboxylesterase-6, chitinase 10, juvenile hormone esterase and multidrug resistance-associated protein 4, and several microRNAs, such as miR-12-3p and miR-750-3p.	Bastarache et al. (2020)

Potato cyst nematode (*Globodera rostochiensis* (*Ro*1))	*S. phureja* (k-11291 and k-9836) and samling at 0, 24 and 72 h after inoculation	Illumina NextSeq500	21,113 genes were used for further analysis. A chain of molecular events initiates the hypersensitive response at the juveniles' invasion sites and provides high-level resistance and reveals resistance mechanisms and candidate R-genes.	Kochetov et al. (2020)
Columbia root-knot nematode (*Meloidogyne chitwoodi*)	Clone PA99N82–4 (resistance from wild *S. bulbocastanum*) and cv. Russet Burbank (susceptible), and sampling at 48 h, 7d, 14d, 21 d post inoculation	Illumina Hiseq3000	Differential gene expression analysis identified 1268, 1261, 1102 and 2753 up-regulated genes in PA99N82–4 at 48 h, seven days, 14 days and 21 days post inoculation respectively, of which 61 genes were common across all the time points. These genes were mapped to plant-pathogen interaction, plant hormonal signaling, antioxidant activity and cell wall re-enforcement pathways annotated for potato.	Bali et al. (2019)
Wart (*Synchytrium endobioticum*)	cv. Qingshu 9 and sampling after a month (L0- L7) of inoculation	Transcriptomics by M/s BioMarker	Identified a total of 214 DEGs for disease resistance involved in pathways, like cell metabolism, cell cycle, biosynthesis, enzyme catalysis, cellular nucleic acid-binding protein, AP2-like transcription factor and E3 ubiquitin-protein ligase.	Li et al. (2021b)
Antimicrobial activity	Chitosan (0.05%) application on abaxial surfaces of the leaves and sampling at 2 and 5 h after treatment	Illumina HiSeq 3000	Identified a total of 83 DEGs mainly enriched in gene modulation associated with electron transfer chains in chloroplasts and mitochondria. Chitosan positively influences plant growth, yield and resistance and an ecofriendly plant-protecting agent.	Lemke et al. (2020)
Early blight (*Alternaria solani*)	cv. Dutch 15 (a.k.a. Favorita) and sampling at 24 h, 36 h and 48 h post-inoculation	RNA-seq Illumina NovaSeq 6000	A total of 6184, 10887 and 8109 DEGs were obtained at 24hpi, 36hpi and 48hpi respectively and analysis revealed that resistance response is mainly mediated by JA and SA signals.	Tian et al. (2020)
Abiotic stress				
Salt stress	cv. Longshu No. 5 was treated with 500 mmol/L NaCl salt and sampled at 0, 24, 48, 72, and 96 h post treatment	Illumina HiSeq X Ten	Identified 5,508 DEG significantly enriched in the categories of nucleic acid binding, transporter activity, ion or molecule transport, ion binding, kinase activity and oxidative phosphorylation.	Li et al. (2020)
Drought stress	In tissue culture cv. Desiree was treated with -1.8 Mpa polyethylene glycol-8000 at 0, 6, 12, 24, and sampled 48 h post treatment	Illumina NextSeq	Identified 5,118 DEGs; drought-tolerance mechanism can mainly be explained by two aspects – the photosynthetic antenna protein and protein processing of the endoplasmic reticulum.	Moon et al. (2018)

Table 2 contd. ...

...Table 2 contd.

Trait	Genotype/stress	Platform	Key findings	References
Drought stress	cv. Jancko Sisu Yari (tolerant) with three conditions: mild (20%), severe (10%), and re-watered (80% soil water content)	Illumina HiSeq 4000	Identified drought-responsive genes (308–3203 DEGs) mainly involved in photosynthesis, signal transduction, lipid metabolism, sugar metabolism, wax synthesis, cell wall regulation and osmotic adjustment. The recovery of rehydration is mainly related to patatin, lipid metabolism, sugar metabolism, flavonoids metabolism and detoxification.	Chen et al. (2019)
Heat stress	cv. Russet Burbank and heat stress (35°C day/28°C night) at 3rd days after treatment	Illumina HiSeq-2000	Identified 1,420 DEGs and *StHsp26-CP* and *StHsp70* were markedly increased in expression under heat stress.	Tang et al. (2020)
Cold stress	*S. commersonii*, *S. tuberosum* (cv. Umatilla) and *A. thaliana* under low temperature (2°C) and sampling at 2, 8, 24, and 168 h post treatment	10K potato cDNA microarray (TIGR)	Identified 53 groups of putative orthologous genes that are cold-regulated in all three species.	Carvallo et al. (2012)
Cold, heat and salt stress	cv. Gilroy under cold (4°C), heat (35°C), salt (100 mM NaCl) stress for up to 27 h post treatment	Potato cDNA microarray (12,000)	Potato cDNA (about 12,000 clones) microarray expression profiling identified genes 2,584 (cold), 1,149 (salt) and 998 (heat) transcription factors, signal transduction factors and heat-shock proteins.	Rensink et al. (2005)
Nitrogen stress	cv. Kufri Jyoti, low N (0.2 mM) and high N (4 mM) under aeroponics	Illumina NextSeq500	Identified 19,730 DEGs (761 up-regulated and 280 down-regulated), such as glutaredoxin, Myb-like DNA-binding protein, WRKY transcription factor 16 in shoots; high-affinity nitrate transporter, protein hosphatise-2c in roots; and glucose-6-phosphate/phosphate translocator 2, BTB/POZ domain-containing protein in stolons.	Tiwari et al. (2020a)
Nitrogen stress	cv. Kufri Gaurav, low N (0.75 mM) and high N (7.5 mM) under aeroponics	Illumina NextSeq500	Identified significant differentially expressed genes, such as 176 (up-regulated) and 30 (down-regulated) in leaves, 39 (up-regulated) and 105 (down-regulated) in roots, and 81 (up-regulated) and 694 (down-regulated) in stolons. The candidate genes associated with improving nitrogen use efficiency in potato variety Kufri Gaurav were superoxide dismutase, GDSL esterase lipase, probable phosphatase 2C, high affinity nitrate transporters, sugar transporter, proline-rich proteins, transcription factors (VQ motif, SPX domain, bHLH), etc.	Tiwari et al. (2020b)

Nitrogen stress	Yanshu 4, Xiabodi and Chunshu 4 under without N, and with N (3.3 kg Urea/100 m^2)	Illumina HiSeq 4000	Identified DEGs varying between 0–1,446 and involved in the processes of nitrate transport, nitrogen compound transport and N metabolism. The key genes were glutamate dehydrogenase (*StGDH*), glutamine synthetase (*StGS*) and carbonic anhydrase (*StCA*), and the Major Facilitator Superfamily (MFS) members, like nitrate transporter 2.4 (*StNRT2.4*), 2.5 (*StNRT2.5*) and 2.7 (*StNRT2.7*).	Zhang et al. (2020)
Quality traits				
Cold-induced sweetening	Eight potato genotypes including cv. Doremi, Lady Claire, Summer Delight and *S. brevicaule* lines stored for 3 months at 4°C and sampling at before and after cold storage	HiSeq 2000 (Illumina)	Cold-induced sweetening high-resistant genotypes (with low glucose after cold) showed increased expression of an invertase inhibitor gene and genes involved in DNA replication and repair after cold storage, while low-resistance cultivars (high glucose after cold storage) showed differential expression of abiotic stress-related genes. A few genes with similar expression patterns in all cultivars were associated in cell wall strengthening and phospholipases.	Tai et al. (2020)
Multi-omics (Transcriptomics/Proteomics/metabolomics/ionomics)				
Biotic stress				
Late blight (*P. infestans*)	cv. Favorita, Mira, and E-malingshu No. 14 and sampling at 5 days after inoculation of *P. infestans* (15000 sporangia/ml)	iTRAQ-based quantitative proteomics	A total of 2,044 were quantified, mediating late blght resistance involved in induction of elicitors, protease inhibitors, serine/threonine kinases, terpenoid, hormone signaling, transport proteins, pathogenesis-related proteins, LRR receptor-like kinases, mitogen-activated protein kinase, WRKY transcription factors, jasmonic acid and phenolic compounds.	Xiao et al. (2020)
Late blight (*P. infestans*)	cv. Sarpo Mira and sampling at 0 h, 48 h, and 120 h post inoculation with *P. infestans* (15000 sporangia/ml)	LC-MS/MS	A total of 1,229 Differentially-Expressed Proteins (DEPs) were identified and which belonged to different expression patterns and functional categories, and enrichment of cell wall-associated defense response proteins during the early stage of infection, whereas the late stage was characterized by cellular protein modification, membrane protein complex and cell death.	Xiao et al. (2019)

Table 2 contd. ...

...Table 2 contd.

Trait	Genotype/stress	Platform	Key findings	References
Late blight (*P. infestans*)	cv. Kuras (moderate), Sarpo Mira (highly resistant) and Bintje (very susceptable), sampling at 4, 16, 65, 120, and 258 h after spraying	LC-MS/MS	Detected and quantified, between 3,248 and 3,529, unique proteins from each cultivar, and upto758 *P. infestans* derived proteins were identified.	Larsen et al. (2016)
Late blight	Desiree, Sarpo Mira and SW93-1015 and sampled at 6, 24, and 72 h post-inoculation	Microarray (JHI *S. tuberosum* 60 k v1), MS	Detected over 17,000 transcripts and 1,000 secreted proteins including many putative hypersensitive and effector-target proteins of large gene families. A few candidate genes for hypersensitive response include a Kunitz-like protease inhibitor, transcription factors and an RCR3-like protein.	Ali et al. (2014)
Potato virus Y (PVY^{NTN})	cv. Désirée and sampling at –1 (1 d before), 0, 1, 2, 3, 4, 5, 6, 7, 8, 9 and 11 d post-infection	Microarray (POCI arrays) and LC/Orbitrap LTQ XL MS	This multiomics analysis provides better insights into the mechanisms leading to tolerant response of potato to viral infection and can be used as a base in further studies of plant immunity regulation.	Stare et al. (2019)
Potato virus A (PVA)	Diplid line v2–108 and sampling at 20 days post-inoculation	LC-MS/MS	Out of a total 807 nuclear proteins identified, 16 unique proteins in each healthy and PVA-infected leaves were identified. The protein Dnajc14 was detected only in healthy leaves whereas different ribosomal proteins, ribosome-biogenesis proteins and RNA splicing–related proteins were over-represented in PVA-infected leaves.	Rajamaki et al. (2020)
Bacterial wilt (*R. solanacearum*)	Wild species *S. chacoense* sampling at 6 days post-inoculation with UW551 and T3SS mutant	iTRAQ	Identified 21 differentially accumulated proteins like miraculin, HBP2 and TOM20 which contribute to immunity, whereas PP1 contributes to susceptibility. Notably, these four proteins were significantly downregulated at the post-transcriptional level.	Wang et al. (2021)
Soft rot and blackleg (*Pectobacterium brasiliense*)	*S. tuberosum* (DM1) and *S. chacoense* (M6)	LC-MS	Study investigated various compounds, such as alkaloids, amines, organic acids, terpenes, peptides, saccharides and lipids. *S. chacoense* extracts did not affect bacterial multiplication rate; however, they did reduce pectinase, cellulase and protease activities. Selected alkaloids, phenolic amines, phenols, amines and peptides which are integrative chemical sources of resistance against bacteria.	Joshi et al. (2021)

Powdery scab (*Spongospora subterranea* f. sp. *subterranean*)	5 tolerant and 5 susceptible varieties	UPLC-Q-TOF/MS	Metabolites belonging to amino acids, organic acids, fatty acids, phenolics, sugars and cell-wall-thickening compounds were abundant in tolerant varieties than in susceptible varieties. Root-exuded compounds belonging to the chemical class of phenolics were also found in abundance in the tolerant cultivars compared to susceptible cultivars.	Lekota et al. (2020)
Tuber black dot (*Colletotrichum coccodes*)	cv. Cheyenne, Erika, Gwenne, Lady Christl, and Lady Felicia grown in greenhouse for 4 months	GC-MS, UHPLC-HRMS/MS, LC-MS/MS	Hydroxycinnamic acids, hydroxycinnamic acid amides and steroidal saponins were found to be biomarkers of resistance under control conditions, while hydroxycoumarins were found to be specifically induced in the resistant cultivars.	Massana-Codina et al. (2020)
Black leg and soft rot (*Dickeya solani*)	Highly resistant: Bea and Humalda, DG00–270) and susceptible (Irys, Katahdin, Ulster Supreme, DG 08–305)	LC-MS-MS/MS	Identified increased expression of proteins, mainly probable inactive patatin-03-Kuras 1 and the proteinase inhibitor PTI in resistant varieties, whereas in the diploid clones, only metallocarboxype ptidase and metallocarboxypeptidase-like inhibitors exhibited much higher fold changes after pathogen invasion.	Lebecka et al. (2019)
Wart (*Synchytrium endobioticum*)	Resistant (Calrose and Humalda) and susceptible (Sebago, Seneca and Wauseon) cultivars	2-DE, LC-MS/MS	Identified 24 proteins associated with resistance mechanism and grouped into four categories, i.e., stress and defence, cell structure, protein turnover and metabolism. Stress and defence genes, mainly heat-shock proteins (HSPs), such as HSP70, HSP60 and HSP20 families play an important role in resistance. Besides others, genes were chaperone factors, *S*-adenosyl-l-homocysteine hydrolase-like, superoxide dismutase [Mn], inactive patatin-3-Kuras 1 and patatin-15.	Szajko et al. (2019)
Verticilium wilt (*V. Dahlia*)	cv Kufri Chipsona-1 and Kufri Sutlej at 0, 72 and 168 h post-inoculation	2-DE, HPLC, Q-TRAP MS	Identified 75 disease-resistance responsive proteins predominantly involved in wall hydration, architecture and redox homeostasis. Study observed that wall crosslinking and salicylic acid signaling significantly altered during patho-stress and increase in reactive oxygen species and scavenging proteins led to cell death and necrosis of the host.	Elagamey et al. (2017)

Table 2 contd. ...

...Table 2 contd.

Trait	Genotype/stress	Platform	Key findings	References
Abiotic stress				
Heat stress	cv. Hezuo 88 grown under high-temperature at 35°C (light)/33°C (dark) and sampling at before and 6 h and 3 days post treatment	Illumina HiSeq 4000, LC-MS	Identified 448 (up) and 918 (down) regulated genes by transcriptomics and 325 (positive) and 219 (negative) ionization mode compounds by proteomics approaches under heat stress. Differentially-expressed genes were enriched in photosynthesis, cell wall degradation, heat response, RNA processing and protein degradation, while differentially-expressed metabolites were involved in amino acid biosynthesis and secondary metabolism was mostly induced during heat stress.	Liu et al. (2021)
Drought stress	Transgenics potatoes (cv. Solara) with Cyanobacterial Fld	Potato Oligo Chip Initiative (POCI) microarray, sugars and amino acids	Identified about 300 genes which potentially contribute to stress acclimation. Improvement of physiological and molecular stress responses by chloroplast Fld presence resulted in increased tuber yield under long-term non-lethal water restriction. Tuber yield losses under chronic water limitation were mitigated in flavodoxin-expressing plants, indicating that the flavoprotein has the potential to improve major agronomic traits in potato.	Karlusich et al. (2020)
Drought stress	cv. Maxi (tolerant) cv. Eurobravo (sensitive), MS medium with/without 0.2 M sorbitol	2D-IEF/SDS-PAGE, MS	Out of a total of 679 distinct protein spots, 118 and 20 spots with differential abundance were found in sensitive and tolerant genotype, respectively under osmotic stress. The chaperone and hydrogen peroxide detoxifying proteins were abundant in the tolerant genotype, while an increase in the abundance of proteinase inhibitors and their precursors, changes in stress responsive proteins and an altered RNA/DNA-binding response were observed in the sensitive genotype.	Bündig et al. (2016)
Drought stress	*S. tuberosum* group Andigenum (clones Sullu and SS2613)	Microarray, HPLC-FLD	Transcriptome analysis revealed pururbation in photosynthesis and carbohydrate-related genes in the cloens. Metabolite analysis indicated that differential accumulation of osmotically active solutes, mainly contents of galactose, inositol, galactinol, proline, and proline analogues were higher upon drought stress.	Evers et al. (2010)

Heat and drought stress	cv. Desiree, Unica, Agria and Russett Burbank *Drought*: Withdrawal of irrigation for 12 days at 24/18°C (day/night) *Heat*: Gradual increase up to 39/27°C for 9 days, and then constant (39/27°C) for 3 days	Microarray, GC-MS	18 of 23 drought-associated transcripts and 45 of 65 heatassociated transcripts were differentially expressed besides highlighting a decrease in the abundance of transcripts encoding proteins associated with PSII light harvesting complex in stress-tolerant cultivars. Stress-tolerant cultivars exhibited stronger expression of genes associated with plant growth and development, hormone metabolism and primary and secondary metabolism than stress-susceptible cultivars. Accumulaiton of proline was recorded in all genotypes under drought stress. On the contrary, sugar alcohols, mainly inositol and mannitol, were strongly accumulated under heat and combined heat and drought stress while galactinol was strongly accumulated under drought. Combined heat and drought also resulted in the accumulation of valine, isoleucine and lysine in all genotypes.	Demirel et al. (2020)
Osmotic and cold stress	cv. Désirée, *S. commersonii*, Osmotic stress (MS medium + 0.21 M sucrose) and chilling (at 6°C)	2-D, MS and MS/MS (5800 MALDI TOF/TOF)	Identified common and differential responses to abiotic stresses, indicating that some pathways could be crucial for tolerance to osmotic stress as well as cryopreservation. Identified proteins and carbohydrates associated with cryopreservation in two potato species – *S. commersonii* and *S. tuberosum*.	Folgado et al. (2014)
Cold stress	cv. Zhengshu 6 and low temperature treatment (4°C day/2°C night) for 1, 3, 5 and 7 days	2-DE, LC-ESI-MS/MS	Identified 52 differentially-expressed proteins associated with metabolic functions, such as defense response, energy metabolism, photosynthesis, protein degradation, ribosome formation, signal transduction, cell movement, nitrogen metabolism and other physiological processes.	Li et al. (2021a)
Cold stress	cv Atlantic and storage at low temp. (0, 4 and 15°C with 85–95% RH) and samling at 0, 3, 10, 20 and 30 days after cold storage	Illumina HiSeq, iTRAQ proteomics	Identified accumulation of soluble sugars under low temperatures and regulated by granule-bound starch synthase 1, beta-amylase, invertase inhibitor and fructokinase. Additionally, 15 Heat Shock Proteins (Hsps), including three Hsp70s, two Hsp80s, one each of Hsp90 and Hsp100 and eight small Hsps, were also induced under low temperatures.	Lin et al. (2019)

Table 2 contd. ...

...Table 2 contd.

Trait	**Genotype/stress**	**Platform**	**Key findings**	**References**
Salt stress	cv BARI-401 (red skin) and Spunta (yellow skin) in MS media supplemented with lithium chloride (LiCl 20 mM) and mannitol (150 mM)	GC-MS and spectrophotometry	Study showed differential response to salt stress in both the varieties. Cv. Spunta was found more salt-tolerant than cv. BARI-401. Because of a greater response to ionic and osmotic stress in cv. Spunta, there was greater accumulation of defensive metabolites by increased POD/PPO activity, reduced ROS production, trehalose accumulation and increased saturated fatty acids composition.	Hamooh et al. (2020)
Nitrogen stress	Topas (tolerant) and Lambada (sensitive) under N sufficient (60 mM) and N deficient 3.75 mM, and root sampling after 7 post-treatment	LC-ESI-Q-TOF-MS/MS	Observed a difference in tolerant and sensitive varieties for genes involved in glutamine synthetase/glutamine oxoglutarate aminotransferase pathway, tricarboxylic acid cycle, glycolysis/gluconeogenesis pathway, protein and amino acid synthesis, protein catabolism and defense mechanisms. N stress tolerance involves increase in nitrogen reassimilation (GDH), carbon metabolism (glycolysis/TCA cycle), cell wall synthesis/lignifications.	Jozefowicz et al. (2017)
Fe, Mn and Zn deficiency	cv. Atlantic, treated with nutrient deficiency Fe (20, 10 and 0 µM Na-Fe-EDTA), Mn (1 and 0 µM MnCl2 $\cdot 4H_2O$), and Zn (0 µM $ZnCl_2$)	2-DE, MALDI-TOF/TOF MS	A total of 146, 55 and 42 protein spots were identified under Fe, Mn and Zn deficiency, respectively and mainly involved in bioenergy and metabolism, photosynthesis, defence, redox homeostasis and protein biosynthesis/degradation, signaling, transport, cellular structure and transcription-related proteins under the metal deficiencies. The signaling cascades involving auxin and NDPKs (Nucleoside Diphosphate Kinases) might also play roles in micronutrient stress signaling.	Cheng et al. (2019)
Quality traits				
Multi-coloured fleshed potato	57 vareities (multi-colored)	LC–MS/MRM, LC-UV/MS, UPLC and UHR-TOF–MS	Identified 21 anthocyanins in the analyzed potato cultivars, including two anthocyanins (pelargonidin feruloyl-xylosyl-glucosyl-galactoside and cyanidin 3-p-coumaroylrutinoside-5-glucoside) which are not described before in potato. Expression of hydroxylases and methyltransferases play key roles in red and blue potato cultivars.	Oertel et al. (2017)
Tuber quality and sugar	*S. tuberosum* diploid F_1 population (98)	2-DE	This study established a relationship between proteins and 26 potato tuber quality traits (e.g., flesh colour, enzymatic discoloration). Detected 1,643 unique protein spots and protein QTLs were identified on chromosomes 3, 5, 8 and 9.	Acharjee et al. (2018)

Cold-induced sweetening	Cold storage for 5 months at 5°C	High pH reverse phase (hpRP) LC and off-gel electrophoresis (OGE), nanoLC–MS/MS (Q-TOF and Orbitrap)	Identified a total of 4,463 potato proteins, of which 46 showed differential expressions during cold storage of potato tubrs. Also identified several key proteins important in controlling starch–sugar conversion during cold-induced sweetening. Study suggests that the hpRP-RP shotgun approach is a feasible and practical workflow for discovering potential protein candidates in plant proteomic analysis.	Yang et al. (2011)
Starch	cv. Mona Lisa	SDS-PAGE LC-MS/MS	Identified presenece of starch proteins, such as PTST1 (Protein Targeting to Starch), ESV1 (Early StarVation1) and LESV (Like ESV) and a new isoform of starch synthase, SS6, containing both K-X-G-G-L catalytic motifs. This finding indicates prospects of SS6 protein to understand storage starch metabolism.	Helle et al. (2018)
Metabolites	cv. Daisy in organic cultivation	HPLC-ESI-QTOF-MS	Identified a total of 109 compounds in leaves, including organic acids, amino acids and derivatives, phenolic acids, flavonoids, iridoids, oxylipins and other polar and semi-polar compounds, of which, quinic acid and its derivatives represented more than 45% of the bioactive compounds.	Rodríguez-Pérez et al. (2018)
Metabolites	13 *Solanum* species (tuber-bearing)	HPLC-DAD and LC-ESI-MS	Ascorbic acid was found to be the most abundant antioxidant in potato and chlorogenic acid was the primary polyphenol; analyzed genomes of wild *S. commersonii* and reference potato for genes associated in ascorbic acid, aromatic amino acid, phenylpropanoid and glycoalkaloid biosynthesis.	Aversano et al. (2017)
Male fertility/sterility	8 tetraploid genotypes, male sterile (cv. Evraziya, Gusar, Sudarynja and 1604/16), and male fertile (cv. Lomonosovskij, 1101/10, 2103/7, 211/9)	GC-MS	Detected 192 compounds by metabolites profiling. Metabolic profiles in the anthers of fertile genotypes were significantly distinguished from male-sterile ones by the accumulation of carbohydrates, while the anthers of sterile genotypes contained a higher amount of amino acids. Also, male-sterile genotypes had undeveloped pollen grain characters and the absence of pollen apertures that might be due to a disorder in the metabolism of carbohydrates and fatty acids. Thus, sterility is characterized by a significant decrease in the carbohydrate pool and an increase in amino acid content in anthers at the stage of mature pollen.	Shishova et al. (2019)

Table 2 contd. ...

...Table 2 contd.

Trait	Genotype/stress	Platform	Key findings	References
Sprouting	cv. Atlantic, Snowden, Toyoshiro and Kitahime, and sampling at sprouting after 7 months of storage	GC/MS	Of the several metabolites identified, sucrose, phosphate and amino acids were verified as robust biochemical markers for prediction of potato sprouting during long-term storage.	Fukuda et al. (2019)
Chip colour	Chipping potato varieties and breeding lines	GC/MS, GC/FID	Of the several metabolites, nine, including alanine, glyceric acid, malic acid, citric acid, allantoin, fructose, glucose and inositol were detected by GC/MS, and further validated by GC/FID. Therefore, this strategy will provide a cost-effective application for potato chip color.	Tomohiko et al. (2020)
Graft potato	cv. Hopehely and White Lady, and sampling of leaves ans tubers at six weeks after grafting	GC-MS	No change in sucrose concentration of tubers was detected after grafting, thus indicating that sucrose and major metabolites are genetically controlled, whereas the galactinol level was highly altered by grafting. Tuberization is higly trigerred by source-derived mobile signals and a positive correlation was observed between the rate of leaf growth and the time of tuber initiation.	Odgerel and Banfalvi (2021)
Carotenoids	36 genotypes (*S. tuberosum*, *S. phureja* and *S. chacoense*)	LC-HRMS	Identified novel *CHY2* and *ZEP* alleles in the collection. Yellow-fleshed potatoes had high levels of epoxy-xanthophylls and xanthophyll esters and *β-Carotene Hydroxylase 2* (*CHY2*) gene; white-fleshed potato was characterized by low carotenoid levels and recessive *chy2* alleles while orange-fleshed potato had high levels of zeaxanthin but low levels of xanthophylls esters, and homozygosity for a *Zeaxanthin Epoxidase* (*ZEP*) recessive allele.	Sulli et al. (2017)
Glycoalkaloids, phenols, flavonoids	Diploid F_1 population 15–1 (*Solanum* hybrid DG 88–89 × *S chacoense*) and leaf sampling from plants grown in greenhouse (May to October)	BGISEQ-500, Nano-LC-MS-MS/MS	The most upregulated transcripts of high phytotoxic (α-solasonine and α-solamargine in leaves) potential were anthocyanin 5-aromatic acyltransferase-like and subtilisin-like protease SBT1.7-transcript variant X2, whereas the most downregulated genes were carbonic anhydrase chloroplastic-like and miraculin-like.	Szajko et al. (2021)

Sprouting	cv. Favorita storage for 50 days and sampling at dormancy, sprouting, camphor inhibition, and recovery sprouting	Illumina HiSeqTM 2000, iTRAQ labelling	Identified 4,000 transcripts and 700 proteins for the roles of camphor inhibition on sprouting. Camphor inhibited particularly gibberellic acid, brassinosteroids and ethylene, leading to dysregulation of physiological processes. Camphor inhibition significantly increased abundance of pathogenesis-related protein PR-10a (or STH-2), the pathogenesis-related P2-like precursor protein and the kirola-like protein as compared to sprouting.	Li et al. (2017)
Nutrients and bioactive compounds	60 genotypes (raw and cooked)	UPLC- and GC-MS, ICP-MS	Detected 2,656 compounds, including 43 bioactive compounds, 42 nutrients, 76 lipids and 23 metals. Most nutrients and bioactives were partially degraded during cooking (52%); however, genotypes with high quantities of bioactives remained highest in the cooked tuber. The yellow-flesh potato had more carotenoids and specially potatoes had more chlorogenic acid as compared to the other market classes. Besides, variations in several molecules were recorded and associated with health benefits.	Chaparro et al. (2018)
K, Fe, Mn, Ca, Mn and Si	cv. Désirée (*Rpi-blb1*, *Rpi-blb2* and *Rpi-vnt1.1* transgenics) and sampling at 24 h, 48 h, 72 h and 120 h post inoculation	ICP-OES (Agilent 5100)	Observed distinctly different distribution patterns of accumulation at the site of inoculation in the resistant lines for calcium (Ca), magnesium, Mn and silicon (Si) compared to the susceptible cultivar. The results reveal different ionomes in diseased plants as compared to resistant plants.	Brouwer et al. (2021)
Heavy metals	cv Roclas	Atomic absorption spectrometry (AAS)	The maximum levels of contamination with heavy metals causing high concern were the ppresence of lead (2.53 mg/kg) and cadmium (0.06 mg/kg) in potato tubers. Study revealed low concentrations of heavy metals in potato tubers.	Edward et al. (2019)

Note: Two-dimensional electrophoresis (2-DE), two-dimensional isoelectric focusing on sodium dodecyl sulphate polyacrylamide gel electrophoresis (2D-IEF/SDS-PAGE), High performance liquid chromatography coupled with a fluorescence detector (HPLC-FLD), liquid chromatography-tandem mass spectrometry (LC–MS/MS), nano-liquid chromatography coupled with tandem mass spectrometry (LC-MS-MS/MS), sodium dodecyl sulfate polyacrylamide gel electrophoresis (SDS-PAGE), high performance liquid chromatography coupled with quadrupole-time of flight mass spectrometry (HPLC-ESI-QTOF-MS), gas chromatography-mass spectrometry (GC-MS), ultra-performance liquid chromatography coupled with quadrupole time-of-flight mass spectrometry (UPLC-Q-TOF/MS), LC–MS/MRM (multiple reaction monitoring) profiling, LC-UV/MS (an ultra-performance liquid chromatography (UPLC), an ultra-high resolution time of flight mass spectrometer (UHR-TOF–MS) and atomic absorption spectrometry (AAS) protein, calcium-transporting ATPase, calmodulin-like protein, dehydrin and protein phosphatase.

Identification of genes and pathways affected by high temperature is crucial for developing thermo-tolerant cultivars. In general, high night temperature beyond 20°C drastically reduces tuber formation in potato (heat tolerant: Kufri Surya vs. heat sensitive: Kufri Chandramukhi). Microarray gene expression analysis showed that a total of 2,500 genes were differentially expressed on 21 days and 4,096 genes on 14 days after stress. This study provided useful information on potato tuberization at elevated night temperatures (24°C) and made available a framework for further investigation into heat stress in potato (Singh et al., 2015). Hancock et al. (2014) investigated physiological, biochemical and molecular responses in potato under moderately high temperature conditions 30/20°C day/night for five weeks and identified 2,190 differentially expressed genes. A large number of genes were over-represented in the study belonging to photosynthesis, lipid metabolism and amino acid biosynthesis pathways and 2,886 genes showed major transcripts changes under heat stress. The genes involved in heat stress belonged to ABA, ethylene, auxin, brassinosteroid and abiotic stress responses. Earlier, Rensink et al. (2005) investigated potato seedlings in response to heat (35°C), cold (4°C) and salt (100 mM NaCl) stress for 27 h and identified a large set of differentially expressed genes, such as transcription factors, DNA binding proteins, transporter proteins, phosphatases and HSPs in response to abiotic stress.

In a recent study, potato varieties Kufri Jyoti (N inefficient) and Kufri Gaurav (N efficient) were grown in aeroponics without N (starvation), low N and high N. Transcriptomes (RNA-seq) were analyzed in leaf, root and stolon tissues to identify genes and regulatory elements controlling NUE in potato. Under N limitation condition, hundreds of differentially up/down regulated genes were identified in comparison to sufficient N in potato cvs. Kufri Jyoti (N inefficient) and Kufri Gaurav (N efficient) (Tiwari et al., 2020a, 2020b) (Fig. 4). In another experiment, sequence variation was analyzed for genes involved in nitrogen (N) metabolism, in potato. Of the total 17 PCR primers tested for N metabolism genes in two contrasting potato varieties, only single, distinct and un-fractioned 12 fragments amplified by six primers, representing five genes (nitrate transporter-NRT, ammonium transporter-AMT, nitrate reductase-NR, nitrite reductase-NIR and asparagines synthetase-AS) involved in N metabolism, were cloned and sequenced. Following the sequence analysis, non-redundant sequences with uninterrupted open reading frames of 12 '*N*-homologous genes' were identified in the known N metabolic pathways genes (Tiwari et al., 2018). Thus, the identified 12 *N*-homologous genes may serve as an important genomic resource for novel genes/markers discovery and prove useful for molecular breeding in potato for improving nitrogen use efficiency.

3.3 Tuber Quality and Other Traits

Potato nutrition and quality are very important parameters from the food security point of view. Recently, Liu et al. (2021) studied transcriptome profiling for reduced cold-induced sweetening traits in contrasting (resistant vs. susceptible) genotypes. They identified genes involved in the pathways of starch degradation, sucrose

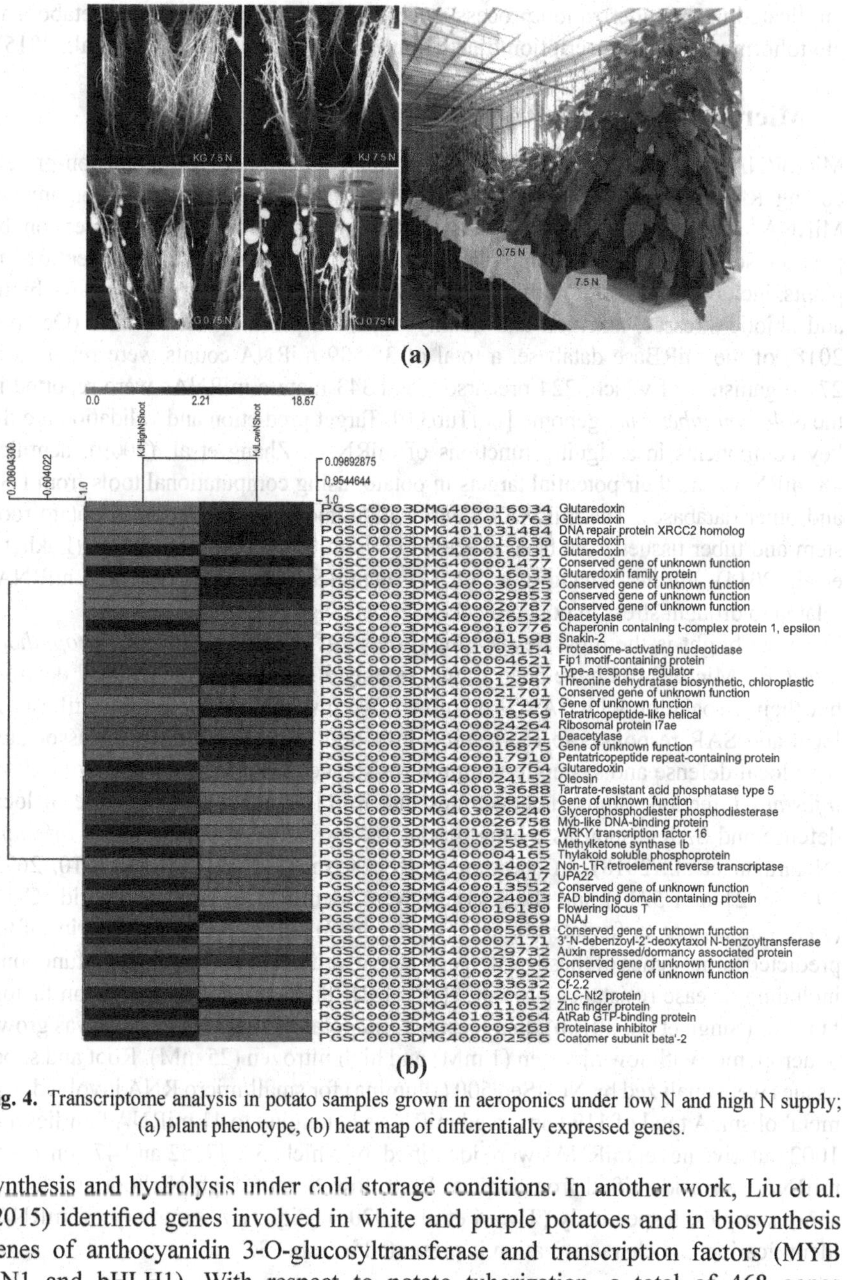

Fig. 4. Transcriptome analysis in potato samples grown in aeroponics under low N and high N supply; (a) plant phenotype, (b) heat map of differentially expressed genes.

synthesis and hydrolysis under cold storage conditions. In another work, Liu et al. (2015) identified genes involved in white and purple potatoes and in biosynthesis genes of anthocyanidin 3-O-glucosyltransferase and transcription factors (MYB AN1 and bHLH1). With respect to potato tuberization, a total of 468 genes (94 up-regulated and 374 down-regulated) was identified that was statistically significant and differentially expressed in tuber-bearing potato somatic hybrid (E1-3: C-13 + *S. etuberosum*) versus control non-tuberous wild species, *S. etuberosum*. This study showed that candidate genes induced in leaves of E1-3 were implicated in the tuberization process, such as transport, carbohydrate metabolism, phytohormones and transcription/translation/binding functions (Tiwari et al., 2015).

4. MicroRNAs

MicroRNAs (miRNAs) are the small (21–24) nucleotides, endogenous and non-protein coding RNA molecules and play important roles in gene regulation in organisms. MiRNAs pair with messenger RNAs (mRNAs) and mediate gene expression by post-transcriptional suppression. A large number of miRNAs have been identified in plants, including potato for multiple traits, such as yield and its attributing traits, biotic and abiotic stresses, nutrition and quality traits. As per the Release 22.1 (October, 2018) of the miRBase database, a total of 38,589 miRNA counts were reported in 271 organisms, of which, 224 precursors and 343 mature miRNAs were reported in the *Solanum tuberosum* genome [SolTub3.0]. Target prediction and validation are the key components in assigning functions of miRNAs. Zhang et al. (2009) identified 48 miRNAs and their potential targets in potato, using computational tools from EST and other databases. The miRNAs identification and characterization in potato root, stem and tuber tissues have been executed by high-throughput sequencing (Lakhotia et al., 2014). Further, Zhang et al. (2014) identified novel and conserved miRNAs related to drought stress in potato by deep sequencing.

Late blight is the most devastating disease of potato caused by *Phytophthora infestans*. MicroRNAs have been shown to play a significant role in local defense, but their association with SAR is unknown. We investigated the role of miR160 in local and SAR responses to *P. infestans* infection in potato. MiR160 is associated with local defense and systemic-acquired resistance of potato against *Phytophthora infestans* infection. Study demonstrates that miR160 plays a crucial role in local defense and SAR responses during the interaction between potato and *P. infestans* (Natarajan et al., 2018). In addition, miRNAs (miR395, 821, 1030, 1510, 2673, 3979, 5021 and 5213) and their targets were identified in somatic hybrid 'C-13' (+) *S. pinnatisectum* for late blight resistance by *in silico* approach. Majority of the predicted target genes of these miRNAs are involved in different biological functions, including disease resistance proteins (NBS-LRR domains) and transcription factors families (Singh et al., 2016b). In a recent work, potato variety Kufri Jyoti was grown in aeroponics with low nitrogen (1 mM) and high nitrogen (25 mM). Root and shoot tissues were analyzed by NextSeq 500 (Illumina) for small/micro RNA involved in N metabolism. A total of 119 conserved miRNAs belonging to 41 miRNA families and 1002 putative novel miRNAs were identified, of which, 52, 47, 52 and 47 conserved miRNAs were identified from high N Root, low N root, high N shoot and low N shoot samples, respectively (Tiwari et al., 2020c) (Fig. 5). A summary of miRNAs identification and characterization is presented in Table 3.

5. Proteomics

Proteomics is the study of proteins in response to various stresses and growth stages in plants. Proteomics includes the integration of technologies for the identification, quantification and characterization of total protein contents of an organism. It

Table 3. Summary of some miRNAs identified in potato and their potential targets for various traits.

Trait	miRNA	Target gene	Key findings	References
Late blight	miR8788	*StLL1*	Demonstrated role of miR8788 in late blight resistance targeting a potato lipase-like membrane protein-encoding gene (*StLL1*).	Hu et al. (2020)
Leaf curvature	miR160	StARF10, StARF16, and StARF17	Role of microRNA (miR160) was confirmed in regulating leaf curvature in potato. The potential targets involve auxin siganling pathway genes, mainly a group of Auxin Response Factors –StARF10, StARF16, and StARF17.	Natarajan and Banerjee (2020)
Late blight	miR160	*StGH3.6* *StARF10*	Role of miR160 was confirmed in local and systemic acquired resistance (SAR) responses to late blight (*P. infestans*) in potato. Analysis of the expression of defense and auxin pathway genes and direct regulation of *StGH3.6* by the miR160 target *St*ARF10 revealed the involvement of miR160 in antagonistic cross-talk between salicylic acid-mediated defense and auxin-mediated growth pathways.	Natarajan et al. (2018)
Osmotic stress	miR164	NAC transcription factor (StNAC262)	Exhibited role of Stu-miR164 in regulation of the target NAC transcription factor StNAC262 under PEG-treated osmotic stress in potato. Real-time PCR analysis of transgenic potato plants under osmotic (PEG) stress showed that potato plants overexpressing Stu-mi164 had reduced expression of StNAC262 and their osmotic resistance decreased.	Zhang et al. (2018)
Potato tuberization	miR399, miR482, miR319, miR8006, miR479, miR477	StGRAS, StTCP2/4 and StPTB6	Identified seven (out of 324) conserved and 12 (out of 311) novel miRNAs showing differential expression in early stolon stages under short day (8 h light, 16 h dark) versus long day (16 h light, 8 h dark) photoperiods. The potential target genes involved in tuberization process were StGRAS, StTCP2/4, StPTB6, StTAS3, StTAS5, StPTB1, POTH1 and StCDPKs.	Kondhare et al. (2018)
Potato virus A	miR482, miR397	Pathogenesis realted (PR) genes	Identifed a total of 2,062 differentially-expressed genes and 201 miRNAs. Small RNA sequencing revealed miRNA-mRNA interactions related to PVA infection. Some of the miRNAs (stu-miR482d-3p, stu-miR397-5p, etc) which target pathogenesis-related (PR) genes showed negative correlations between differentially-expressed miRNAs and genes.	Li et al. (2017)
Colorado potato beetles	miR-100	Cytochrome P450	Identified 33 differentially-expressed miRNAs. Previously miR282 and miR989 have been known to be modulated by imidacloprid in other insects. Here, miR100, an miRNA associated with regulation of cytochrome P450 expression, was significantly modulated in imidacloprid-treated beetles.	Morin et al. (2017)

Table 3 contd. ...

...Table 3 contd.

Trait	miRNA	Target gene	Key findings	References
Potato virus X	miR165/166, miR159	Target mimics (TMs) function	Demonstrated *Potato virus X* (PVX)-based target mimics (TM) expression, causing strong miRNA silencing in *Nicotiana benthamiana*. The PVX-based expression of short tandem target mimics (STTMs) against miR165/166 and 159 caused the corresponding phenotype in all infected plants. Thus, a PVX-based VbMS is a powerful method to study miRNA function and may be useful in high-throughput investigation of miRNA function in *N. benthamiana*.	Zhao et al. (2016)
Potato virus Y	miR162, miR168a miR482	*DCL1, AGO1-2* and *Cc-nbs-lrr*	A novel gene, Ny-DG, conferring resistance to PVY was mapped on chromosome IX in the diploid potato clone, DG 81-68. Study showed that increased expression of stu-miR162, stu-miR168a and stu-miR482 promoted the downregulation of their targets *DCL1*, *AGO1-2* and *Cc-nbs-lrr*, respectively on six-day post-inoculated leaves at 28°C.	Szajko et al. (2019)
Drought	miR811 miR814 miR835 miR4398	MYB transcription factor	Differential expression analysis identified 100 down-regulated and 99 up-regulated miRNAs under drought stress. Additionally, 119 up-regulated and 151 down-regulated novel miRNAs were also identified. Four miRNAs were identified as regulating drought-related genes (miR811, miR814, miR835 miR4398). Their target genes were MYB transcription factor (CV431094), hydroxyproline-rich glycoprotein (TC225721), quaporin (TC223412) and WRKY transcription factor (TC199112), respectively.	Zhang et al. (2014)
Salinity	miR166 miR159	Transcription factors (Myb101)	Established inverse relationships by reverse co-expression between two salinity stress-regulated miRNAs (miR166 and miR159) and their target transcriptional regulators – HD-ZIP-Phabulosa/Phavulota and Myb101, respectively. The miR159-Myb101 network may be important in modulation of vegetative growth under stress conditions.	Kitazumi et al. (2015)
Cold	miR156 miR169	Transcription factors	The wild species *Solanum commersonii* possesses the highest tolerance to low temperatures under both acclimated (ACC) and non-acclimated (NACC) conditions. Of the several miRNAs identified, conserved miR408a and miR408b changed their expression under NACC conditions, whereas miR156 and miR169 were differentially expressed under ACC conditions.	Esposito et al. (2020)
Purple tuber skin and flesh	miR828	MYB transcription factors (MYB12, R2R3-MYB, MYB-36284)	Observed strong associations between the high levels of miR828, TAS4 D4(-) and purple/red color of tuber skin and flesh. MYB transcription factors were predicted as potential targets of miR828 and TAS4 D4(-) and it was confirmed that R2R3-MYB and MYB-36284 are direct targets of the small RNAs.	Bonar et al. (2018)

Nitrogen use efficiency	miR397 miR398	Nitrate transporter, sugar transporter, transcription factor, F-box family protein	A total of 119 conserved miRNAs belonging to 41 miRNAs families and 1,002 putative novel miRNAs were identified under low N stress. The predicted target genes were universal stress protein, heat shock protein, salt tolerance protein, calmodulin binding protein, serine-threonine protein kinsae, Cdk10/11-cyclin dependent kinase, amino acid transporter, nitrate transporter, sugar transporter, transcription factor, F-box family protein and zinc finger protein, etc.	Tiwari et al. (2020c)
Drought and heat	miR156, miR160, miR162, miR172, miR398	Zinc finger family protein, Auxin response factor ARF16, transcription factor (MADS-box GRAS, AP2)	Large numbers of miRNAs were expressed in Unica, whereas Russet Burbank indicated lesser number of changes in miRNA expression in response to drought and heat stress. Identified miR156d-3p, miR160a-5p, miR162a-3p, miR172b-3p and miR398a-5p and their putative targets where results indicate that they may play a vital role at different post-transcriptional levels against drought and heat stresses.	Öztürk Gökçe et al. (2021)
Phosphorous use efficiency	miR399	-	Identified miR399 for increasing phosphorus use efficiency. A large root system might be the most important trait for P acquisition on such soils and therefore in breeding P-efficient crops.	Wacker-Fester et al. (2019)
Potassium phosphate-induced resistance	miR159, miR166 miR167, miR171 miR398, miR482 miR530, miR7985, miR4376	-	Identified 25 differentially-expressed miRNAs (14 known and 11 new) in potato in response to potassium phosphate-induced resistance. The porential target genes were related to pathogen resistance, transcription factors and oxidative stress belonging to transcription factor gene families, like bZip, GRAS, AP2, REV HD-ZipIII, ARF, among others and genes involved in the regulation of plant metabolism and stress responses.	Rey-Burusco et al. (2019)

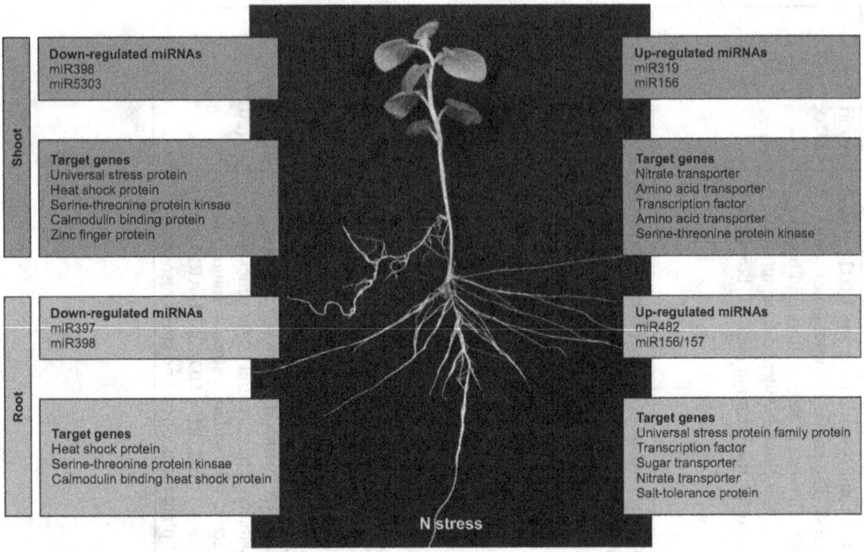

Fig. 5. Identification of miRNA in potato plants grown in aeroponics under N stress.

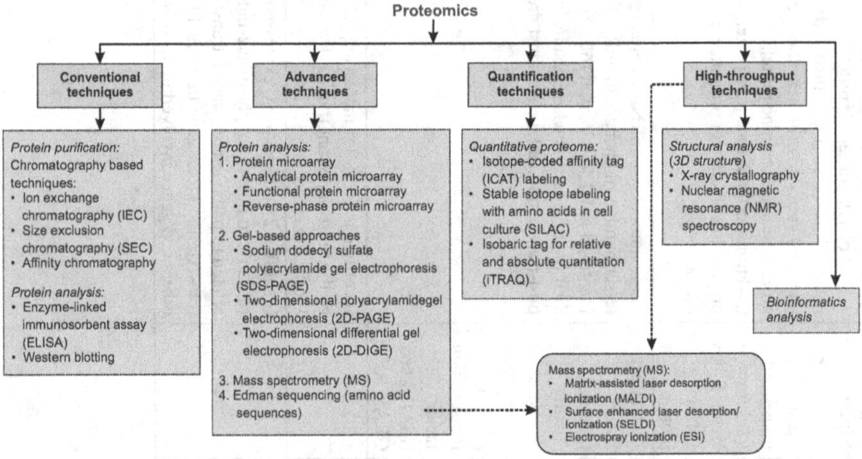

Fig. 6. An overview of proteomics technologies in plants including potato.

adds information on proteins to genomics and transcriptomics technologies and characterizes the structure and functions of proteins. Proteomics technologies are applied in various research fields, such as identification of diagnostic markers, vaccine production, effector proteins/pathogenisity and understanding the mechanism of candidate proteins involved in biotic and abiotic stress resistance/tolerance and quality traits in plants. Proteomics is highly complicated because it involves the analysis and categorization of overall proteins of a genome. Mass spectrometry with LC–MS-MS and MALDI-TOF/TOF has been used widely in current proteomics research. However, utilization of proteomics facilities, including the software for equipment, databases and the requirement of skilled personnel, substantially increase the costs, restricting their wider use, especially in the developing world (Aslam et al., 2017). Furthermore, the proteome is highly dynamic because of complex regulatory systems that control the expression levels of proteins. Advances in proteomics study strengthen the genetic and molecular information for various traits on the network of proteins/genes involved in biosynthesis pathways of disease-pest and quality traits. The techniques of proteomics technologies are depicted in Fig. 6. A summary of proteomics studies is outlined in Table 2 for various biotic and abiotic stresses and tuber quality traits in potato.

5.1 Biotic Stress

Potyviruses (e.g., PVY, PVA) are the largest and most damaging RNA viruses infecting the potato crop at subcellular level. Rajamaki (2020) studied nuclear proteins of virus-infected and healthy potato leaves. They identified 807 nuclear proteins, using gel-free liquid chromatography combined with tandem mass spectrometry, of which 16 proteins in each were specific to infected and healthy samples. In particular, the protein Dnajc14 was only identified in healthy leaf tissues, while various ribosomal proteins, ribosome-biogenesis proteins and RNA splicing-related proteins were observed in infected tissues. Similarly, Xiao et al. (2019) investigated proteins in potato leaves (cv. Sarpo Mira) after challenge inoculation (0, 48 and 120 h) with the pathogen *P. infestans*, causing late blight disease and identified a total of 1,229 differentially expressed proteins. In the early stages, cell-wall-associated proteins were predominant, while at a later stage of infection, cellular protein modification process, membrane protein complex and cell death induction were noticed. Overall, Table 1 summarizes recent proteomics work on biotic stress resistance in potato.

5.2 Abiotic Stress

Aghaei et al. (2008) studied proteome analysis in potato (cvs. Kennebec and Concord) under salt stress (90 mM NaCl) and identified 322 and 305 differentially expressed proteins, respectively. A few important up-regulated proteins were like osmotine-like proteins, TSI-1 protein, heatshock proteins and protein inhibitors so on. In another research, Evers et al. (2012) elucidated a synthetic view of gene network in potato in response to cold and salt (150 mM NaCl) stress by transcriptomic and proteomic analyses. A strong repression in proteins associated with various biological/metabolic pathways was observed under salt stress, such as photosynthesis (e.g., glyceraldehyde-3-phosphate dehydrogenase and triose phosphate isomerise), nitrogen and amino acid metabolisms, particularly polyamine synthesis (e.g., arginine decarboxylase and S-adenosyl-methionine decarboxylase). Proteome analysis in potato was performed in drought-tolerant potato (cv. Ninglang 182) leaves under drought stress and identified 12 differentially expressed proteins, using 2D gel electrophoresis and MALDI-TOF-TOF/MS analyses. Likewise, a number of studies have been done in potato. Table 2 lists recent proteomics work on abiotic stress tolerance in potato.

5.3 Quality and Other Traits

Mitochondrial proteins were analyzed in potato tubers using Blue-Native PAGE and 2D (Eubel et al., 2004) and a total of 18 proteins of super complexes and respiratory complexes were identified. Further, they provided insights into the respiratory chain of plant mitochondria (Eubel et al., 2003, 2004). Mitochondrial proteome was analyzed in potato, applying shotgun proteomics approach containing 1060 non-redundant proteins and identified abundant proteins associated with electron transport chain and TCA (tricarboxylic acid cycle) cycle (Salvato et al., 2014). Recently, genetical genomics of quality-related traits in potato tubers was investigated, using proteomics and established raltionship between proteins and tuber quality traits (Acharjee et al., 2018). Table 1 summarizes recent proteomics work on tuber quality and other traits in potato.

6. Metabolomics

Metabolomics involves identification and quantification of metabolites present in the biological systems of organisms at a certain period of time (Daviss et al., 2005). Metabolites are produced after downstream applications of transcriptomics and proteomics when metabolites are the end products of organisms having a size less than 1 kDa (Samuelsson et al., 2008). A network analysis allows understanding of complex data of plant metabolism (Toubiana et al., 2013). Application of metabolomics research in potato is summarized in Table 1.

6.1 Biotic Stress

Plant defense system allows production of bioactive chemical compounds to defend themselve from attacking pathogens. Lekota et al. (2020) studied metabolmic fingerprinting in potato roots for resistance/susceptibility to *Spongospora subterranea* f. sp. *subterranean*. Metabolomics profiling was performed in potato roots and root exudates, using ultra-performance liquid chromatography coupled with quadrupole time-of-flight mass spectrometry (UPLC-Q-TOF/MS). The metabolites belonging to amino acids, organic acids, fatty acids, phenols, sugards and cell wall thickening compounds were abundandnt in the resistant cultivars than in the susceptible cultivars. Moreover, metabolites belonging to secondary defense metabolism and phenolics compounds were also more in tolerant cultivars than in susceptible ones. Table 1 summarizes recent metabolomics work on biotic stress resistance in potato.

6.2 Abiotic Stress

Metabolomics has emerged as an important tool to investigate plant metabolism for many traits, particularly abiotic stress tolerance in plant lifecycles affecting crop yield. Metabolomics applies the use of advanced chromatography methods and highly precise detection tools to uncover the metabolic spectra of organisms under the influence of internal and external stimuli. A number of metabolimics processes were investigated to understand the mechanisms for tuber yields, nutritional compounds, quality traits and resistance mechanism to insect-pests and diseases

(Puzanskiy et al., 2017). Metabolomics study was applied in potato cultivars/lines to detect changes with respect to environmental factors and also during tuber storage. This finding provides clues to elucidate the underlying mechanism associated with stress tolerance in potato to increase yield. Table 1 outlines the recent metabolomics work on abiotic stress tolerance in potato.

6.3 Quality and Other Traits

Potato contains many nutrional and quality beneficial compounds for human health. Chaparro et al. (2018) investigated comprehensive metabolimics (UPLC- and GC-MS) and ionomics (ICP-MS) study in potato tubers (raw and cooked). They identified 43 bioactive compounds, 42 nutrients, 76 lipids and 23 metals. Most nutrients and bioactive compounds were partially degraded during cooking (52%); however, genotypes with high quantities of bioactive compounds remained highest in cooked tubers and yellow-fleshed potato tubers recorded higher levels of carotenoids. This study provides valuable information on potato breeding for nutritional traits. In another study, Plischke et al. (2012) performed metabololite profiling in GM and non-GM potato leaves in response to leaf age, aphid herbivory and virus infection using the NMR approach. Study observed that only young leaves of healthy plants of GM lines recorded a different response from its parents. Also, aphid performance on excised leaves was influenced by leaf age, whereas no difference was found between GM and non-GM plants. Recent metabolomics work on potato tuber quality and other traits is summarized in Table 1.

7. Ionomics

Ionomics is the term emerging from plant mineral nutrition and metabolomics, revealing a network among various mineral nutrients in an organism (Baxter et al., 2008). Later ionomics was expanded to include all metals, metalloids and non-metals present in an organism (Lahner et al., 2003). Furthermore, the metallome term was expanded to include biologically significant non-metals, such as nitrogen, phosphorus, sulfur, selenium, chlorine and iodine (Outten et al., 2001). Ionomics, metabolmics and proteomics may cover some overlapping nutrients, such as phosphorous, sulphur and nitrogen compounds containing non-metals falling under ionome and metabolomes, and metals like zinc, copper, manganese and iron falling within proteome and metallproteome categories (Szpunar, 2004). Chao et al. (2011) performed an inonomics study and identified sphingolipids in the root playing an important role in regulating the leaf ionome in *Arabidopsis thaliana*. Sha et al. (2012) investigated ionomics to understand the impact of seeds affected by previous cropping with mycorrhizal plant and manure application in soybean. A number of stodie shave were conducted on ionomics, particulalty on silicon. Usually a plant uptakes very small quantities of silicon whilch gets accumulated in very small quantities in their tissues (< 0.5%). Moreover, silicon has been found to be associated with drought tolerance and is beneficial under other stresses in plants. Chaparro et al. (2018) carried out ionomics (ICP-MS) work in raw and cooked tubers of 60 potato genotypes and observed significant mineral variations in the cultivars, particularly for calcium, iron, molybdenum among the market classes. A very limited research is available on ionomics application in potato (Table 1).

8. Multi-Omics System

A large amount of datasets have been generated in potato by applying various omics approaches, such as genomics, transcriptomics, proteomics, metabolomics, phenomics and ionomics. These data can be used to identify key genetic and physiological regulatory elements associated with particular traits Cohen et al. (2017), but there is a need for integration of different omics technologies to understand the complex mechanism of the traits. Moreover, analysis of high throughput data of different omics approaches is one of the biggest challenges to reach a conclusive result. Scholz et al. (2014) detected biological features by independent component analysis through metabolite fingerprinting and suggested development of bioinformatics tools to enable in-depth pathways analysis of omics data for interpretation of biological functions. The ultimate goal of the omics tools is to integrate different components of biological functions through a comprehensive and integrated knowledge on DNA sequences, RNA, genes, proteins, metabolites and trait phenotypes. Cho et al. (2016) investigated an integrated regulatory network analysis to identify genes and metabolites for anthocyanin biosynthesis in three colored potato cvs. Hongyoung (light-red), Jayoung (dark-purple), and Atlantic (white). Transcriptomics and metabolomics approaches were applied by using RNA-sequencing and ultraperformance liquid chromatography quadrupole time-of-flight tandem mass spectrometry (UPLC-Q-TOF-MS), respectively. Correlation analysis of anthocyanin content and transcriptional changes showed 823 high correlations between 22 compounds and 119 transcripts in various categories, such as flavonoid metabolism, hormones, transcriptional regulation and signaling. Thus, based on the correlation studies between colored potato and genes involved therein, this study identified potential candidate genes for breeding for multi-colored potatoes, using the system-based approach. We have also reviewed integrated genomics, physiology and breeding approaches for improving nitrogen use efficiency in potato (Tiwari et al., 2018, 2020). A few numbers of web-based resources are available online to understand the transcriptomes and metabolites data, such as *Kyoto Encyclopedia of Genes and Genomes* (KEGG; http://www.genome.ad.jp/kegg/) contains pathways database (PATHWAY) of genes and metabolic pathways of various biological processes of many organisms. Some of the user-friendly tools are listed below for further use in breeding and biotechnology research on potato (Table 4).

9. Conclusion

A wide range of omics tools has become popular to understand molecular mechanisms in plants. Genotype, phenotype and environmental conditions affect adaptation mechanism of plants in response to various stresses by alteration in gene expression, proteins and metabolites. Hence, it is important to decipher functions of genes, proteins and metabolites for traits applying genomics, transcriptomics,

Table 4. Online tools/resources for omics studies in plants including potato.

Database	**Description**	**Website link**
Potato omics tools		
GoMapMan	GoMapMan is an open-access resource for functional annotation of genes in plants under three sub-sets, viz. proteins, metabolites and smallRNA.	http://www.gomapman.org/
NCBI	The NCBI genome database contains the representative genome sequences of potato (*Solanum tuberosum*, assembly SolTub_3.0),	https://www.ncbi.nlm.nih.gov/genome/400
PlantGDB	Web resources for plant genome browsers, sequences, annotation, assemblies, BLAST analysis, alignments, etc.	http://www.plantgdb.org
PoMaMo	PoMaMo (Potato Maps and More) is a GABI primary database for potato maps, sequences, SNPs, tools, etc.	http://www.gabipd.org/projects/Pomamo/#Tools
POTATOCYC 4.0	Metabolic pathway prediction.	https://plantcyc.org/databases/potatocyc/4.0
Sol Genomics Network	The Solanaceae Genomics Network database is for Solanaceous crops (tomato, pepper, eggplant, potato and tobacco).	https://solgenomics.net/
SolCAP	The Solanaceae Coordinated Agricultural Project (SolCAP) database includes genotype and phenotype data of Solanaceous crops.	http://solcap.msu.edu/index.shtml
Spud DB	The Spud DB: Potato Genomics Resource is web browser of the potato genome sequence and its updates and data mining tool.	http://solanaceae.plantbiology.msu.edu/index.shtml
Transcriptomics database		
ArrayExpress	ArrayExpress is a gene expression database based on arrays for *Solanum tuberosum* hosted by European Nucleotide Archive (ENA) and imports datasets from the GEO.	https://www.ebi.ac.uk/arrayexpress/search.html?query=Solanum+tuberosum
Expression Atlas	Tissue specific gene expression database (microarray and RNA-seq) are hosted by EMBL-EBI and display functional enrichment analysis.	https://www.ebi.ac.uk/gxa/home
Gene Expression Omnibus (GEO Datasets)	Gene Expression Omnibus (GEO Datasets) is the NCBI transcriptomics database of microarrays and RNA-sequences.	https://www.ncbi.nlm.nih.gov/gds/?term=
Genevestigator	It is a manually curated transcriptomics database.	https://genevestigator.com/
NONCODE	NONCODE is an integrated knowledge database of non-coding RNAs, excluding tRNA and rRNA.	http://www.noncode.org/
RefEx	Database of human, mouse and rat transcriptomes from 40 different organs hosted by DDBJ.	https://www.ddbj.nig.ac.jp/index-e.html

Table 4 contd. ...

...Table 4 contd.

Database	Description	Website link
Multi-omics database		
BioCyc	BioCyc is a collection of 19,534 Pathways/ Genome Databases (PGDBs) and integrates genome data of metabolites, regulatory networks, proteins, orthologs, genes and atom mappings.	https://biocyc.org
COBRA	The COBRA (COnstraint-based Reconstruction and Analysis) is a MATLAB tool for quantitative prediction of cellular and multicellular biochemical networks.	https://opencobra.github.io/ cobratoolbox
COVAIN	COVAIN (Covariance inverse) is a tool for uni- and multivariate statistics, time-series and correlation network analysis of metabolomics covariance data.	http://www.univie.ac.at/mosys/ software.html
GIM3E	GIM3E is a tool for condition-specific models of cellular metabolism, developed from metabolomics and expression data.	https://github.com/ brianjamesschmidt/gim3e
IMPaLA	IMPaLA (Integrated Molecular Pathway Level Analysis) is a tool for pathway over-representation and enrichment analysis with expression and metabolite data.	http://impala.molgen.mpg.de
KaPPA-View4	KaPPA-View4 is a metabolic pathway database for better understanding of metabolic regulation.	http://kpv.kazusa.or.jp/kpv4/
KBCommons	KBCommons (Knowledge Base Commons) is a multi-omics web-based data integration framework for biological discoveries.	https://kbcommons.org
KEGG PATHWAY	KEGG PATHWAY is a collection of metabolic pathway maps representing knowledge on molecular interaction, reaction and relation networks.	https://www.genome.jp/kegg/ pathway.html
Mapman	The Mapman Site of Analysis the Gabi primary database used for enrichment analysis and visualization of gene expression data.	https://mapman.gabipd.org
MetaCyc	MetaCyc metabolic pathways database includes experimentally elucidated 2,937 pathways from 3295 different organisms.	https://metacyc.org
mixOmics	The mixOmics tool offers a wide range of multivariate methods for the exploration and integration of omics datasets and similarity relationships.	http://www.mixOmics.org
ModelSEED	ModelSEED is a resource for the reconstruction, exploration, comparison and analysis of metabolic models.	https://modelseed.org/
Omicade4	A multivariate approach to the integration of multi-omics datasets.	http://bioconductor.org/packages/ release/bioc/html/omicade4.html
OmicKriging	OmicKriging is a poly-omic prediction tool for complex traits.	https://github.com/hakyimlab/ OmicKriging

Table 4 contd. ...

...Table 4 contd.

Database	Description	Website link
OmicsAnalyzer	OmicsAnalyzer is useful for vidualization and integrated analysis of large-scale omics data.	https://apps.cytoscape.org/apps/omicsanalyzer
OmicsPLS	OmicsPLS is useful for integrating two high dimensional datasets and visualizing relationship, particularly in life sciences.	https://github.com/selbouhaddani/OmicsPLS
PaintOmics	PaintOmics v4.5 is a web tool for integrative visualization of multiple omics datasets on to KEGG pathways.	http://www.paintomics.org
PATHVIEW	PATHVIEW allows pathway-based data integration and visualization	https://pathview.uncc.edu/
PathVisio	PathVisio is a pathway analysis and drawing software which allows drawing, editing and analyzing of biological pathways.	https://pathvisio.github.io/
Phytozome	Phytozome (v13) is a comprehensive platform for green plant genomics and hosts 261 assembled and annotated genomes.	https://phytozome-next.jgi.doe.gov/
Plant Regulomics	Plant regulomics is a data-driven interface for retrieving upstream regulators from plant multi-omics data.	http://bioinfo.sibs.ac.cn/plant-regulomics/
VANTED	VANTED (Visualization and Analysis of Network) containing experimental data for metabolic mappings and correlation networks analysis.	https://www.cls.uni-konstanz.de/software/vanted/
Proteomics database		
BLAST	NCBI basic local alignment search tool.	https://blast.ncbi.nlm.nih.gov/Blast.cgi
BLASTP	NCBI BLASTP programs search protein database using protein query.	https://blast.ncbi.nlm.nih.gov/Blast.cgi
CATH/Gene3D	Classifies protein structure (3D) from Protein Data Bank.	https://www.cathdb.info/
CDD	The Conserved Domain Databse (CDD) is a protein annotation resource that consists of a collection of well-annotated multiple sequence alignment models for ancient domains and full-length proteins.	https://www.ncbi.nlm.nih.gov/Structure/cdd/cdd.shtml
ConSurf	The ConSurf server is used for the identification of fucnitonal regions in proteins.	http://consurf.tau.ac.il/
DELTA-BLAST	Domain Enhanced Lookup Time Accelerated BLAST.	https://blast.ncbi.nlm.nih.gov/Blast.cgi
DIALIGN	DIALIGN is a software program for multiple sequence alignment.	http://bibiserv.techfak.uni-bielefeld.de/dialign/
EMBL-EBI	Tools and data resources maintained by the European Bioinformatics Institute (EMBL-EBI).	https://www.ebi.ac.uk/services/all
EMBL-EBI	Resources for various tools, including nucleotide and proteins analysis.	https://www.ebi.ac.uk/Tools/structure/
ExPASy	UniProtKB/Swiss-Prot is the curated component of UniProtKB containing thousands of protein descriptions.	https://www.expasy.org/resources/uniprotkb-swiss-prot

Table 4 contd. ...

...Table 4 contd.

Database	Description	Website link
GenBank	NCBI protein database.	https://www.ncbi.nlm.nih.gov/protein/
Gibbs	Gibbs Motif Sampler (Gibbs) is a motif-finding tool in protein sequence.	http://bayesweb.wadsworth.org/gibbs/gibbs.html
InterPro	InterPro provides functional analysis of proteins by classifying them into families and predicting domains and important sites.	https://www.ebi.ac.uk/interpro/
MEME	Motif-based alignment server.	https://meme-suite.org/meme/tools/meme
MMDB	Molecular modeling database (MMDB) is a part of Entrez system to facilitate 3-dimensional structure of biomolecules.	http://www.ncbi.nlm.nih.gov/Structure/
Motif Scan	Motif scan allows finding of all known motifs that occur in a protein sequence.	https://myhits.sib.swiss/cgi-bin/PFSCAN
NetPhos 3.1	The NetPhos 3.1 server predicts serine, threonine or tyrosine phosphorylation sites in eukaryotic proteins.	http://www.cbs.dtu.dk/services/NetPhos
OWL	A composite protein sequence database of four publicly available primary resources: SWISS-PROT, PIR (1-3), GenBank (translation) and NRL-3D.	http://www.bioinf.man.ac.uk/dbbrowser/OWL/ http://130.88.97.239/OWL/
Pfam	Pfam database is a large collection of protein families.	http://pfam.xfam.org/
PHI-BLAST	NCBI Pattern Hit Initiated BLAST.	https://blast.ncbi.nlm.nih.gov/Blast.cgi
PIR	Protein Information Resource is an integrated protein informtaics resource for genomic, proteomic and system biology research.	http://pir.georgetown.edu/
PRABI Lyon Gerland	PRABI Lyon Gerland is a tool for protein sequence structure prediction.	https://prabi.ibcp.fr/htm/site/web/home
PROSITE	PROSITE consists of protein domains, families and functions.	https://prosite.expasy.org/
PSI-BLAST	NCBI Position Specific Iterated BLAST.	https://blast.ncbi.nlm.nih.gov/Blast.cgi
RCSB PDB	Research Collaboratory for Structural Bioinformatics (RSCB) Protein Data Bank (PDB) is the protein structure-alignment database.	http://www.rcsb.org/pdb/
RefSeq	NCBI reference sequence database including genomic, transcript and protein.	https://www.ncbi.nlm.nih.gov/refseq/
SCOP	Structural Classification of Proteins (SCOP) is a protein structure database.	http://scop.mrc-lmb.cam.ac.uk/scop/
SIRW	Simple Indexing and Retrieval System (SIRW) is a web interface to search protein/nucleotide databases and a sequence motif.	http://sirw.embl.de/index.html
SwissModel	SwissModel is the automated protein structure homology-modeling platform.	https://www.expasy.org/resources/swiss-model

Table 4 contd. ...

...*Table 4 contd.*

Database	Description	Website link
SWISS-MODEL	The SWISS-MODEL Repository is a database of annotated 3D protein structure models.	http://swissmodel.expasy.org/repository/
TBLASTN	NCBI BLASTN programs search nucleotide database using a nucleotide query.	https://blast.ncbi.nlm.nih.gov/Blast.cgi
TrEMBL	UniProtKB/TrEMBL is a computer-annotated protein sequence database complementing the UniProtKB/Swiss-Prot Protein Knowledge Base.	http://www.bioinfo.pte.hu/more/TrEMBL.htm
UniParc	The UniProt Archive (UniParc) is a comprehensive and non-redundant database.	https://www.uniprot.org/help/uniparc
UniProt	Universal Protein Resources (UniProt) is a comprehensive and high-quality tool of protein sequence and functional information.	https://www.uniprot.org/
UniRef	The UniProt Reference Cluster (UniRef) sets of sequences.	https://www.uniprot.org/help/uniref

Source: Omics tools data are modified from Jamil et al. (2020) and Aslam et al. (2017)

metabolomics, ionomics and phenomics tools. A cascade of pathways linked to trait phenotyping has been devised to understand genetic makeup of plants. Although, genomics and transcriptomics have been applied to a great extent, and to some extent in proteomics and metabolics also, however, works on ionomics and phenomics are yet to be seen. Taken together, integration of various omics is needed to relate the data in order to identify functions of genes for biotic/abotic stress and quality traits.

Acknowledgement

I am thankful to the Director, ICAR-Central Potato Research Institute, Shimla, and scientists/technicians/research fellows and other colleagues of the institute for their support under the institute research projects on biotechnology, germplasm, breeding, and seed research on aeroponics. I am also grateful to the funding agencies for support under the externally funded projects (CABin, ICAR-IASRI, New Delhi; ICAR-LBS Young Scientist Award Project, and DBT, Government of India).

References

Acharjee, A., Chibon, P.Y., Kloosterman, B., America, T., Renaut, J. et al. (2018). Genetical genomics of quality related traits in potato tubers using proteomics. *BMC Plant Biol.*, 18: 20.

Aghaei, K., Ehsanpour, A.A. and Komatsu, S. (2008). Proteome analysis of potato under salt stress. *J. Proteome Res.*, 7(11): 4858–4868.

Ali, A., Alexandersson, E., Sandin, M., Resjö, S., Lenman, M. et al. (2014). Quantitative proteomics and transcriptomics of potato in response to phytophthora infestans in compatible and incompatible interactions. *BMC Genomics*, 15(1): 497.

Aslam, B., Basit, M., Nisar, M.A., Khurshid, M. and Rasool, M.H. (2017). Proteomics: technologies and their applications. *J. Chromatographic Sci.*, 55: 182–196.

Aversano, R., Contaldi, F., Adelfi, M.G., D'Amelia, V., Diretto, G. et al. (2017). Comparative metabolite and genome analysis of tuber-bearing potato species. *Phytochemistry*, 137: 42–51.

Bali, S., Vining, K., Gleason, C., Majtahedi, H., Brown, C.R. and Sathuvalli, V. (2019). Transcriptome profiling of resistance response to *Meloidogyne chitwoodi* introgressed from wild species *Solanum bulbocastanum* into cultivated potato. *BMC Genomics*, 20(1): 907.

Barra, M., Meneses, C., Riquelme, S., Pinto, M., Lagüe, M. et al. (2019). Transcriptome profiles of contrasting potato (*Solanum tuberosum* L.) genotypes under water stress. *Agronomy*, 9(12): 848.

Bastarache, P., Wajnberg, G., Dumas, P., Chacko, S., Lacroix, J. et al. (2020). Transcriptomics-based approach identifies spinosad-associated targets in the Colorado potato beetle, *Leptinotarsa decemlineata*. *Insects*, 11(11): 820.

Baxter, I.R., Vitek, O., Lahner, B., Muthukumar, B., Borghi, M. et al. (2008). The leaf ionome as a multivariable system to detect a plant's physiological status. *Proc. Natl. Acad. Sci., U.S.A.*, 105: 12081–12086.

Bonar, N., Liney, M., Zhang, R., Austin, C., Dessoly, J. et al. (2018). Potato miR828 is associated with purple tuber skin and flesh color. *Front. Plant Sci.*, 9: 1742.

Brouwer, S.M., Lindqvist-Reis, P., Pergament, D.P., Marttila, S., Grenville-Briggs, L.J. and Andreasson, E. (2021). Visualising the ionome in resistant and susceptible plant-pathogen interactions. *The Plant J.* Doi: https://doi.org/10.1111/tpj.15469.

Bündig, C., Jozefowicz, A.M., Mock, H.P. and Winkelmann, T. (2016). Proteomic analysis of two divergently responding potato genotypes (*Solanum tuberosum* L.) following osmotic stress treatment *in vitro*. *J. Proteomics*, 143: 227–241.

Cao, W., Gan, L., Shang, K., Wang, C., Song, Y. et al. (2020). Global transcriptome analyses reveal the molecular signatures in the early response of potato (*Solanum tuberosum* L.) to *Phytophthora infestans*, *Ralstonia solanacearum* and Potato virus Y infection. *Planta*, 252(4): 57.

Carvallo, M.A., Pino, M.T., Jeknić, Z., Zou, C., Doherty, C.J. et al. (2011). A comparison of the low temperature transcriptomes and CBF regulons of three plant species that differ in freezing tolerance: *Solanum commersonii*, *Solanum tuberosum*, and *Arabidopsis thaliana*. *J. Exp. Bot.*, 62(11): 3807–3819.

Chao, D.Y., Gable, K., Chen, M., Baxter, I., Dietrich, C.R. et al. (2011). Sphingolipids in the root play an important role in regulating the leaf ionome in *Arabidopsis thaliana*. *The Plant Cell*, 23(3): 1061–1081.

Chaparro, J.M., Holm, D.G., Broeckling, C.D., Prenni, J.E. and Heuberger, A.L. (2018). Metabolomics and ionomics of potato tuber reveals an influence of cultivar and market class on human nutrients and bioactive compounds. *Front. Nutr.*, 5: 36.

Chen, Y., Li, C., Yi, J., Yang, Y., Lei, C. and Gong, M. (2019). Transcriptome response to drought, rehydration and re-dehydration in potato. *Inter. J. Mol. Sci.*, 21(1): 159.

Cheng, L., Zhang, S., Yang, L., Wang, Y., Yu, B. and Zhang, F. (2019). Comparative proteomics illustrates the complexity of Fe, Mn and Zn deficiency-responsive mechanisms of potato (*Solanum tuberosum* L.) plants *in vitro*. *Planta*, 250(1): 199–217.

Cho, K., Cho, K.S., Sohn, H.B., Ha, I.J., Hong, S.Y. et al. (2016). Network analysis of the metabolome and transcriptome reveals novel regulation of potato pigmentation. *J. Exp. Bot.*, 67(5): 1519–1533.

Cohen, H., Aharoni, A., Szymanski, J. and Dominguez, E. (2017). Assimilation of 'omics' strategies to study the cuticle layer and suberin lamellae in plants. *J. Exp. Bot.*, 68: 5389–5400.

Daviss, B. (2005). Growing pains for metabolomics. *Scientist*, 19: 25–28.

Demirel, U., Morris, W.L., Ducreux, L., Yavuz, C., Asim, A. et al. (2020). Physiological, biochemical and transcriptional responses to single and combined abiotic stress in stress-tolerant and stress-sensitive potato genotypes. *Front. Plant Sci.*, 11: 169.

Edward, M., Muntean, N., Duda, M. and Michalski, R. (2019). Heavy metals' uptake from soil in potato tubers: an ionomic approach. *ProEnvironment*, 12: 159–162.

Elagamey, E., Sinha, A., Narula, K., Abdellatef, M., Chakraborty, N. and Chakraborty, S. (2017). Molecular dissection of extracellular matrix proteome reveals discrete mechanism regulating *Verticillium dahliae* triggered vascular wilt disease in potato. *Proteomics*, 17: 1600373.

Esposito, S., Aversano, R., Bradeen, J.M., Matteo, A.D., Villano, C. and Carputo, D. (2020). Deep-sequencing of *Solanum commersonii* smallRNA libraries reveals ribo regulators involved in cold stress response. *Plant Biol. (Stuttg)*, 22 Suppl., 1: 133–142.

Eubel, H., Heinemeyer, J. and Braun, H.P. (2004). Identification and characterization of respirasomes in potato mitochondria. *Plant Physiol.*, 134: 1450–1459.

Eubel, H.L., Heinemeyer, J. and Braun, H.P. (2003). New insights into the respiratory chain of plant mitochondria. Supercomplexes and a unique composition of complex II. *Plant Physiol.*, 133: 274–286.

Evers, D., Lefevre, I., Legay, S., Lamoureux, D., Hausman, J.F. et al. (2010). Identification of drought-responsive compounds in potato through a combined transcriptomic and targeted metabolite approach. *J. Exp. Bot.*, 61(9): 2327–2343.

Evers, D., Legay, S., Lamoureux, D., Hausman, J.F., Hoffmann, L. and Renaut, J. (2012). Towards a synthetic view of potato cold and salt stress response by transcriptomic and proteomic analyses. *Plant Mol. Biol.*, 78(4-5): 503–514.

Fofana, B., Somalraju, A., Fillmore, S., Zaidi, M., Main, D. and Ghose, K. (2020). Comparative transcriptome expression analysis in susceptible and resistant potato (*Solanum tuberosum*) cultivars to common scab (*Streptomyces scabies*) revealed immune priming responses in the incompatible interaction. *PLoS ONE*, 15(7): e0235018.

Folgado, R., Sergeant, K., Renaut, J., Swennen, R., Hausman, J.F. and Panis, B. (2014). Changes in sugar content and proteome of potato in response to cold and dehydration stress and their implications for cryopreservation. *J. Proteomics*, 98: 99–111.

Fukuda, T., Takamatsu, K., Bamba, T. and Fukusaki, E. (2019). Gas chromatography-mass spectrometry metabolomics-based prediction of potato tuber sprouting during long-term storage. *J. Biosci. Bioeng.*, 128(2): 249–254.

Gong, L., Zhang, H., Gan, X., Zhang, L., Chen, Y. et al. (2015). Transcriptome profiling of the potato (*Solanum tuberosum* L.) plant under drought stress and water stimulus conditions. *PLoS ONE*, 10(5): e0128041.

Goyer, A., Hamlin, L., Crosslin, J.M., Buchanan, A. and Chang, J.H. (2015). RNA-seq analysis of resistant and susceptible potato varieties during the early stages of potato virus Y infection. *BMC Genomics*, 16(1): 472.

Hamooh, B.T., Sattar, F.A., Wellman, G. and Mousa, M. (2021). Metabolomic and biochemical analysis of two potato (*Solanum tuberosum* L.) cultivars exposed to *in vitro* osmotic and salt stresses. *Plants (Basel)*, 10(1): 98.

Hancock, R.D., Morris, W.L., Ducreux, L.J., Morris, J.A., Usman, M. et al. (2014). Physiological, biochemical and molecular responses of the potato (*Solanum tuberosum* L.) plant to moderately elevated temperature. *Plant Cell Environ.*, 37: 439–450.

Helle, S., Bray, F., Verbeke, J., Devassine, S., Courseaux, A. et al. (2018). Proteome analysis of potato starch reveals the presence of new starch metabolic proteins as well as multiple protease inhibitors. *Front. Plant Sci.*, 9: 746.

Hu, X., Hodén, K.P., Liao, Z., Dölfors, F., Åsman, A.K. and Dixelius, C. (2020). Phytophthora infestans *Ago1-bound miRNA Promotes Potato Late Blight Disease.* BioRxiv doi: https://doi.org/10.1101/2020.01.28.924175.

Jamil, I.N., Remali, J., Azizan, K.A., Muhammad, N.A.N., Arita, M. et al. (2020). Systematic multi-omics integration (MOI) approach in plant systems biology. *Front. Plant Sci.*, 11: 944.

Jeevalatha, A., Sundaresha, S., Kumar, A., Kaundal, P., Guleria, A. et al. (2017). An insight into differentially regulated genes in resistant and susceptible genotypes of potato in response to *Tomato leaf curl New Delhi virus-[potato]* infection. *Virus Res.*, 232: 22–33.

Joshi, J.R., Yao, L., Charkowski, A.O. and Heuberger, A.L. (2021). Metabolites from wild potato inhibit virulence factors of the soft rot and blackleg pathogen *Pectobacterium brasiliense*. *Mol. Plant-Microbe Interact.*, 34(1): 100–109.

Jozefowicz, A.M., Hartmann, A., Matros, A., Schum, A. and Mock, H.P. (2017). Nitrogen deficiency induced alterations in the root proteome of a pair of potato (*Solanum tuberosum* L.) varieties contrasting for their response to low N. *Proteomics*, 17: 1700231.

Karlusich, J., Arce, R.C., Shahinnia, F., Sonnewald, S., Sonnewald, U. et al. (2020). Transcriptional and metabolic profiling of potato plants expressing a plastid-targeted electron shuttle reveal modulation of genes associated to drought tolerance by chloroplast redox poise. *Inter. J. Mol. Sci.*, 21(19): 7199.

Kitazumi, A., Kawahara, Y., Onda, T.S., De Koeyer, D. and de los Reyes, B.G. (2015). Implications of miR166 and miR159 induction to the basal response mechanisms of an andigena potato (*Solanum tuberosum* subsp. *andigena*) to salinity stress, predicted from network models in *Arabidopsis*. *Genome*, 58(1): 13–24.

Kloosterman, B., De Koeyer, D., Griffiths, R., Flinn, B., Steuernagel, B. et al. (2008). Genes driving potato tuber initiation and growth: identification based on transcriptional changes using the POCI array. *Funct. Integr. Genomics*, 8: 329–340.

Kochetov, A.V., Egorova, A.A., Glagoleva, A.Y., Strygina, K.V., Khlestkina, E.K. et al. (2020). The mechanism of potato resistance to *Globodera rostochiensis*: comparison of root transcriptomes of resistant and susceptible *Solanum phureja* genotypes. *BMC Plant Biol.*, 20(Suppl 1): 350.

Kondhare, K.R., Malankar, N.N., Devani, R.S. and Banerjee, A.K. (2018). Genome-wide transcriptome analysis reveals small RNA profiles involved in early stages of stolon-to-tuber transitions in potato under photoperiodic conditions. *BMC Plant Biol.*, 18(1): 284.

Lahner, B., Gong, J., Mahmoudian, M., Smith, E.L. and Abid, K.B. (2003). Genomic scale profiling of nutrient and trace elements in *Arabidopsis thaliana*. *Nat. Biotechnol.*, 21: 1215–1221.

Lakhotia, N., Joshi, G. and Bhardwaj, A.R. (2014). Identification and characterization of miRNAome in root, stem, leaf and tuber developmental stages of potato (*Solanum tuberosum* L.) by high-throughput sequencing. *BMC Plant Biol.*, 14: 6.

Larsen, M.K., Guldstrand, J., Malene, M., Bennike, T.B. and Stensballe, A. (2016). Time-course investigation of *Phytophthora infestans* infection of potato leaf from three cultivars by quantitative proteomics. *Data Brief*, 6: 238–248.

Lebecka, R., Kistowski, M., Dębski, J., Szajko, K., Murawska, Z. and Marczewski, W. (2019). Quantitative proteomic analysis of differentially expressed proteins in tubers of potato plants differing in resistance to *Dickeya solani*. *Plant Soil*, 441: 317–329.

Lekota, M., Modisane, K.J., Apostolides, Z. and van der Waals, J.E. (2020). Metabolomic fingerprinting of potato cultivars differing in susceptibility to *Spongospora subterranea* f. sp. *subterranea* root infection. *Int. J. Mol. Sci.*, 21(11): 3788.

Lemke, P., Moerschbacher, B.M. and Singh, R. (2020). Transcriptome analysis of *Solanum tuberosum* genotype RH89-039-16 in response to chitosan. *Front. Plant Sci.*, 11: 1193.

Li, H., Luo, W., Ji, R., Xu, Y., Xu, G., Qiu, S. and Tang, H. (2021a). A comparative proteomic study of cold responses in potato leaves. *Heliyon*, 7(2): e06002.

Li, L., Zou, Q., Deng, X., Peng, M.S., Huang, J., Lu, X.L. and Wang, X. (2017). Comparative morphology, transcription, and proteomics study revealing the key molecular mechanism of camphor on the potato tuber sprouting effect. *Int. J. Mol. Sci.*, 18(11): 2280.

Li, P., Fan, R., Peng, Z., Qing, Y. and Fang, Z. (2021b). Transcriptome analysis of resistance mechanism to potato wart disease. *Open Life Sci.*, 16(1): 475–481.

Li, Q., Qin, Y. and Hu, X. (2020). Transcriptome analysis uncovers the gene expression profile of salt-stressed potato (*Solanum tuberosum* L.). *Sci. Rep.*, 10: 5411.

Lin, Q., Xie, Y., Guan, W., Duan, Y., Wang, Z. and Sun, C. (2019). Combined transcriptomic and proteomic analysis of cold stress induced sugar accumulation and heat shock proteins expression during postharvest potato tuber storage. *Food Chem.*, 297: 124991.

Liu, B., Kong, L., Zhang, Y. and Liao, Y. (2021). Gene and metabolite integration analysis through transcriptome and metabolome brings new insight into heat stress tolerance in potato (*Solanum tuberosum* L.). *Plants* (Basel), 10(1): 103.

Liu, X., Chen, L., Shi, W., Xu, X., Li, Z., Liu, T. et al. (2021). Comparative transcriptome reveals distinct starch-sugar interconversion patterns in potato genotypes contrasting for cold-induced sweetening capacity. *Food Chem.*, 334: 127550.

Liu, Y., Lin-Wang, K., Deng, C., Warran, B., Wang, L. and Yu, B. (2015). Comparative transcriptome analysis of white and purple potato to identify genes involved in anthocyanin biosynthesis. *PLoS ONE*, 10(6): e0129148.

Lowe, R., Shirley, N., Bleackley, M., Dolan, S. and Shafee, T. (2017). Transcriptomics technologies. *PLoS Comput Biol.*, 13(5): e1005457.

Massa, A.N., Childs, K.L., Lin, H., Bryan, G.J., Giuliano, G. and Buell, C.R. (2011). The transcriptome of the reference potato genome *Solanum tuberosum* Group Phureja clone DM1-3 516R44. *PLoS ONE*, 6(10): e26801.

Massana-Codina, J., Schnee, S., Allard, P.M., Rutz, A., Boccard, J. et al. (2020). Insights on the structural and metabolic resistance of potato (*Solanum tuberosum*) cultivars to tuber black dot (*Colletotrichum coccodes*). *Front. Plant Sci.*, 11: 1287.

Moon, K.B., Ahn, D.J., Park, J.S., Jung, W.Y., Cho, H.S. et al. (2018). Transcriptome profiling and characterization of drought-tolerant potato plant (*Solanum tuberosum* L.). *Mol. Cells*, 41(11): 979–992.

Morin, M.D., Lyons, P.J., Crapoulet, N., Boquel, S. and Morin, P.J. (2017). Identication of differentially expressed miRNAs in Colorado potato beetles (*Leptinotarsa decemlineata* (Say)) exposed to imidacloprid. *Int. J. Mol. Sci.*, 18: 2728.

Natarajan, B., Kalsi, H.S., Godbole, P., Malankar, N., Thiagarayaselvam, A. et al. (2018). MiRNA160 is associated with local defense and systemic acquired resistance against *Phytophthora infestans* infection in potato. *J. Exp. Bot.*, 69(8): 2023–2036.

Natarajan, B. and Banerjee, A.K. (2020). MicroRNA160 regulates leaf curvature in potato (*Solanum tuberosum* L. cv. Désirée). *Plant Signal. Behav.*, 15(5): 1744373.

Odgerel, K. and Banfalvi, Z. (2021). Metabolite analysis of tubers and leaves of two potato cultivars and their grafts. *PLoS ONE*, 16(5): e0250858.

Oertel, A., Matros, A., Hartmann, A., Arapitsas, P., Dehmer, K.J. et al. (2017). Metabolite profiling of red and blue potatoes revealed cultivar and tissue specific patterns for anthocyanins and other polyphenols. *Planta*, 246(2): 281–297.

Osmani, Z., Sabet, M.S., Shams-Bakhsh, M., Moieni, A., Vahabi, K. and Wehling, P. (2019). Virus-specific and common transcriptomic responses of potato against PVY, PVA and PLRV using microarray meta-analysis. *Plant Breed.*, 138: 216–228.

Outten, C.E. and O'Halloran, T.V. (2001). Femtomolar sensitivity of metalloregulatory proteins controlling zinc homeostasis. *Science*, 292: 2488–2492.

Öztürk Gökçe, Z.N., Aksoy, E., Bakhsh, A., Demirel, U., Çalışkan, S. and Çalışkan, M.E. (2021). Combined drought and heat stresses trigger different sets of miRNAs in contrasting potato cultivars. *Funct. Integ. Genomics*, 21(3-4): 489–502.

Plischke, A., Choi, Y.H., Brakefield, P.M., Klinkhamer, P.G.L. and Bruinsma, M. (2012). Metabolomic plasticity in gm and non-GM potato leaves in response to aphid herbivory and virus infection. *J. Agric. Food Chem.*, 60(6): 1488–1493.

Puzanskiy, R.K., Yemelyanov, V.V., Gavrilenko, T.A. and Shishova, M.F. (2017). The perspectives of metabolomic studies of potato plants. *Russ. J. Genet. Appl. Res.*, 7(7): 744–756.

Rajamaki, M.L., Sikorskaite-Gudziuniene, S., Sarmah, N., Varjosalo, M. and Valkonen, J.P.T. (2020). Nuclear proteome of virus-infected and healthy potato leaves. *BMC Plant Biol.*, 20(1): 355.

Regan, S., Gustafson, V., Rothwell, C., Sardana, R., Flinn, B. et al. (2006). Finding the perfect potato: using functional genomics to improve disease resistance and tuber quality traits. *Can. J. Plant Pathol.*, 28: S247–S255.

Rensink, W.A., Iobst, S., Hart, A., Stegalkina, S., Liu, J. and Robin, B.C. (2005). Gene expression profiling of potato responses to cold, heat, and salt stress. *Funct. Integ. Genomics*, 5: 201–207.

Rey-Burusco, M.F., Daleo, G.R. and Feldman, M.L. (2019). Identification of potassium phosphite responsive miRNAs and their targets in potato. *PLoS ONE*, 14(9): e0222346.

Rodríguez-Pérez, C., Gómez-Caravaca, A.M., Guerra-Hernández, E., Cerretani, L., García-Villanova, B. and Verardo, V. (2018). Comprehensive metabolite profiling of *Solanum tuberosum* L. (potato) leaves by HPLC-ESI-QTOF-MS. *Food Res. Inter.*, 112: 390–399.

Salvato, F.J.F., Havelund, M., Chen, R.S.P., Rao, A., Rogowska-Wrzesinska, O.N. et al. (2014). The potato tuber mitochondrial proteome. *Plant Physiol.*, 164: 637–653.

Samuelsson, L.M. and Larsson, D.G. (2008). Contributions from metabolomics to fish research. *Mol. Biosyst.*, 4(10): 974.

Scholz, M., Gatzek, S., Sterling, A., Fiehn, O. and Selbig, J. (2014). Metabolite fingerprinting: detecting biological features by independent component analysis. *Bioinformatics*, 20: 2447–2454.

Sha, Z., Oka, N., Watanabe, T., Tampubolon, B.D., Okazaki, K. et al. (2012). Ionome of soybean seed affected by previous cropping with mycorrhizal plant and manure application. *J. Agric. Food Chem.*, 60: 9543–9552.

Shishova, M., Puzanskiy, R., Gavrilova, O., Kurbanniazov, S., Demchenko, K. et al. (2019). Metabolic alterations in male-sterile potato as compared to male-fertile. *Metabolites*, 9(2): 24.

Singh, A., Sundaresha, S., Bhardwaj, V., Singh, B., Kumar, D. and Singh, B.P. (2015). Expression profile of potato cultivars with contrasting tuberization at elevated temperature using microarray analysis. *Plant Physiol. Biochem.*, 97: 108–116.

Singh, R., Tiwari, J.K., Rawat, S., Sharma, V. and Singh, B.P. (2016a). Monitoring gene expression pattern in somatic hybrid of *Solanum tuberosum* and *S. pinnatisectum* for late blight resistance using microarray analysis. *Plant Omics*, 9: 99–105.

Singh, R., Tiwari, J.K., Rawat, S., Sharma, V. and Singh, B.P. (2016b). *In silico* identification of candidate microRNAs and their targets in potato somatic hybrid *Solanum tuberosum* (+) *S. pinnatisectum* for late blight resistance. *Plant Omics*, 9: 159–164.

Stare, T., Ramšak, Ž., Križnik, M. and Gruden, K. (2019). Multiomics analysis of tolerant interaction of potato with potato virus Y. *Sci. Data*, 6(1): 250.

Sulli, M., Mandolino, G., Sturaro, M., Onofri, C., Diretto, G. et al. (2017). Molecular and biochemical characterization of a potato collection with contrasting tuber carotenoid content. *PLoS ONE*, 12(9): e0184143.

Sundaresha, S., Tiwari, J.K., Sindhu, R., Sharma, S., Bhardwaj, V. et al. (2014). *Phytophthora infestans* associated global gene expression profile in a late blight resistant Indian potato cv. Kufri Girdhari. *Australian J. Crop Sci.*, 8: 215–222.

Szajko, K., Plich, J., Przetakiewicz, J., Sołtys-Kalina, D. and Marczewski, W. (2019). Comparative proteomic analysis of resistant and susceptible potato cultivars during *Synchytrium endobioticum* infestation. *Planta*, 251(1): 4.

Szajko, K., Yin, Z. and Marczewski, W. (2019). Accumulation of miRNA and mRNA targets in potato leaves displaying temperature-dependent responses to potato virus Y. *Potato Res.*, 62: 379–392.

Szajko, K., Ciekot, J., Wasilewicz-Flis, I., Marczewski, W. and Sołtys-Kalina, D. (2021). Transcriptional and proteomic insights into phytotoxic activity of interspecific potato hybrids with low glycoalkaloid contents. *BMC Plant Biol.*, 21(1): 60.

Szpunar, J. (2004). Metallomics: A new frontier in analytical chemistry. *Anal. Bioanal. Chem.*, 378: 54–56.

Tai, H.H., Lagüe, M., Thomson, S., Aurousseau, F., Neilson, J. et al. (2020). Tuber transcriptome profiling of eight potato cultivars with different cold-induced sweetening responses to cold storage. *Plant Physiol. Biochem.*, 146: 163–176.

Tang, R., Gupta, S.K., Niu, S., Li, X.Q., Yang, Q. et al. (2020). Transcriptome analysis of heat stress response genes in potato leaves. *Mol. Biol. Rep.*, 47(6): 4311–4321.

The Potato Genome Sequencing Consortium. (2011). Genome sequence and analysis of the tuber crop potato. *Nature*, 475: 189–195.

Tian, M., Wang, J., Li, Z., Wang, C., Zhang, D. et al. (2020). Transcriptome analysis of potato leaves in response to infection with the necrotrophic pathogen Alternaria solani. *bioRxiv*. Doi: https://doi.org/10.1101/2020.09.21.307314.

Tiwari, J.K., Devi, S., Sundaresha, S., Chandel, P., Ali, N. et al. (2015). Microarray analysis of gene expression patterns in the leaf during potato tuberization in the potato somatic hybrid *Solanum tuberosum* and *Solanum etuberosum*. *Genome*, 58: 305–313.

Tiwari, J.K., Devi, S., Ali, N., Dua, V.K., Singh, R.K. and Chakrabarti, S.K. (2018). Cloning and sequence variation analysis of candidate genes involved in nitrogen metabolism in potato (*Solanum tuberosum*). *Indian J. Agric. Sci.*, 88: 751–756.

Tiwari, J.K., Plett, D., Garnett, T., Chakrabarti, S.K. and Singh, R.K. (2018). Integrated genomics, physiology and breeding approaches for improving nitrogen use efficiency in potato: Translating knowledge from other crops. *Funct. Plant Biol.*, 45: 587–605.

Tiwari, J.K., Buckseth, T., Zinta, R., Saraswati, A., Singh, R.K. et al. (2020a). Transcriptome analysis of potato shoots, roots and stolons under nitrogen stress. *Sci. Rep.*, 10: 1152.

Tiwari, J.K., Buckseth, T., Devi, S., Varshney, S., Sahu, S. et al. (2020b). Physiological and genome-wide RNA-sequencing analyses identify candidate genes in a nitrogen-use efficient potato cv. Kufri Gaurav. *Plant Physiol. Biochem.*, 154: 171–183.

Tiwari, J.K., Buckseth, T., Zinta, R., Saraswati, A., Singh, R.K. et al. (2020c). Genome-wide identification and characterization of microRNAs by small RNA sequencing for low nitrogen stress in potato. *PLoS ONE*, 15(5): e0233076.

Tomohiko, F., Kiyofumi, T., Takeshi, B. and Eiichiro, F. (2020). Potato tuber metabolomics-based prediction of chip color quality and application using gas chromatography/flame ionization detector. *Biosci. Biotechnol. Biochem.*, 84(11): 2193–2198,

Toubiana, D., Fernie, A.R., Nikoloski, Z. and Fait, A. (2013). Network analysis: tackling complex data to study plant metabolism. *Trends in Biotechnol.*, 31: 29–36.

Wacker-Fester, K., Uptmoor, R., Pfahler, V., Dehmer, K.J., Bachmann-Pfabe, S. and Kavka, M. (2019). Genotype-specific differences in phosphorus efficiency of potato (*Solanum tuberosum* L.). *Front. Plant Sci.*, 10: 1029.

Wang, B., He, T., Zheng, X., Song, B. and Chen, H. (2021). Proteomic analysis of potato responding to the invasion of *Ralstonia solanacearum* UW551 and its type III secretion system mutant. *Mol. Plant-Microbe Interact.*, 34(4): 337–350.

Xiao, C., Gao, J., Zhang, Y., Wang, Z., Zhang, D. et al. (2019). Quantitative proteomics of potato leaves infected with phytophthora infestans provides insights into coordinated and altered protein expression during early and late disease stages. *Int. J. Mol. Sci.*, 20(1): 136.

Xiao, C., Huang, M., Gao, J., Wang, Z., Zhang, D. et al. (2020). Comparative proteomics of three Chinese potato cultivars to improve understanding of potato molecular response to late blight disease. *BMC Genomics*, 21(1): 880.

Yang, X., Guo, X., Yang, Y., Ye, P., Xiong, X. et al. (2018). Gene profiling in late blight resistance in potato genotype SD20. *Int. J. Mol. Sci.*, 19(6): 1728.

Yang, Y., Qiang, X., Owsiany, K., Zhang, S., Thannhauser, T.W. and Li, L. (2011). Evaluation of different multidimensional LC–MS/MS pipelines for isobaric tags for relative and absolute quantitation (iTRAQ)-based proteomic analysis of potato tubers in response to cold storage. *J. Proteome Res.*, 10: 4647–4660.

Zhang, J., Wang, Y., Zhao, Y., Zhang, Y., Zhang, J. et al. (2020). Transcriptome analysis reveals Nitrogen deficiency induced alterations in leaf and root of three cultivars of potato (*Solanum tuberosum* L.). *PLoS ONE*, 15(10): e0240662.

Zhang, L., Yao, L., Zhang, N., Yang, J., Zhu, X. et al. (2018). Lateral root development in potato is mediated by Stu-mi164 regulation of NAC transcription factor. *Front. Plant Sci.*, 9: 383.

Zhang, N., Yang, J., Wang, Z., Wen, Y., Wang, J. et al. (2014). Identification of novel and conserved micrornas related to drought stress in potato by deep sequencing. *PLoS ONE*, 9(4): e95489.

Zhang, W., Luo, Y., Gong, X., Zeng, W. and Li, S. (2009). Computational identification of 48 potato microRNAs and their targets. *Comput. Biol. Chem.*, 33: 84–93.

Zhao, J., Liu, Q., Hu, P., Jia, Q., Liu, N. et al. (2016). An efficient Potato virus X-based microRNA silencing in *Nicotiana benthamiana*. *Sci. Rep.*, 6: 20573.

Zuluaga, A.P., Solé, M., Lu, H., Góngora-Castillo, E., Vaillancourt, B. et al. (2015). Transcriptome responses to *Ralstonia solanacearum* infection in the roots of the wild potato *Solanum commersonii*. *BMC Genomics*, 16(1): 246.

Chapter 7
Phenomics in Potato

1. Introduction

Potato is an autotetraploid crop with tetrasomic inheritance, high heterozygosity and acute inbreeding depression. Hence, potato breeding through conventional breeding is a long process, time consuming and labour intensive. Conventional method includes hybridization between selected parents, followed by recurrent selection of elite clones for tuber and agronomic traits over multiple years (> 10 years) in different environments. Nearly 40 traits in potato are believed to be important during the varietal development process (Gebhardt, 2013). Hence, it is a very big challenge to handle a large population to select elite clones with desirable agronomic traits under the climate change scenario (Dahal et al., 2019). Manual collection of huge phenotypic data is laborious, time consuming, erroneous and requires a lot of manpower. Although, conventional breeding method is quite slow, there is a need to utilize high throughput phenotyping platform for precise measurement of traits. This new phenotyping system can screen thousands of lines in a short time and thus accelerate potato breeding. Increasing genomics information puts pressure on the breeders to provide rapid and accurate phenotypic analyses. Precise and efficient phenotypic platform is important for potato breeders for developing new improved varieties. Various breeding techniques, such as marker-assisted selection, quantitative trait loci mapping, mutants population analysis and association mapping require assistance of proper trait phenotyping. It is very difficult to manage and analyze huge datasets. Thus, the major challenge for phenotyping is to develop tools that can collect, manage, access, organize, integrate and analyze huge phenotypic data set of breeding program. New advanced imaging techniques and bioinformatics/computational tools are now available to assist high-throughput phenotyping.

Numerous traits have been reported to be associated with agronomic and stress resistance, of which a few have been used in germplasm screening conventional methods. Present phenotyping methods are costly, labour intensive and slow and moreover, utilize destructive sampling and allow analysis of a limited number of traits at a time (Virlet et al., 2017). Recently progress has been made in crop phenotyping techniques in a non-destructive manner to analyze traits for plant breeders, biotechnologists and geneticists to meet the world's increasing food production demands (Sankaran et al., 2017). These non-destructive high-throughput phenotyping technologies focus on various traits that directly or indirectly show plant water content, chlorophyll content, biomass and growth potential and also offer a new dimension that increases precision, speed and analysis of captured data (Zhao et al., 2019; Danilevicz et al., 2021). Development of high-yielding potato cultivars with improved tolerance to biotic and abiotic stresses requires precise phenotyping for various morphological, structural, physiological, biochemical and molecular traits (Zia et al., 2017). Therefore, breeders must be assisted by high-throughput phenotyping for functional analysis of specific genes, forward and reverse genetic studies and generation of new improved varieties with desired traits. Thus breeders can manage many trials in different growth conditions with the use of different lines in mapping populations, breeding populations, mutant populations and germplasm pool. Further advancement in high throughput and accurate phenotyping, modeling and association with potato breeders is a major challenge, particularly when developing new potato varieties in target environments (York, 2019).

2. Phenomics

Phenomics includes the study of plant phenotype with large datasets of multiple environments. The major aim of genetics is to understand genotype and phenotype and interaction with environmental factors, where phenotype is the manifestation of genotype and environment (Araus et al., 2018). Phenomics includes translation of genes or whole genome-level information into plant phenotypes through genomics application for target traits. To understand the whole plant system, integration of genotype and phenotype is important to enhance genetics and breeding of plants (Rebetzke et al., 2019). As compared to the genomes, phenomics is lagging behind and therefore there is a need for rapid phenotypic characterization to integrate phenomics with other omics approaches.

Precision phenotyping is important to understand the performance of genotypes in various environments, over many years/seasons. High-throughput data of genotypes for traits, such as agronomical, physio-biochemical, phenotype, quality and nutrition, and biotic and abiotic stresses are essential for genetic dissection of traits (Moreira et al., 2020; Kim et al., 2021). This would enhance accuracy in association among genotype, phenotype, traits and functions when coupled with molecular data (Pasala and Pandey, 2020). Identifying a large number of genes (genomics), proteins (proteomics), metabolites (metabolomics) and ions (inonomics) is now popular to dissect genetic variations in plants. However, there is a lack of infrastructure and accuracy in measurement of these parameters. Moreover, High Throughput Phenotyping (HTP) measurements of physiological processes, such as photosynthesis, nutrient uptake and transport with precision and accuracy in a large number of samples are gaining importance through phenomics (http:/www.lemnatec.com/; http:/www.plantaccelerator.org.au/). The recent advances in automated high throughput phenotyping platform through spectral imaging and sensor technologies allow measurement of numerous parameters, using computational tools (Yang et al., 2020).

Now with increasingly genomics resources and databases in many crops, including potatoes, phenomics has limited success. This could be due to lack of costly automated phenotyping facilities for precise measurement of various biotic/abiotic stress resistance/tolerance and quality traits of multi-populations in multiple environmental conditions, coupled with integrated computations tools to analyze the huge datasets (Großkinsky et al., 2015). The integrated HTP, analytical and computational tools for the large datasets would be the keys to link plant phenomes with genomes while exploring complex biological phenomenon. Knowledge about the different components of peonotypic variation, genotype and its performance in different environments is essential for selection of genotypes. Several studies have reported that genotype × environment interaction plays a key role in the yield performance of crops. Thus, HTP allows large-scale measurement of phenotype data with higher precision, accuracy and resolution, with reduced labour cost than conventional phenotyping methods. This would further augment linkage with plant genomes data for precision and targeted potato breeding in future. HTP technology allows various advantages like:

- *Automatic phenotyping of traits*: HTP platform enables automatic and non-destructive imaging by sensors/high-resolution cameras in controlled and/or in-field conditions and image analysis, using advanced computing tools/software.
- *High accuracy*: HTP allows rapid and accurate traits measurement of a large set of genotypes in less time on both foliage and roots traits.
- *High precision*: HTP is enabled with advanced sensors/cameras and software and therefore allows precise measurement of traits of complete plant lifecycle.
- *Increase selection efficiency and reduce breeding cycle*: HTP increases selection efficiency and reduces breeding cycles in less time through selection of desirable traits.
- *Integration of genomics and phenomics approaches*: HTP has the capacity to deliver genomics and phenomics research advancement via rapid genes and markers discovery for speedy breeding of potato.

3. High-Throughput Phenotyping (HTP) Platforms

HTP is essential to phenotype a genotype to utilize its ultimate potential. As mentioned before, conventional phenotyping method is slow, time-taking, laborious and less precise, often leading to destructive sampling and has limited phenotyping ability. High-throughput phenotyping (HTP) platform is essential for precision phenotyping and further application in breeding new cultivars (Tables 1 and 2). HTP platforms are usually based on automation, sensors, high resolution imaging and robotics, which have been applied in a wide range of crops, including potato (Atefi et al., 2021). In potato, a few technologies have been applied to both underground and aboveground plant parts. For example, Phenofab and Keytrack System (KeyGene, The Netherlands) have been developed, using multiple imaging systems and thermal sensors with automated handling under controlled environments for measuring plant growth traits.

HTP platforms have been applied for non-destructive analysis of whole plant system in different crops under controlled or field conditions, which utilize advanced automation and robotics, imaging (2D/3D) techniques, unique sensors, hardware and

Table 1. Summary of high-throughput phenotyping (HTP) work in crops including potato.

Sr. No.	Crop	Trait	HTP system	Key findings	References
1.	Potato	Heat and drought stress	X-ray Computed Tomography (CT)	CT analysis revealed that tuber growth is inhibited under stress within a week and resumed when stress was terminated.	Harsselaar et al. (2021)
2.	Potato	Plant height, canopy cover and volume	Unmanned Aerial Vehicle (UAV) system	Demonstrated use of UAV imaging for accurate phenotyping of plant canopy architecture traits.	Colwell et al. (2021)
3.	Potato	Tuber shape	Commercially available camera (Canon EOS Rebel T7i)	Developed an image analysis pipeline to extract tuber shape statistics from images taken by using normal cameras up to ~ 76–86% accuracy.	Neilson et al. (2021)
4.	Potato	Drought stress	Canopy spectral reflectance using spectroradiometer FieldSpec 4 Hi-Res (ASD Inc., USA)	Spectral information was correlated with physiological variables, such as foliar area, total water content, relative growth rate of potato tubers, leaf area ratio and foliar area index.	Romero et al. (2017)
5.	Potato	Crop emergence and canopy cover	RGB imaging using unmanned aerial vehicle (UAV)	Developed semi-automated image analysis software to estimate crop emergence from high-resolution RGB ortho-images captured from an Unmanned Aerial Vehicle (UAV).	Li et al. (2019)
6.	Potato	Crop emergence	UAV integrated with multispectral imaging using octocopter ARF OktoXL 6S12 (HiSystems GmbH, Germany)	Revealed significant correlations between multispectral image-based features, such as plant count, SUM-NDVI and SUM-BINARY, and manual plant counts.	Sankaran et al. (2017)
7.	Potato	Stomatal behaviour	Infra-red thermal images using ThermaCAM P25 infrared camera (FLIR systems, USA)	Developed as an easy, rapid and non-destructive screening method, using infra-red thermography (IRT) imaging for high throughput field phenotyping in potato.	Prashar et al. (2014)
8.	Wheat	Plant height	UAV-based RGB imaging	Provided the use of UAV-based high-resolution RGB images to obtain time-series estimates of plant height suitable for breeding trials.	Volpato et al. (2021), Juliana et al. (2019)
9.	Wheat	Complex traits	Deep learning (DL) algorithms: multilayer perceptron (MLP), and convolutional neural network (CNN)	DL models gave 0–5% higher prediction accuracy than rrBLUP model under both cross and independent validations for the traits studied. MLP produces 5% higher prediction accuracy than CNN for grain yield and grain protein content.	Sandhu et al. (2021)

Table 1 contd. ...

...Table 1 contd.

Sr. No.	Crop	Trait	HTP system	Key findings	References
10.	Potato	Late blight	Deep learning: Mask Region-based Convolutional Neural Network (Mask R-CNN)	Developed enhanced field-based detection method of potato blight in complex backgrounds.	Johnson et al. (2021)
11.	Grape	Canopy volume estimation and bunch detection and counting.	RGB camera with agricultural vehicle using deep learning modules (AlexNet,VGG16, VGG19 and GoogLeNet)	Proposed methods of correct detection of fruits, with a maximum accuracy of 91.52%, obtained by the VGG19 deep neural network.	Milella et al. (2021)
12.	Potato	Drought stress	RGB camera and LED light system (5500 K ± 500 K)	Developed methods for multi-scale characterization of plant morphology using high definition RGB color cameras (top and side views) and three-dimensional images of the underground organs, using a 1.5 T MRI scanner in response to water stress.	Musse et al. (2021)
13.	Rice	Drought stress	PlantScreen™ Robotic XYZ System (Photons Systems Instruments, Czech Republic); 3D scanalyzer with a 6576 × 4384 resolution RGB camera (LemnaTec, GmbH, Germany); NIR and IR images	Developed high-throughput phenotyping methods for short breeding cycles as well as for functional genetic studies of tolerance to drought stress. Red-green-blue (RGB) images recorded plant area, color and compactness; near-infrared (NIR) images determined water content of rice; and infrared (IR) images measured plant temperature and fluorescence was used to examine photosynthesis efficiency. DroughtSpotter technology was used to determine water use efficiency, plant water loss rate and transpiration rate.	Kim et al. (2020)
14.	Soybean	Seed traits	Digital camera images	Enabled the measurement and analysis of large amounts of plant seed phenotype data in a short time.	Baek et al. (2020)
15.	Arabidopsis	Seedling hypocotyl	Deep learning methods	Developed high-throughput plant phenotyping methods using a simple flatbed scanner or smartphone cameras applying a deep-learning–based approach. https://github.com/biomag-lab/hypocotyl-UNet	Dobos et al. (2019)
16.	Arabidopsis	Plant phenotype	RGB camera (Logitech C920) with machine learning tools	Developed a new, automated, high-throughput plant phenotyping system, using simple and robust hardware consisting of machine-learning- based plant segmentation.	Lee et al. (2018)
17.	Potato	Plant phenotype	Various cameras/sensors mounted on ground/aerial vehicles	Reviewed application of high-throughput phenotyping in automated glasshouse systems equipped with imaging sensors and in-field systems with sensors mounted on ground- or aerial-based vehicles.	Slater et al. (2017)

Table 2. An overview of phenotyping platforms.

HTP	Platform	Sensors	Plants	Traits	Country
A. Indoor HTP system					
i. Conveyor type	LemnaTec Scanalyzer 3D	RGB, NIR, FLUO	Barley	Biomass, plant height, width, compactness, drought stress.	Germany
	LemnaTec Scanalyzer 3D	RGB, NIR, FLUO, hyperspectral	Sorghum, maize, barley	Biomass, leaf water content.	USA
	LemnaTec Scanalyzer 3D	RGB, NIR, FLUO, hyperspectral	Chickpea, Wheat	Nutrient stress, salt stress, water content, nitrogen content.	Australia
	Bellwether	RGB, NIR, FLUO	Setaria	Plant height, biomass, water-use efficiency, water content.	USA
	-	Hyperspectral	Maize	PLA, NDVI, perimeter, major axis length, minor axis length, eccentricity.	USA
	HRPF	RGB, CT	Rice	Drought stress, tiller number.	China
ii. Benchtop type	Phenovator	Monochrome	Arabidopsis	PLA, PSII efficiency.	Netherlands
	Phenoscope	RGB	Arabidopsis	Rosette size, expansion rate, evaporation.	France
	-	RGB	Arabidopsis	Radiation dosage stress, projected area, convex hull area, perimeter length.	Korea
	Phenoarch	RGB	Maize	Growth rate of ear and silk.	France
	Glyph	RGB	Soybean	Water use efficiency, drought stress.	Argentina
	LemnaTec Scanalyzer HTS	RGB, FPUO, NIR	Arabidopsis	Water stress.	USA
B. Field HTP system					
i. Pole/tower-based	-	Terrestrial laser scanner	Maize, soybean, wheat	Canopy height.	Switzerland
	CropQuant	RGB, NIR	Crop	Crop growth rate.	UK
	CropQuant	RGB	Ecosystem	Canopy greenness.	USA
	-	Hemispherical video camera	Wheat, oat, barley	Crop lodging.	USA
	-	Laser-Induced Fluorescence Transient (LIFT)	Barley, sugar beet	Photosynthesis.	Germany
	-	RGB, NIR	Rice	Shoot biomass, panicle number, grain weight.	Colombia

Table 2 contd. ...

...Table 2 contd.

HTP	Platform	Sensors	Plants	Traits	Country
ii. Mobile	-	LiDAR, RGB, thermal IR, IR thermometer, hyperspectral	Wheat	Canopy height, leaf angular distribution, leaf area, leaf volume, spike number, VIs, canopy transpiration.	Australia
	-	Ultrasonic, NDVI, thermal IR, spectrometers, RGB	Soybean, wheat	Canopy height, NDVI, canopy temperature.	USA
	Phenomobile Lite	LiDAR, RGB, NDVI	Wheat	Plant height, biomass, ground cover.	Australia
	GPhenoVision	RGB-D, thermal, hyperspectral	Cotton	Canopy height, width, growth rate, projected leaf area, volume and yield.	USA
	-	LiDAR	Maize	Plant height.	China
	-	Ultrasonic, RGB, spectrometer, IR radiometer	Soybean	Canopy height, canopy coverage, NDVI.	USA
iii. Gantry-based	LeasyScan	Planteye	Peanut, cowpea, pearl millet, maize	Canopy transpiration, plant height, 3D leaf area, water use efficiency.	India
	Phénofieldr	RGB, VIS-NIR, LiDAR	Wheat	Water stress resistance, nitrogen stress resistance.	France
	Field Scanalyzer	RGB, FLUO, thermal IR, hyperspectral, 3D laser scanner	Wheat	Canopy height, spike number, canopy closure, canopy temperature, NDVI, photosynthesis.	UK
	Mini-Plot	Hyperspectral	Barley	Disease severity.	Germany
iv. Cable-suspended	NU-Spidercam	Multispectral, thermal IR, LiDAR, VIS-NIR spectrometer	Soybean, maize	Plant height, ground cover, canopy Temperature.	USA
	FIP	Spectrometer, ultrasonic, DSLR, thermal, laser scanner, operator camera	Winter wheat, maize, soybean	Canopy cover, canopy height.	Switzerland

C. Aerial HTP system					
i. UAP	Phantom 4	RGB	Lettuce	Carotenoid content.	
	3DR Solo quadcopter	Multispectral	Maize	NDVI, chlorophyll red-edge index (CHL), hemispherical-conical reflectance factors (HCRF).	
	Customized	Hyperspectral	Winter barley	NDVI yield.	
	Self-developed octorotor	RGB, multispectral	Rice	Canopy height, VIs, canopy Coverage.	
	Matrice 600 Pro	RGB, multispectral camera, infrared thermal	Cotton	Yield.	
	Tuffwing Mapper	RGB	Sorghum	Plant height.	
	Ebee	Multispectral	Wheat	Yield.	
	Self-developed	Multispectral, thermal	Maize	Low-nitrogen stress Resistance.	
	Anaconda	RGB, multispectral	Sorghum, maize	Plant height, Vis.	
ii. MAP	Robinson R44 Raven helicopter	Radiometrically calibrated thermal	Wheat	Canopy temperature.	
	Air Tractor AT-402B	RGB	Crop	Pest severity.	
	-	LiDAR	Maize	Biomass.	
iii. Satellite	GeoEye-1	Multispectral	Turfgrasses	Nitrogen content.	
	RapidEye	Multispectral	Wheat	Nitrogen stress.	
	Sentinel-1 and RADARSAT-2	Synthetic aperture radar (SAR)	Wheat	Crop height, angle of inclination.	
	Fluorescence explorer (FLEX)	Fluorescence Imaging Spectrometer (FLORIS)	Terrestrial vegetation	Photosynthesis.	

Source: Adapted in modified form from Li et al. (2021) and Nguyen and Kant (2018) software to monitor various traits of plant growth and development (Prashar et al., 2013; Zia et al., 2017). HTP is based on automated real-time measurement of plant growth and developmental stages and physiological and biochemical responses in smart glasshouses (or field conditions by unmanned aerial vehicles, i.e., UAV) equipped with imaging system. The visible imaging system measures phenotypes like plant growth rates, biomass accumulation, architecture, canopy, phenology, pathogen lesion area, senescence and chlorophyll content, etc., whereas hyperspectral imaging system allows measurements of internal traits, such as sugar, starch, protein, moisture content and several stress-related parameters (Slater et al., 2017). HTP platforms take multiple images at various time intervals at different wavelengths to generate data for software-based analysis. These imagery data is processed into a desired readable format, using image analysis software. HTP platforms utilize application of visible light (RGB: red-blue-green) imaging system for estimation of various traits, such as crop phenology, senescence, shoot biomass, plant height, leaf area, germination rate, flowering time, yield contributing traits, growth-related parameters, disease and pest symptoms (Sugiura et al., 2016), thermal imaging for analyzing leaf/canopy temperature, leaf senescence, transpiration rate, heat dissipation, pathogen and disease detection, fluorescence imaging for analyzing photosynthesis status, non-photochemical quenching, plant health, shoot architecture drought and heat stress (Li et al., 2014). Near-Infrared (NIR) hyperspectral and multispectral imaging techniques are used for analyzing leaf area index, transpiration rate, water content, Normalized Difference Vegetation Index (NDVI), heat dissipation, plant health status, nutrient status, photosynthetic efficiency, canopy temperature (Deery et al., 2016) and 3D laser imaging is mostly used for assessment of root system architecture (Smith et al., 2018, 2021). Machine learning computer-based methods are widely used in HTP (Mochida et al., 2019) along with root omics studies (Kumar et al., 2021). There are various image analysis software programs for high throughput phenotyping, namely LAMINA, LEAFPROCESSOR, RootReader2D and 3D, PlaRoM, LeafAnalyzer, Gia-Roots, GROWSCEEN 3D, LemnaTec 3D Scanalyzer and TraitMill (Cobb et al., 2013). A study in Australia shows application of HTP technologies in field-grown crops by applying LiDAR, multispectral and hyperspectral sensors, thermal sensors, ultrasonic sensors, data loggers and RTK-GNSS receivers (Slater et al., 2017). Another new application of HTP demonstrates use of RGB camera coupled with unmanned aerial vehicles (UAV) for detection of late blight disease in potato (Chawade et al., 2019; Choudhury et al., 2019). Overall, HTP technologies have been successfully applied in various crops for automated precision phenotyping to assist crop breeding (Tables 3 and 4). Application of various crop sensor and imaging technologies, HTP platforms and HTP data management systems are outlined in Tables 1, 2, 3 and 4. Although, limited applications are available in potato crop but there is a great scope to integrate HTP in precision phenotyping of traits in potato breeding in near future.

Table 3. Sensors available to monitor traits.

Sensor	Morphological traits				Physiological traits				Component content traits		
	Plant height	Biomass	Canopy	Lodging	FAPAR	Stay green/ senescence	Light use efficiency	Disease-pest	Chlorophyll content	Nitrogen content	Water content
RGB camera	Yes	Yes	Yes	Yes	Yes	Yes	-	Yes	Yes	Yes	Yes
Multi/Hyperspectral camera	Yes	Yes	Yes	Yes	Yes	Yes	-	Yes	Yes	Yes	Yes
Thermal camera	-	-	Yes	Yes	-	-	-	Yes	-	-	Yes
Photosynthesis sensor	-	Yes	-	Yes	-	-	Yes	Yes	-	-	-
Fluorescence sensor	-	Yes	-	Yes	-	-	Yes	Yes	-	-	-
Stereo camera	Yes	Yes	Yes	Yes	-	-	-	-	-	-	-
LiDAR	Yes	Yes	Yes	Yes	-	-	-	-	-	-	-

Table 4. HTP platforms and data management systems and networks available worldwide.

HTP platform	Country	Description	Reference/web link
Laboratory of Plant Ecophysiological Responses to Environmental Stresses (LEPSE)	France	Study plant responses and adaptation to variable environmental stresses	http://www1.montpellier.inra.fr/ibip/lepse/english/
The Julich Plant Phenotyping Centre (JPPC)	Germany	JPPC is one of the leading institutes to generate phenomics data on crops and model plants. It has a unique infrastructure for non-invasive phenotyping methods.	http://www.fz-juelich.de/ibg/ibg-2/EN/Research/Phenotyping/Phenotyping_article.html?nn=548814
Biotron Experimental Climate Change Research Facility	Canada	Famous for its imaging of confocal, fluorescence, transmission and scanning electron microscopy	http://www.thebiotron.ca/
Green Crop Network (GCN)	Canada	GCN research program aimed at increasing understanding of genetic, physiological and ecological processes of crop plants	http://www.greencropnetwork.com/
IBERS, Aberystwyth University	U.K.	HTP facility at Aberystwyth makes use of automated non-destructive image analysis techniques by LemnaTec.	http://www.aber.ac.uk/en/ibers/facilities/new_builds_at_ibers/
International Plant Phenomics Network (IPPN)	Many countries	IPPN aims to provide novel technologies to analyze phenotypes and gene/QTL functions. Includes countries like Australia, France, Germany, Canada, etc.	http://www.plantphenomics.com/
German Plant Phenotyping Network (DPPN)	Germany	DPPN promotes use of non-invasive new technologies for basic and applied research on plants and plant breeding	http://www.dppn.de/SharedDocs/Pressemitteilungen/UK/EN/2013/13-01-23DPPN.html
European Plant Phenotyping Network (EPPN)	22 partners institutions	The network offers access to dozens of different plant phenotyping facilities worldwide and includes countries like Germany, France, Denmark, U.K., The Netherlands, Hungary, Australia, etc.	http://www.plant-phenotyping-network.eu/eppn/home
Mutant genotype and phenotype dataset	U.S.A.	This dataset comprises a set of 2,400 genes with a loss-of-function mutant phenotype in *Arabidopsis*.	http://www.plantphysiol.org/content/early/2012/01/13/pp.111.192393.short?rss=1
New European Ecotron of Montpellier	France	This phenotypic platform aims to study ecosystems and organisms in response to environmental changes.	http://www.ecotron.cnrs.fr/

PhenoPhyte	U.S.A.	PhenoPhyte allows recording of plant area from images of *Arabidopsis*.	https://vphenodbs.rnet.missouri.edu/PhenoPhyte/index.php
SciNetS	Japan	This database provides phenotypic information about activation tagging lines, etc.	https://database.riken.jp/
The Australian Plant Phenomics Facility	Australia	This is high resolution plant phenomics centre. The Plant Accelerator® focuses on deep phenotyping and reverse phenomics offering state-of-the-art HTP system.	http://www.plantphenomics.org.au/
BRC Plant Stress Diagnostics Laboratory	Hungary	This is engaged in testing for drought tolerance using modern imaging technologies.	http://www.buzapr.hu/#
Phenom-Networks	Israel	It provides efficient platform for measuring complex phenotypic data.	http://www.phenome-networks.com/aboutus
KeyTrack® by Keygene Inc.	The Netherlands	It is a robust imaging phenotyping platform with automated and non-destructive system.	http://www.keygene.com/services/services_KeyTrack.php
PhenoFab	Germany and The Netherlands	It provides high-throughput, noninvasive phenotyping facility and data analysis.	http://www.phenofab.com/index.php?id=the-companies
RadiMax	Denmark	The facility is meant for phenotyping of plant roots to select drought-tolerant plants.	https://www.dlf.com/about-dlf/news-and-press-releases/article/radimax—new-research-project-sheds-light-on-plant-roots?Action=1&PID=1905
Aeroponics	Europe (EPPN)	This UCL aeroponics platform aims to root phenotyping system.	http://www.uclouvain.be/aeroponics
CropSight	Norwich Research Institute, UK	Plant phenotyping and IoT-based crop management system.	https://github.com/Crop-Phenomics-Group/cropsight/releases
Fraunhofer-PhenoCT	Germany	High-throughput phenotyping facility under controlled environments for roots.	https://www.iis.fraunhofer.de/de/ff/zfp/projekte/hotpot.html
MP3	Canada	McGill plant phenomics platform.	http://mp3.biol.mcgill.ca/mcgill_mp3_summary.html

Table 4 contd. ...

...Table 4 contd.

HTP platform	Country	Description	Reference/web link
PGP Repository	Germany	A plant phenomics and genomics data facility.	http://edal.ipk-gatersleben.de/repos/pgp/
Phenotyper	-	Plant phenotyping data centre.	http://www.bioinformatics.org/groups/?group_id=1210
PHIS	France	Plant phenomics centre used under field and controlled conditions.	http://www.phis.inra.fr/
Planteome	-	Integrated resources for reference ontologies, plant genomics and phenomics.	http://www.planteome.org
RAP	China	Rice automatic phenotyping platform for HTP of rice, maize and other crops.	http://plantphenomics.hzau.edu.cn/checkiflogin_en.action
SensorDB		A virtual laboratory aims to integrate, visualize and analyse varied biological sensor data.	
Nanaji Deshmukh Plant Phenomics Centre	India	HTP platform for phenotyping of crops (ICAR-IARI, New Delhi)	https://www.iari.res.in/files/Latest-News/PlantPhenomicsCentre_inauguration_News_13102017.pdf
Plant Phenomics National Facility	India	HTP platform for phenotyping of crops (ICAR-IIHR, Bangalore).	https://www.iihr.res.in/inauguration-plant-phenomics-national-facility-icar-indian-institute-horticultural-research

Source: Modified adapted from Mir et al. (2019) and Song et al. (2021)

4. Application of HTP in Potato

4.1 Agronomic Traits

A large number of traits have been considered in potato, like resistance/tolerance to biotic/abiotic stresses, yield contributing traits, nutritional and tuber quality traits. For example, in Indian conditions, important traits are early tuber bulking, short duration, plant canopy, tolerance to high temperature and drought stresses, nutrient use efficiency, resistance to diseases (late blight, viruses, bacterial wilt, etc.) and pests (aphid, white fly, mite, tuber moth, potato cyst nematode, etc.). Tuber quality traits are very important for consumers and processing industries, such as for tuber dry matter content, storage, nutritional components, tuber traits (shape, size and color) and processing traits.

HTP platforms are deployed on a great scale in many plant species, such as Arabidopsis, rice, wheat, maize, soybean, legumes, beans, tomato, sugar beet and potato, etc. There is a need to analyze various agronomic traits for rapid development of varieties. To develop high-yielding potato varieties with enhanced resistance/tolerance to biotic/abiotic stresses, nutritional superior, high nutrient use efficiency and consumer-driven desirable traits require precision phenotyping or HTP of various morphological, physiological, biochemical and molecular traits for molecular breeding (Zia et al., 2017). A large number of conventional phenotyping methods have been applied to germplasm/genotypes screening for numerous traits in potato. Monneveux et al. (2013) and Wishart et al. (2014) screened potato genotypes under drought stress conditions. Recently, high-resolution aerial imaging HTP system has been applied for estimation of crop emergence in potato (Sankaran et al., 2017; Li et al., 2019). Unlike manual detection of a large number of genotypes, HTP is helpful in phenotyping of several potato genotypes, particularly for early maturing varieties via analysis of plant growth and biomass digital images while detecting a fast growing and early vigor genotypes. High throughput visible near infrared (NIR) and infrared (IR) range imaging of potato under controlled conditions are shown in Figs. 1 and 2.

4.2 Crop Canopy

Crop canopy cover is an important trait in potato. Crop emergence and canopy-cover-associated traits are important for varietal screening, field management and tuber yield in potato. These traits are affected by various factors, like seed quality, dormancy, and soil moisture, nutrient status and soil temperature. Therefore, consistent monitoring of traits is important for efficient crop management. Crop canopy is one of the most commonly used traits monitored by using remote sensing technology at the early growth stage. Amount of sunlight interception is determined by crop canopy cover and thus affects photosynthetic efficiency. Crop emergence rate and uniformity are crucial for field-scale phenotyping. Crop emergence is conventionally estimated by time-consuming and laborious manual counting practice, while canopy cover is assessed by subjective and inaccurate manual scoring method (Li et al., 2019). The repeatability and reproducibility of these manual assessment techniques are very low

Fig. 1. Potato plants grown in the high-throughput phenotyping facility of LemnaTec platform at ICAR-National Institute of Abiotic Stress Management, Baramati, Maharashtra, India (*Courtesy: Dr. Sushil S. Changan*).

and it is very difficult to execute these practices in thousands of trial plots of large agronomical and breeding experiments.

Non-destructive phenotyping of potato can be done in HTP platforms under controlled conditions in pots. However, there are some problems in correlating these results with field conditions, as the plant canopy structure differs between field and controlled conditions. Potato crop contains a large canopy and shows restricted growth and development in pots under controlled conditions, whereas large canopy is found in fields. Thus, there is need to develop a highly automated and non-destructive HTP platform for field conditions. As breeding and genetic analysis of most crops comprising potato is generally carried out under natural environment, the field phenotyping approach provides better understanding of crop behavior (Prashar et al., 2013). A fully automated, high-throughput fixed site phenotyping platform, Field Scanalyzer, has been developed; it is equipped with multiple imaging sensors for non-destructive observation of plant growth and development (Virlet et al., 2017). The information generated by Field Scanalyzer may be utilized by potato breeders to screen a large set of germplasm based on desired physiological and morphological traits. In potato, infrared thermography (IRT) was used to estimate stomatal conductance and canopy temperature (Prashar et al., 2013). Even under well-watered conditions, the potato genotypes showed significant differences in canopy temperature. There was a negative correlation between canopy temperature and tuber yield (Prashar et al., 2013). These observations can be utilized to identify SNP (Single Nucleotide Polymorphism) that controls stomatal conductance and canopy temperature (Zia et al., 2017). Estimation of canopy temperature using thermal imagery was done to assess severity of water stress in maize. Thus, stomatal

Fig. 2. High-throughput phenotyping images using LemnaTec Phenomics platform at ICAR-NIASM, Baramati, Maharashtra, India. Visible range imaging side (a) and top (b) view; near infrared imaging side (c) and top (d) view, infra-red imaging side view (e, f) (*Courtesy: Dr. Sushil S. Changan*).

conductance and canopy temperature traits can be used to screen potato germplasm for drought stress-tolerance breeding.

The camera sensor phenotyping approach has been used to monitor green canopy coverage of potato and its correlation was established with fresh plant biomass and leaf area index (LAI) (Dammer et al., 2016). Relative vegetation index and NDVI (Normalized Difference Vegetation Index) were used to estimate leaf area index and biomass in potato. Crop canopy images can be used to detect disease occurrence, severity and level of infestation and also provides information to develop disease forecasting models and management practices (Zia et al., 2017). Thus, potato breeders can utilize this approach to identify biotic stress (late blight of potato) tolerant genotype.

4.3 Biotic and Abiotic Stresses

Chemical fertilizers and pesticides have been used widely to increase potato yield, which leads to potential economic waste and environmental pollution. Strategies for optimization of nutrient doses are required for potato cultivation. Heat and drought stress tolerance and nutrient use efficiency are the major abiotic factors in potato and these are complex phenomena governed by various physiological, biochemical and molecular factors. Among the biotic factors diseases, like late blight, viruses, bacterial wilt and storage disease, and pests like aphid, white fly, mites, potato cyst nematodes, etc. are the devastating causes of yield losses in potato. HTP could be utilized to detect foliar disease and pest symptoms, and also their level of infestation and time of arrival in the potato field. This digital images-driven information can be used for effective management of diseases and insect-pests. Also, this phenotyping will assist in screening out susceptible genotypes in disease/pest resistance breeding program. Application of HTP could be possible to phenotype the traits for these above disease-pests. In potato, abiotic stresses experiments are easily possible under HTP platforms under normal pot conditions by manipulation of temperature and photoperiod regulation of heat stress, soil water (moisture) control for drought stress and nutrients doses for nutrient use efficiency studies. Moreover, HTP facility needs to be standardized for biotic stresses study, where challenging inoculations of pathogens or pests are applied to measure resistance/tolerance in plants that can be investigated by HTP imaging systems. The recently introduced high throughput phenotyping technique provides advanced tools for precision screening by incorporating innovative screening strategies that can assist the selection and pyramiding of drought-responsive genes, appropriate for specific environmental conditions. Several methods for assessment of drought tolerance are available and used in cereal crops, and that could be also applicable to potato. These methods comprise fluorescence, thermometry and reflectance. Chlorophyll fluorescence imaging provides a reliable technique for the study of changes in the photosynthesis rate of potato under water-deficit stress, whereas the ratio of Fv and Fm and the differences in the canopy temperature may be used to screen drought tolerance among potato genotypes (Prashar et al., 2013; Prashar and Jones, 2014). Multispectral imaging has been used for determination of chlorophyll content in potato leaves (Borhan et al., 2004). In the case of potato, genotypes with higher canopy temperature under irrigated conditions were more tolerant to drought stress as compared to genotypes with lower canopy temperature (Stark et al., 1991). Reflectance indices, calculated from the visible and near-infrared light reflected by vegetation, have been used in several crops to estimate biomass and changes in leaf water content. These indices have proved accurate to assess drought-associated traits.

4.4 Root Traits

Roots are an important underground plant part because plant's performance mainly depends on the healthy root system. Thus root phenotyping is as important as shoot phenotyping (Wasaya et al., 2018). Root system architecture is largely associated with drought tolerance and nutrient uptake as compared to the above-ground plant parts and plays an important role in maintaining crop yield under drought stress. A non-invasive and non-destructive phenotyping techniques warrants special attention to more accurately assess the response of various traits under drought conditions. Different techniques have been developed for root phenotyping under controlled as well as field conditions. Root growth, development, architecture and its functionality under water-deficit stress must be a part of potato breeding programs (Iwama, 2008). *In situ* root imaging technique is used to study root system in several crops and in potato also (Richner et al., 2000). Han et al. (2009) successfully used X-ray computed tomography (CT) technique to extract the architecture of first order potato roots. Magnetic Resonance Imaging (MRI) technique can be used to assess root system architecture in early stages of potato development (Monneveux et al., 2013). Several root characters, i.e., morphological plasticity, primary root length, length and number of lateral roots, crown root number, root tip diameter, root hair density, root angles, root tissue density and gravitropism help the plants to adapt and respond under various stress conditions and they might be important for improving water use efficiency in crop species (Wasaya et al., 2018). Root phenotyping techniques comprise some degree of automation with imaging, image analysis and processing. Various imaging and its analysis techniques/software have been used as reliable tools for root phenotyping and they include WinRhizo, Smart Root, EZ-Rhizo, Image J, Root System Analyzer, Root Nav, IJ_Rhizo, and Root Trace (Wasaya et al., 2018). Advancement in sensor technology could also enable to unravel the underground traits like root, stolon and tuber initiation, and their growth and development. These traits could not be evaluated manually. Use of HTP at harvesting stage will help to assess tuber characteristics, viz. tuber shape, size, skin color, texture and number. This information will be used to predict performance of each genotype in terms of yield and tuber distribution. Hyperspectral and multispectral imaging could be used to assess the tuber quality parameters, viz. carbohydrate, starch, protein, reducing sugar and water content which are important for cooking performance.

4.5 Aeroponics Technology

Aeroponic is a soil-less crop cultivation system where nutrient solution is supplied through mist from to the plant roots under controlled chamber. Aeroponic technology has been applied mostly for production of healthy minitubers. Application of aeroponic in seed potato production is recent in India; however, a decade back, it was limited to countries like China and Korea for commercial production of potato quality seeds. Nowadays, aeroponic is being applied in most parts of the potato-growing countries. Thus, aeroponic is an important technique of soil-less culture under controlled conditions for healthy quality seed potato production. Here, we have standardized phenotyping by manual method (semi-automated for nutrient supply) of potato for various plant parts, like roots, shoots and stolons under aeroponic (Tiwari et al., 2020). Further, genomics approaches could be applied to study genes controlling nitrogen use efficiency in potato under controlled supply of nutrients (Tiwari et al., 2018) (Fig. 3). Our recent studies indicate that aeroponic can be applied to screening of genotypes by measurement of root- and shoot-related traits and genes discovery. This technology allows dissection of full root system architecture of plants

Fig. 3. Precision phenotyping of potato plants parts (root, shoot, stolon and tubers) grown in aeroponics with different N supply at ICAR-CPRI, Shimla, HP, India.

for various traits related to root, stolon and minitubers. Further, it has advantages of year-round cultivation of potato and independent from the crop season. Moreover, temperature and photoperiod are the key environmental cues determining potato tuberization, which can be regulated under fully controlled environmental conditions (i.e., automated phenotyping platform or HTP). Further, data generated from aeroponic-HTP systems could be analyzed on a similar pattern like earlier. Until now, HTP has been mostly applied through pot culture experiments in various crops. Hence, with the advancement in technologies, we propose that aeroponic could be integrated with HTP platform and combined with multiple imaging systems and sensors for real time monitoring of plant phenotypes. This would assist in measurement of above-ground parts (shoot/leaves/foliage) and under-ground parts (roots, stolons and tubers) for a wide range of traits, such as abiotic stresses like nutrient, heat and drought, and biotic stresses like disease and pests. This aeroponic-HTP platform would be a novel discovery for precision phenotyping in potato. In particular, nutrient stress-related study could be easily designed in aeroponic where the amount of macro- and micro-nutrients can be monitored in the solution. Nevertheless, integration of aeroponic with HTP platforms would require designing of technologies equipped with imaging systems for whole plant image capture at different growth and developmental stages, including tubers and further data recording and analysis system.

5. Conclusion

In the current post-genomics era, HTP is becoming a high priority research area due to its automation, precision, sensitivity, accuracy, repeatability and reproducibility. Advanced plant breeding is a key factor to address the worldwide food-security issue, keeping in mind climate change, water scarcity and diminishing land. HTP in combination with advanced breeding approaches, viz. genomic selection and genome editing are being used as next-generation breeding strategies for crop improvement. In potato, HTP using digital images at regular intervals of plant growth would enable early selection of genotypes with desirable traits. This high throughput phenotyping can detect foliar diseases with different lesion areas, using multispectral, hyperspectral and thermal sensors on aerial vehicles without manual intervention. Further, these sensors are equipped to measure underground plant parts, like tuber initiation, tuber characters (shape, size, number, color, etc.) and quality traits (starch, reducing sugar and moisture content). Moreover, RGB camera combined with UAV and robotics would be a potential application of HTP in field trials (Atefi et al., 2021). Such methods should maximize reproducibility and reliability of phenotyping experiments for enhancing precision in quantifying variations of plant trait expression. In order to do so, it is important to quantify features of both the plant as well as its growth environment that allows expression of desirable traits. High throughput genotyping and phenotyping can assist in faster, cheaper and more effective potato breeding and can be useful in capturing genetic variations for several traits in breeding programs. For precisely extracting the desired information about the potato plant phenotype from these techniques, protocols need to be optimized for monitoring plant growth. Application of HTP techniques outlined in this chapter provide a roadmap for future rapid improvement in potato breeding. However, for maximum utilization of such HTP platforms, screening protocols should be standardized for different crop species and under different stress conditions.

Acknowledgement

I am thankful to the Director, ICAR-Central Potato Research Institute, Shimla, and scientists/technicians/research fellows and other colleagues of the institute for their support under the institute research projects on biotechnology, germplasm, breeding and seed research on aeroponics. I am also grateful to the funding agencies for support under the externally funded projects (CABin, ICAR-IASRI, New Delhi; ICAR-LBS Young Scientist Award Project, and DBT, Government of India).

References

Araus, J.L., Kefauver, S.C., Zaman-Allah, M., Olsen, M.S. and Cairns, J.E. (2018). Translating high-throughput phenotyping into genetic gain. *Trends Plant Sci.*, 23: 451–466.

Atefi, A., Ge, Y., Pitla, S. and Schnable, J. (2021). Robotic technologies for high-throughput plant phenotyping: contemporary reviews and future perspectives. *Front. Plant Sci.*, 12: 611940.

Baek, J., Lee, E., Kim, N., Kim, S.L., Choi, I. et al. (2020). High throughput phenotyping for various traits on soybean seeds using image analysis. *Sensors*, 20(1): 248.

Borhan, M.S., Panigrahi, S., Lorenzen, J.H. and Gu, H. (2004). Multispectral and color imaging techniques for nitrate and chlorophyll determination of potato leaves in a controlled environment. *Am. Soc. Agric. Eng.*, 47(2): 599–608.

Chawade, A., van Ham, J., Blomquist, H., Bagge, O., Alexandersson, E. and Ortiz, R. (2019). High-throughput field-phenotyping tools for plant breeding and precision agriculture. *Agronomy*, 9: 258.

Choudhury, S.D., Samal, A. and Awada, T. (2019). Leveraging image analysis for high-throughput plant phenotyping. *Front. Plant Sci.*, 10: 508.

Cobb, J.N., Declerck, G., Greenberg, A., Clark, R. and McCouch, S. (2013). Next-generation phenotyping: requirements and strategies for enhancing our understanding of genotype-phenotype relationships and its relevance to crop improvement. *Theor. Appl. Genet.*, 126(4): 867–887.

Colwell, F.J., Souter, J., Bryan, G.J., Compton, L.J., Boonham, N. and Prashar, A. (2021). Development and validation of methodology for estimating potato canopy structure for field crop phenotyping and improved breeding. *Front. Plant Sci.*, 12: 612843.

Dahal, K., Li, X.Q., Tai, H., Creelman, A. and Bizimungu, B. (2019). Improving potato stress tolerance and tuber yield under a climate change scenario—A current overview. *Front. Plant Sci.*, 10: 563.

Dammer, K.H., Dworak, V. and Selbeck, J. (2016). On-the-go phenotyping in field potatoes, using camera vision. *Potato Res.*, 59: 113–127.

Danilevicz, M.F., Bayer, P.E., Nestor, B.J., Bennamoun, M. and Edwards, D. (2021). Resources for image-based high-throughput phenotyping in crops and data sharing challenges. *Plant Physiol.*, 187(2): 699–715.

Deery, D.M., Rebetzke, G.J., Jimenez-Berni, J.A., James, R.A., Condon, A.G. et al. (2016). Methodology for high-throughput field phenotyping of canopy temperature using airborne thermography. *Front. Plant Sci.*, 7: 1808.

Dobos, O., Horvath, P., Nagy, F., Danka, T. and Viczián, A. (2019). A deep learning-based approach for high-throughput hypocotyl phenotyping. *Plant Physiol.*, 181(4): 1415–1424.

Gebhardt, C. (2013). Bridging the gap between genome analysis and precision breeding in potato. *Trends Genet.*, 29: 248–256.

Großkinsky, D.K., Svensgaard, J., Christensen, S. and Roitsch, T. (2015). Plant phenomics and the need for physiological phenotyping across scales to narrow the genotype-to-phenotype knowledge gap. *J. Exp. Bot.*, 66(18): 5429–5440.

Han, L., Dutilleul, P., Prashar, S.O., Beaulieu, C. and Smith, D.L. (2009). Assessment of density effects of the common scab-inducing pathogen on the seed and peripheral organs of potato during growth using computed tomography scanning data. *Transact. ASABE*, 52(1): 305–311.

Harsselaar, J.K.V., Claußen, J., Lübeck, J., Wörlein, N., Uhlmann, N. et al. (2021). X-ray CT phenotyping reveals bi-phasic growth phases of potato tubers exposed to combined abiotic stress. *Front. Plant Sci.*, 12: 613108.

Iwama, K. (2008). Physiology of the potato: new insights into root system and repercussions for crop management. *Potato Res.*, 51: 333.

Johnson, J., Sharma, G., Srinivasan, S., Masakapalli, S.K., Sharma, S. et al. (2021). Enhanced field-based detection of potato blight in complex backgrounds using deep learning. *Plant Phenomics*, 2021: 9835724.

Juliana, P., Montesinos-López, O.A., Crossa, J., Mondal, S., González Pérez, L. et al. (2019). Integrating genomic-enabled prediction and high-throughput phenotyping in breeding for climate-resilient bread wheat. *Theor. Appl. Genet.*, 132(1): 177–194.

Kim, M., Lee, C., Hong, S., Kim, S.L., Baek, J.H. and Kim, K.H. (2021). High-throughput phenotyping methods for breeding drought-tolerant crops. *Int. J. Mol. Sci.*, 22(15): 8266.

Kim, S.L., Kim, N., Lee, H., Lee, E., Cheon, K.S. et al. (2020). High-throughput phenotyping platform for analyzing drought tolerance in rice. *Planta*, 252(3): 38.

Kumar, J., Sen Gupta, D., Djalovic, I., Kumar, S. and Siddique, K. (2021). Root-omics for drought tolerance in cool-season grain legumes. *Physiol. Plant.*, 172(2): 629–644.

Lee, U., Chang, S., Putra, G.A., Kim, H. and Kim, D.H. (2018). An automated, high-throughput plant phenotyping system using machine learning-based plant segmentation and image analysis. *PLoS ONE*, 13(4): e0196615.

Li, B., Xu, X., Han, J., Zhang, L., Bian, C. et al. (2019). The estimation of crop emergence in potatoes by UAV RGB imagery. *Plant Methods*, 15(1): 15.

Li, D., Quan, C., Song, Z., Li, X., Yu, G. et al. (2021). High-throughput plant phenotyping platform (HT3P) as a novel tool for estimating agronomic traits from the lab to the field. *Front. Bioeng. Biotechnol.*, 8: 623705.

Li, L., Zhang, Q. and Huang, D. (2014). A review of imaging techniques for plant phenotyping. *Sensors*, 14(11): 20078–20111.

Milella, A., Marani, R., Petitti, A. and Reina, G. (2021). In-field high throughput grapevine phenotyping with a consumer-grade depth camera. *Comput. Electron. Agric. arXiv*:2104.06945v1. Doi: 10.1016/j.compag.2018.11.026.

Mir, R.R., Reynolds, M., Pinto, F., Khan, M.A. and Bhat, M.A. (2019). High-throughput phenotyping for crop improvement in the genomics era. *Plant Sci.*, 282: 60–72.

Mochida, K., Koda, S., Inoue, K., Hirayama, T., Tanaka, S. et al. (2019). Computer vision-based phenotyping for improvement of plant productivity: A machine learning perspective. *GigaScience*, 8(1): giy153.

Monneveux, P., Ramírez, D.A. and Pino, M.T. (2013). Drought tolerance in potato (*S. tuberosum* L.): Can we learn from drought tolerance research in cereals? *Plant Sci.*, 205: 76–86.

Moreira, F.F., Oliveira, H.R., Volenec, J.J., Rainey, K.M. and Brito, L.F. (2020). Integrating high-throughput phenotyping and statistical genomic methods to genetically improve longitudinal traits in crops. *Front. Plant Sci.*, 11: 681.

Musse, M., Hajjar, G., Ali, N., Billiot, B., Joly, G. et al. (2021). A global non-invasive methodology for the phenotyping of potato under water deficit conditions using imaging, physiological and molecular tools. *Plant Methods*, 17(1): 81.

Neilson, J.A.D., Smith, A.M., Mesina, L., Vivian, R., Smienk, S. and De Koyer, D. (2021). Potato tuber shape phenotyping using RGB imaging. *Agronomy*, 11: 1781.

Nguyen, G.N. and Kant, S. (2018). Improving nitrogen use efficiency in plants: Effective phenotyping in conjunction with agronomic and genetic approaches. *Funct. Plant Biol.*, 45(6): 606–619.

Pasala, R. and Pandey, B.B. (2020). Plant phenomics: High-throughput technology for accelerating genomics. *J. Biosci.*, 45: 111.

Prashar, A., Yildiz, J., McNicol, J.W., Bryan, G.J. and Jones, H.G. (2013). Infra-red thermography for high throughput field phenotyping in *Solanum tuberosum*. *PLoS ONE*, 8: e65816.

Prashar, A. and Jones, H.G. (2014). Infra-red thermography as a high-throughput tool for field phenotyping. *Agronomy*, 4: 397–417.

Rebetzke, G.J., Jimenez-Berni, J., Fischer, R.A., Deery, D.M. and Smith, D.J. (2019). Review: High-throughput phenotyping to enhance the use of crop genetic resources. *Plant Sci.*, 282: 40–48.

Richner, W., Liedgens, M., Bürgi, H., Soldati, A. and Stamp, P. (2000). Root image analysis and interpretation. *In*: Smit, A.L. et al. (eds.). *Root Methods*. Springer, Berlin, Heidelberg.

Romero, A.P., Alarcón, A., Valbuena, R.I. and Galeano, C.H. (2017). Physiological assessment of water stress in potato using spectral information. *Front. Plant Sci.*, 8: 1608.

Sandhu, K.S., Lozada, D.N., Zhang, Z., Pumphrey, M.O. and Carter, A.H. (2021). Deep learning for predicting complex traits in spring wheat breeding program. *Front. Plant Sci.*, 11: 613325.

Sankaran, S., Quirós, J.J., Knowles, N.R. and Knowles, L.O. (2017). High-resolution aerial imaging-based estimation of crop emergence in potatoes. *Am. J. Potato Res.*, 94: 658–63.

Slater, A.T., Cogan, N.O.I., Rodoni, B.C., Daetwyler, H.D., Hayes, B.J. et al. (2017). Breeding differently the digital revolution: High-throughput phenotyping and genotyping. *Potato Res.*, 60: 337–352.

Smith, D.T., Potgieter, A.B. and Chapman, S.C. (2021). Scaling up high-throughput phenotyping for abiotic stress selection in the field. *Theor. Appl. Genet.*, 134: 1845–1866.

Smith, L.N., Zhang, W., Hansen, M.F., Hales, I.J. and Smith, M.L. (2018). Innovative 3D and 2D machine vision methods for analysis of plants and crops in the field. *Comput. Ind.*, 97: 122–131.

Song, P., Wang, J., Guo, X., Yang, W. and Zhao, C. (2021). High-throughput phenotyping: Breaking through the bottleneck in future crop breeding. *The Crop Journal*, 9: 633–645.

Stark, J.C., Pavek, J.J. and McCann, I.R. (1991). Using canopy temperature measurements to evaluate drought tolerance of potato genotypes. *J. Am. Soc. Hortic. Sci.*, 116: 412–415.

Sugiura, R., Tsuda, S., Tamiya, S., Itoh, A., Nishiwaki, K. et al. (2016). Field phenotyping system for the assessment of potato late blight resistance using RGB imagery from an unmanned aerial vehicle. *Biosyst. Eng.*, 148: 1–10.

Tiwari, J.K., Plett, D., Garnett, T., Chakrabarti, S.K. and Singh, R.K. (2018). Integrated genomics, physiology and breeding approaches for improving nitrogen use efficiency in potato: translating knowledge from other crops. *Funct. Plant Biol.*, 45: 587–605.

Tiwari, J.K., Devi, S., Buckesth, T., Ali, N., Singh, R.K. et al. (2020). Precision phenotyping of contrasting potato (*Solanum tuberosum* L.) varieties in a novel aeroponics system for improving nitrogen use efficiency: in search of key traits and genes. *J. Integr. Agric.*, 19: 51–61.

Virlet, N., Sabermanesh, K., Sadeghi-Tehran, P. and Hawkesford, M.J. (2017). Field Scanalyzer: An automated robotic field phenotyping platform for detailed crop monitoring. *Funct. Plant Biol.*, 44(1): 143–153.

Volpato, L., Pinto, F., González-Pérez, L., Thompson, I.G., Borém, A. et al. (2021). High throughput field phenotyping for plant height using UAV-based RGB imagery in wheat breeding lines: feasibility and validation. *Front. Plant Sci.*, 12: 591587.

Wasaya, A., Zhang, X., Fang, Q. and Yan, Z. (2018). Root phenotyping for drought tolerance: A review. *Agronomy*, 8: 241.

Wishart, J., George, T.S., Brown, L.K., White, P.J., Ramsay, G. et al. (2014). Field phenotyping of potato to assess root and shoot characteristics associated with drought tolerance. *Plant Soil*, 378: 351–363

Yang, W., Feng, H., Zhang, X., Zhang, J., Doonan, J.H. et al. (2020). Crop phenomics and high-throughput phenotyping: past decades, current challenges, and future perspectives. *Mol. Plant*, 13(2): 187–214.

York, L.M. (2019). Functional phenomics: an emerging field integrating high-throughput phenotyping, physiology, and bioinformatics. *J. Exp. Bot.*, 70(2): 379–386.

Zhao, C., Zhang, Y., Du, J., Guo, X., Wen, W. et al. (2019). Crop phenomics: current status and perspectives. *Front. Plant Sci.*, 10: 714.

Zia, M.A.B., Naeem, M., Demirel, U. and Caliskan, M.E. (2017). Next generation breeding in potato. *Ekin. J.*, 3: 1–33.

Chapter 8

Genome Editing (CRISPR/Cas) Technology in Potato

1. Introduction

Potato is the third most important food crop after rice and wheat for human consumption. The increasing world population from 7.7 billion to 9.7 billion by 2050 has created a huge demand for foods (United Nations, 2019). However, potato suffers from various pathogens, insect-pests and environmental abiotic stresses, which are becoming more severe under the climate change scenario. Conventional breeding has shown key roles in varietal development coupled with marker-assisted selection to fast breeding and later transgenics have also been developed in potato for many traits. Now, with the availability of the potato genome (Potato Genome Sequencing Consortium, 2011), it is possible to modulate genes to accelerate crop breeding by applying genome editing tools.

Genome editing is an advance breeding tool which can be deployed for crop improvement by manipulation of target genes through gene knock-out and insertion/deletion mutagenesis (Hameed et al., 2018). Genome editing is the targeted alteration in the genome with new allelic variation. CRISPR is an advanced and improved molecular breeding technique in terms of creation of mutagenesis for better crop improvement. Sequence editing is done either by deleting or modifying genes individually and then studying the subsequent mutant phenotypes to understand the gene function. Sequence Specific Nucleases (SSNs) have been applied for genetic manipulation in the genome through genome editing. The SSNs technology is rapidly becoming important in plants where three major nucleases are used—Zinc Finger Nucleases (ZFNs), Transcription Activator-Like Effector Nucleases (TALENs), and Clustered Regularly Interspaced Short Palindromic Repeats/CRISPR-associated proteins (CRISPR/Cas9) (Nadakuduti et al., 2018).

CRISPR/Cas9 is the most widely used genome editing technology due to its simplicity, multiplexing capability, cost-effectiveness and high efficiency. The CRISPR/Cas9 (*Streptococcus pyogenes* Cas9, i.e., SpCas9) is a RNA-guided method to target DNA sequence. For precise DNA manipulation, the new genome editing system is able to induce Double-Stranded Breaks (DSBs) at a specific site in the genome and repair naturally via DNA repair mechanisms, such as error-prone Non-Homologous End-Joining (NHEJ) or precise Homologous Recombination (HR) pathways. However, the issue involved with this technology is the off-target mutants due to mismatch base pairing between gRNA and DNA. Unlike GM technology, which creates stable integration in line by cisgene or transgene, CRISPR/Cas, a new breeding tool, offers crop improvement where no foreign gene is introduced and the selected line is not likely to be treated as GMOs. Gene editing, particularly CRISPR/Cas9, has been applied for many traits in different crops, including potato for tuber quality, disease resistance (late blight and potato virus Y), phenotype and other traits, particularly cold-induced sweetening, glycoalkaloids (solanine and chaconine), acetochalactate synthase, granule bound starch synthase, etc. (Dangol et al., 2019; Hameed et al., 2020; Hofvander et al., 2021). Indeed, the application of gene editing technology based on CRISPR/Cas has emerged as an alternative and efficient technology (Andersson et al., 2018) and tremendous progress has been reported worldwide on its utility in crops.

2. CRISPR/Cas Genome Editing

Initially, mutagenesis was performed with radiation or chemical agents to induce genetic modifications, but thousands of individuals carrying random mutations need to be screened to identify the desired or the aberrant phenotypes. Then Zinc Finger Nucleases (ZFNs) and transcription activator, like effector nucleases (TALENS), were used to produce double-stranded DNA breaks at the precision location of the genome of plants. But it could be successfully applied to a few plant species; not in potato and other horticultural crops. The reason could be due to an erratic or lesser mutation rate and high off-targets (Hameed et al., 2018; Dangol et al., 2019). The high cost and complexity of synthesizing DNA-binding proteins have also limited their use for the study and genetic improvement of plants (Koltun et al., 2018). Compared to ZFNs and TALENs, CRISPR-Cas9 system is an efficient and easier system that has been widely used in recent years (Khatodia et al., 2016).

CRISPR-Cas is a prokaryotic adaptive molecular immune system found in bacteria or archaea to confront the invading viruses or phages. CRISPR/Cas is the most powerful biological tool to create specific targeted modifications in the genome. This allows ease in designing and construction of gene-specific single-guide RNA (sgRNA) vectors for target traits. Then sgRNA vectors are easily reprogrammable to direct *Streptococcus pyogenes* Cas9 (SpCas9) to generate DSBs and subsequently repair endogenously by the error-prone NHEJ or HR mechanisms, as mentioned above. CRISPR-Cas systems are separated into two distinct classes, based on the differences in sequence, structure and function characteristics of Cas proteins. The class 1 of CRISPR-Cas systems utilizes a multi-protein effector complex and contains types I, III, and IV. In contrast, class 2 of CRISPR-Cas system achieves target editing only with a single effector protein and includes II, V and VI types. Among them, Types II and V Cas proteins are utilized to edit DNA and Type VI Cas proteins are applied for editing RNA (reviewed in Khatodia et al., 2016; Cao et al., 2020) (Fig. 1). Currently, CRISPR/Cas9 has revolutionized plant research due to its ease, multiplexing, cost-effectiveness, high efficiency and minimum off targets.

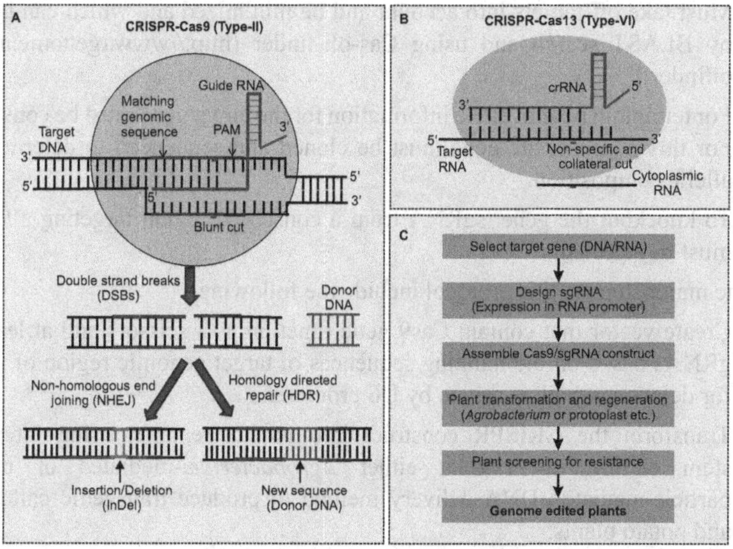

Fig. 1. A schematic layout of CRISPR/Cas genome editing in plants. (a) Cas9, (b) Cas13 and (c) genome edited plant regeneration.

A large number of complex traits (> 40) are needed to be manipulated to develop the new variety. The multigenic controlled biotic and abiotic stresses are difficult to improve through conventional breeding methods. Hence, there is a need for comprehensive CRISPR/Cas to address these issues. Nevertheless, gene knockout mechanisms of susceptibility genes have been applied widely in potato to regenerate desirable mutants.

2.1 Steps Involved in CRISPR/Cas Construct Designing

Procedures for designing sgRNAs, protocols to clone sgRNAs for CRISPR/Cas9 constructs to generate knockouts, design of donor repair templates are described below.

- *Define target sequence/gene or promoter*: Selection of 20-bp target (gene of interest target region, named as spacer) sgRNA Sequence 5'-NNNNN(20 bases)-NGG-3' (targeting template strand) can be selected, using web-based tools, such as CRISPR design tools (http://www.rgenome.net/) or CRISPR-P 2.0 (http://crispr.hzau.edu.cn/CRISPR2/) or CCTOP (https://crispr.cos.uni-heidelberg.de/) which have the *Solanum tuberosum* Group Phureja (PGSC v4.03) potato genome sequence.
- For manual selection, the sgRNA target site:
  - Must be specific/unique to the gene of interest.
  - Must immediately precede the 5'-NGG PAM.
  - Must be in 50 exonic regions, encoding a functional domain so as to disrupt gene function.

- Must take off-targets into account and be minimized and which can be done by BLAST search and using Cas-off-finder (http://www.rgenome.net/cas-offinder/).
- For tetraploid potato, allele information for the target gene must be considered. For this purpose, the gene must be cloned and sequenced to determine the allelic composition.
- To knockout the gene, sgRNA from a conserved region targeting all alleles must be selected.

➢ The major steps in this protocol include the following:

- Create vector that contain Cas9 gene function as a scissors and at least two gRNAs based on the flanking sequences of target genomic region of interest for deletion, which is driven by U6 promoter.
- Transform the CRISPR construct into potato genome, using intermodal stem cuttings as explant either *Agrobacterium*-mediated or biolistic particle-mediated DNA delivery method to produce transgenic callus lines and potato plants.
- The CAS 9 gene will be integrated randomly to produce protein that will be directed by the single-guide RNA to the target site in the genome and cut repair by natural DNA repair mechanism (non-homologous and homologous end joining).
- Selecting plants with mutation or replacement or insertion in the target region.
- Screening the clonally propagated plants for Cas9, random locus containing the T-DNA single-guide RNA.
- Analyze inheritance of the deletions and select plants with only variants.

2.2 CRISPR-Cas Transformation Systems

Various transformation methods have been applied frequently for efficient genome editing, such as *Agrobacterium*-mediated, particle bombardment or biolistic method, floral-dip and PEG-mediated protoplast (Sandhya et al., 2020). *Agrobacterium*-mediated transformation is a most widely used method in plants. In potato, *Agrobacterium*-based transformation with either the classical T-DNA or a geminiviral-based T-DNA vector was used to test in callus and also to induce stable events in diploid MSX914-10 (X914-10) and tetraploid potato cv. Desiree (Butler et al., 2016). Some events retained mutations across clonal generations. Tetraploid mutants were selfed and the diploid mutants were crossed to make the self-compatible diploid M6 line (Butler et al., 2015). Zhan et al. (2019) and Veillet et al. (2020a) also developed transgenic plants of *S. tuberosum* cv. Desiree using *Agrobacterium tumefaciens* and further selection by using antibiotics. But *Agrobacterium*-mediated method cannot be used to deliver ribonucleoprotein complexes. PEG-mediated delivery is widely used to deliver Cas9/gRNA ribonucleoproteins. Establishment of suspension culture, protoplast isolation and regeneration of protoplast into whole plants are the main hurdles in this method (Sandhya et al., 2020). However, Andersson et al. (2017) achieved knockout of all the four alleles of granule-bound starch synthase I (GBSSI) in 2% of regenerated lines using transient transfection of protoplasts through the PEG-mediated transformation method. Johansen et al. (2019) also used protoplasts to edit GBSS gene in potato. In another study, *Arabidopsis* plants, subjected to 37°C heat stress, showed an increased rate of induced mutation by CRISPR-Cas9 system and similar results were observed with citrus plants. So, similar heat treatment at 37°C could improve mutation in potato also (Dangol et al., 2019).

2.3 Gene Knockout Mechanism

An ideal CRISPR-Cas system can be applied to gene knockout (KO) in potato. Appropriate sgRNA promoters derived from dicots along with plant codon-optimized Cas gene should be used. Further use of a single RNA pol II promoter for both Cas gene and sgRNA is beneficial (reviewed in Dangol et al., 2019). In an effort to develop a more robust system, Kusano et al. (2018) used a modified CRISPR-Cas9 system with dMac3 translational enhancer and multiple guide RNAs to target the potato-granule-bound starch synthase I (GBSSI) and found up to 25% of potato transformants with mutation in all the four alleles. However, the efficiency of SaCas9 in potato needs to be further investigated by targeting several other loci (Veillet et al., 2020a).

2.4 DNA-free Genome Editing

Transfection of plasmid DNA has led to unintended inserts of fragments originated from the plasmid DNA. The value of eliminating CRISPR-Cas9 components inserted in the genome via selfing or backcrossing is more complicated in genetically complex and vegetative propagated species, such as potato (Koltun et al., 2018). Of late, preassembled Cas9-gRNA ribonucleoproteins (RNPs) have been directly delivered into the cells to induce mutations in plants (Park et al., 2019). This DNA free editing system is completely free of DNA and thus the risk of integration of DNA into the genome is avoided (Metje-Sprink et al., 2019; Ma et al., 2020). This system is used at least in 14 plant species and is particularly useful for plant species which are vegetatively propagated, such as potato (Andersson et al., 2018; Metje-Sprink et al., 2019). It has several advantages: (i) it circumvents the need for backcrossing and screening of the progeny, (ii) no off-target effects have been reported so far, (iii) many editors can be controlled in a better way, and (iv) the editors are ready to introduce mutations directly after transfection (no lagging phase) (Kim et al., 2017). As the Cas9 nuclease and guide RNA or the isolated RNPs (preassembled) are directly delivered into the cells, the system has two major problems: (i) delivery through the plant cell wall and (ii) regeneration of plants from tissue or cell-wall free cells (Metje-Sprink et al., 2019). However, protoplast was successfully used to alter all the four copies of a single gene in potato, using DNA free genome editing with 2–3% regeneration (Andersson et al., 2018). Other methods, such as protoplast microinjection, biolistic delivery and infiltration with cell-penetrating peptides (CPPs) are to be exploited for the delivery of RNPs (Metje-Sprink et al., 2019). Recently, Liu et al. (2020) demonstrated lipofection-mediated DNA-free delivery of the Cas9/gRNA ribonucleoprotein into plant cells, while in another study, chitoson nanoparticles were used to deliver pre-assembled Cas9-sgRNA ribonucleoproteins into dissected potato shoot apical meristems to produce plants with mutated coilin gene (Makhotenko et al., 2019).

2.5 Virus-induced Genome Editing

A strategy, known as Virus Induced Genome Editing (VIGE) involving plant virus derived vectors for fast and efficient delivery of sgRNAs in Solanaceous plants, was developed by researchers. Butler et al. (2016) used Gemini Virus Replicon (GVR) to deliver sequence-specific nucleases to edit a gene in potato. However, the insert size of GVR is restricted; in that case, PVX has no limitation and can be used for the delivery of large size Sequence Specific Nuclease (SSN) genes (Ariga et al., 2020). Ariga et al. (2020) mechanically inoculated viruses comprising the PVX vector expressing Cas9 in *Nicotiana benthamiana* to develop DNA-free genome-edited plants. Uranga et al. (2021a) engineered PVX to build a vector able to express unspaced gRNA arrays in *N. benthamiana* and the results revealed that it can be successfully applied for multiplex CRISPR/Cas genome editing, particularly in important plants of the *Solanaceae* family. The VIGE system bypasses the requirement of transformation and regeneration of plants as it is time consuming and tedious. But large size of Cas9 challenges the use of virus vector system as the length of foreign insert negatively correlates with the stability of the vector.

2.6 Base Editing

Base editing is an upgraded approach for efficient genome editing, thereby enabling precise nucleotide substitutions without a donor DNA or the induction of a double-strand break (DSB). In this, catalytically inactive CRISPR-Cas9 domain (Cas9 variants, dCas9 or Cas9-nickase) is fused with cytosine or adenosine deaminase domain to introduce desired point mutations (C to T or A to G) in the target region (Mishra et al., 2020). Veillet et al. (2019) employed *S. aureus*-cytosine base editor (CRISPR-SaCas9 CBE) to edit StDMR 6-1 in potato. Ariga et al. (2020) used PVX vector to express a base editor consisting of modified Cas9 fused with cytidine-deaminase to introduce targeted nucleotide substitution in *Nicotiana benthamiana.* However, the size of the base editor is larger than Cas9 and that hindered the delivery into cells by viral vectors. Additionally, CBEs or Cytidine-base editors have been reported to create numerous off targets both at DNA and RNA levels and ABEs have been preferred instead of CBEs in several crops.

3. Application of CRISPR/Cas in Potato Improvement

3.1 Biotic Stress

CRISPR/Cas genome editing has emerged as an alternative and efficient technology to accelerate potato improvement. Its applications in potato for various traits, such as potato virus Y (PVY) and late blight resistance are summarized in Table 1. In potato, Cas13a protein was deployed to confer resistance to three PVY strains (RNA virus) by targeting *P3*, *CI*, *Nib*, and *CP* viral genes (Zhan et al., 2019). Host genes like the

Table 1. Successful examples of CRISPR/Cas in potato for biotic and abiotic stress resistance, and quality traits improvement.

Gene	Trait	CRISPR/	Transformation	Genotype	Key findings	References
Biotic stress resistance						
P3, CI, NIb or *CP* (RNA virus genes)	PVY, PVS and PVA resistance	LshCas13a	*Agrobacterium*	Desiree	Multiple PVY strain-resistant mutants.	Zhan et al. (2019)
Coilin (host gene)	PVY resistance	Cas9	*Agrobacterium*	Chicago	Editing at least one allele of the coilin gene increased resistance to PVY (and also tolerance to salt and osmotic stress).	Makhotenko et al. (2019)
StDND1, *StCHL1* and *StDMR6*-1 (S-genes: Susceptibility genes)	Late blight resistance	Cas9	*Agrobacterium*	Désirée	Tetra-allelic mutants by knockout of *StDMR6-1* and *StCHL1* genes.	Kieu et al. (2021)
Caffeoyl-CoA O-methyltransferase (*StCCoAOMT*)	Late blight resistance	Cas9	*Agrobacterium*	Russet Burbank	Increase in late blight resistance than control.	Hegde et al. (2021)
Abiotic stress tolerance						
StMYB44 (MYB transcription factor)	Phosphate transport (roots)	Cas9	*Agrobacterium*	Désirée	Mutants (84%), *StMYB44* negatively regulates Pi transport by suppressing *StPHO1* gene expression.	Zhou et al. (2017)
Tuber quality traits						
GBBS	Starch quality	Cas9	Protoplast (PEG: *polyethylene glycol*)	Kuras	Multiple allele mutants (67%) and amylopectin rich and waxy potato.	Andersson et al. (2017)
GBBS	Starch quality	Cas9 (RNP: Ribonucleo protein)	Protoplasts	Kuras	Regenerants without transgenes (9%).	Andersson et al. (2018)
GBBS	Starch quality	Cas9	Protoplasts	Desirée and Wotan	Mutants (35%).	Johansen et al. (2019)
GBSS I	Starch quality	Cas9	*Agrobacterium*	Sayaka	Mutants with all four alleles (25%), low amylose starch.	Kusano et al. (2018)

Table 1 contd. ...

...Table 1 contd.

Gene	**Trait**	**CRISPR/**	**Transformation**	**Genotype**	**Key findings**	**References**
GBBS I	Starch quality	Cas9	*Agrobacterium*	Désirée	Tetra-allelic mutants by knockout of amylose-producing *StGBSSI* gene.	Veillet et al. (2019)
Starch synthase gene (*StSS6*)	Starch biosynthesis	Cas9	*Agrobacterium*	Désirée	Specific gRNA design and successful knock-out *SS6*.	Sevestre et al. (2020)
Starch-branching enzymes (SBEs) genes *SBE1*, *SBE2*	Starch quality	Cas9	*Agrobacterium* Protoplasts (PEG)	Désirée	Mutants with valuable starch properties.	Tuncel et al. (2019)
SBE1, *SBE2*	Starch quality	Cas9 RNP	Protoplasts	Désirée	Tree-four allele mutants (72%) with amylase starch with no branching.	Zhao et al. (2021)
PHYTOENE DESATURASE (*PDS*)	Carotenoid biosynthesis	Cas9	*Agrobacterium*	Désirée	Mutants (2–10%).	Bánfalvi et al. (2020)
StPDS	Carotenoid biosynthesis	Cas9	*A. rhizogenes*	Diploid, self-compatible F_1 hybrid DMF1 (DM1-3 x M6)	Transgenic hairy roots mutants (64%–98%).	Butler et al. (2020)
PDS and *coilin*	Carotenoid biosynthesis	Cas9	*In vitro* study without delivery	Chicago	Stimulated activity *in vitro.*	Khromov (2018)
St16DOX	Glycoalkaloids	Cas9	*A. rhizogenes* (electroporation)	Mayqueen	Full knockout of steroidal glycoalkaloids.	Nakayasu et al. (2018)
Sterol side chain reductase 2 (*StSSR2*)	Steroidal glycoalkaloids (SGAs)	Cas9	*Agrobacterium*	Atlantic	Mutants (64%) with significantly reduced SGAs.	Zheng et al. (2021)
Polyphenol Oxidases (PPOs) gene (*StPPO2*)	Enzymatic browning	Cas9 (RNP)	Protoplasts	Désirée	Mutants (69% in four alleles) with 73% reduction in PPO activity than control.	González et al. (2020)
Other traits						
StDMR6-1 and *StGBSSI*	Phenotype	Cas9	*Agrobacterium*	Désirée	SpCas9-NG application in genome editing.	Veillet et al. (2020a)

StIAA2	Phenotype	Cas9	*Agrobacterium*	*S. tuberosum* Gp Phureja double monoploid	Mono- and bi-allelic homozygous mutants (83%).	Wang et al. (2015)
Acetolactate synthase1 (*StALS1*)	Herbicide tolerance	Cas9	*Agrobacterium* Geminivirus replicon (GVR)	Désirée, diploid (MSX914-10)	Targeted mutants (87–100%).	Butler et al. (2015)
StALS	Herbicide tolerance	Cas9	*Agrobacterium* GVR	Désirée, diploid (MSX914-10)	Improved homozygous recombinants (HR) but no change in NHEJ.	Butler et al. (2016)
StALS1 and *StALS2*	Herbicide tolerance	Cas9 (CBE: Cytidine base editing)	*Agrobacterium*	Désirée	Transgene-free mutants (10%).	Veillet et al. (2019)
StALS1 and *StALS2*	Herbicide tolerance	Cas9 Prime editing	*Agrobacterium*	Désirée	Successful prime editing in potato with nucleotide transition/transversion.	Veillet et al. (2020b)
Stylar ribonuclease gene (*S-RNase*)	Self-incompatibility	Cas9	*Agrobacterium*	DRH-195 and DRH-310 F1	Stable self-compatible mutants through *S-RNase* gene knockout.	Enciso-Rodriguez et al. (2019)
S-RNase	Self-incompatibility	Cas9	*Agrobacterium*	*S. tuberosum* Gp Phureja S15-65	Knockout of *S-RNase* gene resulted in self-compatibility.	Ye (2018)
NbFT, *NbPDS3* and *NbXT2B*	Virus induced genome editing (VIGE)	Cas9	*Agrobacterium*	Solanaceous plants	Heritable mutants expressing multiple sgRNAs in *Nicotiana benthamiana*/potato.	Uranga et al. (2021a, 2021b)

GBBS: Granule-bound starch synthase gene

eukaryotic translation initiation factor *eIF4E* and *coilin* have also been found very effective in PVY resistance (Makhotenko et al., 2019). Recently, late blight resistance was demonstrated in potato by knockout of susceptibility genes, *StDMR6-1* and *StCHL1* (Kieu et al., 2021) and *Caffeoyl-CoA O-methyltransferase* (*StCCoAOMT*) (Hegde et al., 2021). To illustrate, strategies for virus resistance are described below.

3.1.1 Targeting DNA Virus Genome

The capacity of CRISPR-Cas9 to create resistance against Gemini virus was demonstrated by targeting coding and non-coding sequences of viral regions (*IR*, *Rep* and *CP*) (Table 1). In potato, resistance to ToLCNDV, a Begomo virus causing apical leaf curl disease was achieved by targeting *IR* gene region or multiple genes, including *CP* and *Rep* genes, by utilizing CRISPR/Cas system (Ali et al., 2016). Hence, modification in the *IR* gene was found to be the most effective target showing a significant reduction in viral accumulation and disease symptoms and confer resistance to Begomo virus tomato yellow leaf curl virus in tobacco (Ali et al., 2015, 2016). The *IR* gene interferes with viral multiplication, limits the generation of viral variants, reduces systemic infections and confers plant resistance.

3.1.2 Targeting RNA Virus Genome

The novel CRISPR-FnCas9-mediated resistance has been elucidated against ssRNA viruses, such as tobacco mosaic virus and cucumber mosaic virus in *Arabidopsis* and tobacco (Zhang et al., 2018). The utility of Cas13a protein has been demonstrated to confer high resistance to three PVY strains in potato via targeting conserved coding regions of *P3*, *CI*, *Nib*, and *CP* genes (Zhan et al., 2019). Similarly, Hameed et al. (2019) suggest that editing in *IR* than *CP* genes provides stable resistance to PVY in potato, because the *CP* gene may induce mutation and break resistance by NHEJ repair mechanism. Also, guide RNA from the conserved regions of the *P3* and *CP* genes of multiple PVY strains shows less off-targets. Though PVY is capable of diversifying fast due to a high mutation rate and multiple strains like PVY^{O}, PVY^{N}, PVY^{NTN} and $PVY^{N:O}$, the FnCas9 and Cas13a nucleases could be applied to confer PVY resistance in potato.

3.1.3 Targeting Host Gene

The eukaryotic translation initiation factor *eIF4E* or isoform *eIF(iso)4E* is essential for the multiplication of VPg of Poty viruses, such as PVY. The Cas9 has been utilized to disrupt the function of the *eIF4E/eIF(iso)4E* to abort host-virus interaction to provide resistance to Poty viruses, such as turnip mosaic virus in *Arabidopsis* (Pyott et al., 2016). Cas9 has also been deployed to edit the *coilin* gene in potato for PVY resistance (Makhotenko et al., 2019). In fact, *coilin* is an essential structural and functional protein of the cajal bodies in plants and which are involved in nucleolus, RNA metabolism and plant-virus interaction. Overall, Cas9 could be applied to edit the *eIF4E/eIF(iso)4E* and *coilin* genes to confer resistance against RNA viruses in potato.

3.1.4 Multiplexing Approach

Most potato viruses are insect-transmitted; for example, aphids transmit PVY and PLRV and whiteflies transmit ToLCNDV. In aphids, virus transmission genes are dynamin, serine protease inhibitors, vesicle transport and endocytosis/exocytosis proteins. Further, aphid virulence genes, like defensive/detoxification genes, salivary genes, secretory proteins and chemoreceptors also play important roles in viral infections. CRISPR-Cas could be explored to edit vector virulence genes to inhibit viral transmission and confer plant resistance. Moreover, simultaneous targeting of multiple plant viral genomes, insect vectors and host genes can be attempted, using a multiplexing approach in an appropriate CRISPR-Cas system to confer virus resistance (Najera et al., 2019).

3.2 Abiotic Stress

Given the importance of abiotic stress tolerance, such as heat, drought, salinity, cold and others, a meagre report is available on the use of CRISPR/Cas in potato till date. Zhou et al. (2017) developed mutants (84%) by manipulating potato MYB transcription factor gene *StMYB44*, which negatively regulates phosphate transport in potato by suppressing *StPHO1* gene expression (Table 1). We have also proposed the use of CRISPR/Cas to increase nitrogen use efficiency in potato by manipulating N metabolisms genes (Tiwari et al., 2020). Manipulation of nitrate transporter gene *NRT1.1B* has been proven by CRISPR/Cas9 (Li et al., 2018) and base-editing (Lu and Zhu, 2017) to increase NUE in rice. Our recent research identified several genes and regulatory elements associated with potato under N stress (Tiwari et al., 2020). Like other plants, high-affinity nitrate transporters are the key candidate genes for manipulation in N uptake and transport in potato roots. Moreover, genes like ferric chelate reductase, protein phosphatase 2 C, glutaredoxin, GDSL esterase/lipase, cytochrome P450 hydroxylase and TFs also appear significantly in roots. In stolon, nitrate transporter, urea active transporter and sodium/proline symporter facilitate N transport. We have also shown candidate miRNAs, like up-regulated (miR156/157 and miR482) and down-regulated (miR397 and miR398), in roots under low N. These candidate genes, transcription factors and miRNAs could be targeted to improve NUE in potato by applying CRISPR/Cas9 and base-editing technologies. Gene expression (over expression/knockout) of particularly N transporters in roots and assimilatory genes of carbohydrate and amino acids metabolism in shoot/stolons, and TFs (Myb and WRKY) could be manipulated by constitutive or tissue-specific promoters and gene knockdown, by applying RNAi via microRNAs (miR156, miR397, miR398, miR319 and miR482) targeting N pathways genes for improving NUE in potato (Tiwari et al., 2020).

3.3 Tuber Quality, Phenotype and Other Traits

Most CRISPR/Cas studies evidence on tuber quality, phenotype and other traits in potato via *Agrobacterium*, protoplast-based Cas9/ribonucleoprotein (RNP) or Virus-Induced Genome Editing (VIGE), like Gemini Virus Replicon (GVR) delivery mechanisms for several traits, such as improved tuber starch quality (Andersson et al., 2017, 2018; Kusano et al., 2018; Johansen et al., 2019; Tuncel et al., 2019; Veillet et al., 2019; Sevestre et al., 2020; Zhao et al., 2021), carotenoid biosynthesis (Khromov et al., 2018; Bánfalvi et al., 2020; Butler et al., 2020), glycoalkaloids (Nakayasu et al., 2018; Zheng et al., 2021) and enzymatic browning (González et al., 2020) (Table 1). In addition, functional mutants were developed for phenotypic traits (Wang et al., 2015; Veillet et al., 2020a) and herbicide tolerance (Butler et al., 2015, 2016). Self-compatible potato regenerants were produced using Cas9 via *Agrobacterium* (Ye et al., 2018; Enciso-Rodriguez et al., 2019) or VIGE (Uranga et al., 2021a, 2021b). Interestingly, Cas9-based base editing and prime editing technology were also generated herbicide tolerant mutants (Veillet et al., 2019, 2020b).

4. CRISPR/Cas Challenges in Tetraploid Potato

High heterozygosity, tetrasomic inheritance, severe inbreeding depression and vegetative propagation cause difficulties in successful application of CRISPR/Cas in potato. Selection of suitable single-guide RNA (sgRNA), robust CRISPR/Cas nucleases, efficient transformation protocols and phenotype without off targets are the main decisive factors. The online tools have been demonstrated to design sgRNA, such as CRISPick (https://portals.broadinstitute.org/gppx/crispick/public) and CCTop-CRISPR/Cas9 target online predictor (https://crispr.cos.uni-heidelberg.de/). PAM limitation is one of the drawbacks of CRISPR/Cas and therefore, more diversity in CRISPR/Cas toolbox is necessary (Veillet et al., 2020). Despite this, some success has been achieved in potato, as mentioned above. Gene knockout is a preferred mechanism in plants and even all four alleles were mutated in potato for *StGBSS* gene (Andersson et al., 2017). The use of multiple sgRNAs along with dMac3 translational enhancer produced mutants (25%) with all four alleles of *GBSSI* gene in potato (Kusano et al., 2018).

The most common *Agrobacterium*-mediated transformation and recently protoplasts have been successfully deployed in CRISPR/Cas in potato. Also, Cas9 and sgRNA, under the control of single RNA pol II promoter, are recommended. *Agrobacterium*-mediated method cannot be used to deliver ribonucleoprotein (RNP) complexes. Moreover, elimination of CRISPR-Cas9 assembly from the plant genome via selfing or backcrossing is more complicated in genetically-complex and vegetatively-propagated potato (Belhaj et al., 2013; Koltun et al., 2018). Hence, DNA-free genome editing is a novel approach. Protoplast (PEG: polyethyle glycol)-mediated transformation is an excellent alternative for efficient delivery of Cas9/gRNA RNPs in potato. DNA-free preassembled Cas9-gRNA RNPs were directly delivered into the plant cells to induce mutations (Park et al., 2019) and also demonstrated in lipofection-mediated DNA-free delivery (Liu et al., 2020). But the establishment of suspension culture, protoplast isolation and regeneration into whole plants are the associated problems of protoplast system (Sandhya et al., 2020).

VIGE is another approach for fast and efficient delivery of sgRNAs in potato (Butler et al., 2016). This VIGE system bypasses the requirement of transformation and regeneration of plants as it is a time consuming and tedious process, but the large size of Cas9 challenges the use of virus vector as the length of foreign insert negatively correlates with the stability of the vector. Very recently, base editing and prime editing are the upgraded and efficient approaches of CRISPR/Cas. The programmable base-editing technology, like adenine base editor (ABE) coverts A.T. to G.C. without DNA cleavage, has emerged as a boon for crop improvement

(Gaudelli et al., 2017). Veillet et al. (2020c) deployed *S. aureus*-cytosine base editor (CRISPR-SaCas9 CBE) to edit *StDMR6-1* in potato. However, the size of base editor is larger than Cas9 and that hinders the delivery into cells by viral vectors.

5. Conclusion

The availability of robust CRISPR/Cas array, target selection, efficient plant transformation protocols and minimum off targets are the major challenges in tetraploid potato. It is a fact that improvement of multigenic traits is difficult than monogenic, especially in potato, due to polyploidy and vegetative propagation. Despite this, considerable success has been achieved for some traits and mostly through gene knockout or insertion/deletion mutagenesis. The unprecedented progress in CRISPR-Cas9 provides tremendous opportunities to develop disease-pests resistance in potato via interference in the pathogen genomes and/or host genes. The use of multiplexing CRISPR-Cas system can handle single or multiple sgRNA/RNPs, targeting conserved sequences combined with protoplast-mediated transformation in potato. Besides, awareness among people would be necessary about transgene-free genome editing research. Since CRISPR-Cas assembly is eliminated from the mutants in the subsequent generations, genome edited plants could be accepted as non-GM by regulatory bodies. Thus, CRISPR-Cas provides an effective next-generation toolbox for fast breeding of varieties to achieve sustainable crop yield.

Acknowledgement

I am thankful to the Director, ICAR-Central Potato Research Institute, Shimla, and scientists/technicians/research fellows and other colleagues of the institute for their support under the institute research projects on biotechnology, germplasm, breeding, and seed research on aeroponics. I am also grateful to the funding agencies for support under the externally funded projects (CABin, ICAR-IASRI, New Delhi; ICAR-LBS Young Scientist Award Project, and DBT, Government of India).

References

Ali, Z., Abulfaraj, A., Idris, A., Ali, S., Tashkand-i, M. et al. (2015). CRISPR/Cas9-mediated viral interference in plants. *Genome Biol.*, 16: 238.

Ali, Z., Ali, S., Tashkandi, M., Zaidi, S.S. and Mahfouz, M.M. (2016). CRISPR/Cas9-mediated immunity to Gemini viruses: Differential interference and evasion. *Sci. Rep.*, 6: 26912.

Andersson, M., Turesson, H., Nicolia, A., Fält, A.S., Samuelsson, M. et al. (2017). Efficient targeted multiallelic mutagenesis in tetraploid potato (*Solanum tuberosum*) by transient CRISPR-Cas9 expression in protoplasts. *Plant Cell Rep.*, 36: 117–128.

Andersson, M., Turesson, H., Olsson, N., Fält, A.S., Ohlsson, P. et al. (2018). Genome editing in potato via CRISPR-Cas9 ribonucleoprotein delivery. *Physiol. Plant.*, 164(4): 378–384.

Ariga, H., Toki, S. and Ishibashi, K. (2020). Potato Virus X vector-mediated DNA-free genome editing in plants. *Plant Cell Physiol.*, 61(11): 1946–1953.

Bánfalvi, Z., Csákvári, E., Villányi, V. and Kondrák, M. (2020). Generation of transgene-free PDS mutants in potato by Agrobacterium-mediated transformation. *BMC Biotechnol.*, 20(1): 25.

Belhaj, K., Chaparro-Garcia, A., Kamoun, S. and Nekrasov, V. (2013). Plant genome editing made easy: Targeted mutagenesis in model and crop plants using the CRISPR/Cas system. *Plant Method.*, 9: 39–47.

Butler, N.M., Atkins, P.A., Voytas, D.F. and Douches, D.S. (2015). Generation and inheritance of targeted mutations in potato (*Solanum tuberosum* L.) using the CRISPR/Cas system. *PLoS ONE*, 10: e0144591.

Butler, N.M., Baltes, N.J., Voytas, D.F. and Douches, D.S. (2016). Gemini virus-mediated genome editing in potato (*Solanum tuberosum* L.) using sequence-specific nucleases. *Front. Plant Sci.*, 7: 1045.

Butler, N.M., Jansky, S.H. and Jiang, J. (2020). First-generation genome editing in potato using hairy root transformation. *Plant Biotechnol. J.*, 18(11): 2201–2209.

Cao, Y., Zhou, H., Zhou, X. and Li, F. (2020). Control of plant viruses by CRISPR/Cas system-mediated adaptive immunity. *Front. Microbiol.*, 11: 593700.

Dangol, S.D., Barakate, A., Stephens, J., Çalıskan, M.E. and Bakhsh, A. (2019). Genome editing of potato using CRISPR technologies: Current development and future prospective. *Plant Cell Tiss. Organ Cult.*, 139: 403–416.

Enciso-Rodriguez, F., Manrique-Carpintero, N.C., Nadakuduti, S.S., Buell, C.R., Zarka, D. and Douches, D. (2019). Overcoming self-incompatibility in diploid potato using CRISPR-Cas9. *Front. Plant Sci.*, 10: 376.

Gaudelli, N.M., Komor, A.C., Rees, H., Packer, M.S., Badran, A.H. et al. (2017). Programmable base editing of A.T to G.C in genomic DNA without DNA cleavage. *Nature*, 551: 464–471.

González, M.N., Massa, G.A., Andersson, M., Turesson, H., Olsson, N. et al. (2020). Reduced enzymatic browning in potato tubers by specific editing of a polyphenol oxidase gene via ribonucleoprotein complexes delivery of the CRISPR/Cas9 system. *Front. Plant Sci.*, 10: 1649.

Hameed, A., Zaidi, S.S., Shakir, S. and Mansoor, S. (2018). Applications of new breeding technologies for potato improvement. *Front. Plant Sci.*, 9: 925.

Hameed, A., Zaidi, S.S, Sattar, M.N., Iqbal, Z. and Tahir, M.N. (2019). CRISPR technology to combat plant RNA viruses: A theoretical model for Potato virus Y (PVY) resistance. *Microb. Pathog.*, 133: 103551.

Hameed, A., Mehmood, M.A., Shahid, M., Fatma, S., Khan, A. and Ali, S. (2020). Prospects for potato genome editing to engineer resistance against viruses and cold-induced sweetening. *GM Crops Food*, 11(4): 185–205.

Hegde, N., Joshi, S., Soni, N. and Kushalappa, A.C. (2021). The caffeoyl-CoA O-methyltransferase gene SNP replacement in Russet Burbank potato variety enhances late blight resistance through cell wall reinforcement. *Plant Cell Rep.*, 40(1): 237–254.

Hofvander, P., Andreasson, E. and Andersson, M. (2021). Potato trait development going fast-forward with genome editing. *Trends Genet.*, S0168-9525(21)00288-2.

Johansen, I.E., Liu, Y., Jørgensen, B., Bennett, E.P., Andreasson E. et al. (2019). High efficacy full allelic CRISPR/Cas9 gene editing in tetraploid potato. *Sci. Rep.*, 9: 17715.

Khatodia, S., Bhatotia, K., Passricha, N., Khurana, S.M. and Tuteja, N. (2016). The CRISPR/Cas genome-editing tool: Application in improvement of crops. *Front. Plant Sci.*, 7: 506.

Khromov, A.V., Gushchin, V.A., Timberbaev, V.I., Kalinina, N.O., Taliansky, M.E. and Makarov, V.V. (2018). Guide RNA design for CRISPR/Cas9-mediated potato genome editing. *Dokl. Biochem. Biophys.*, 479(1): 90–94.

Kieu, N.P., Lenman, M., Wang, E.S., Petersen, B.L. and Andreasson, E. (2021). Mutations introduced in susceptibility genes through CRISPR/Cas9 genome editing confer increased late blight resistance in potatoes. *Sci. Rep.*, 11(1): 4487.

Kim, H., Kim, S.T., Ryu, J., Kang, B.C. et al. (2017). CRISPR/Cpf1-mediated DNA-free plant genome editing. *Nat. Commun.*, 8: 14406.

Koltun, A., Corte, L.E.D., Mertz-Henning, L.M. and Gonçalves, L.S.A. (2018). Genetic improvement of horticultural crops mediated by CRISPR/Cas: A new horizon of possibilities. *Hortic. Brasil.*, 36: 290–298.

Kusano, H., Ohnuma, M., Mutsuro-Aoki. H., Asahi. T., Ichinosawa, D. et al. (2018). Establishment of a modified CRISPR/Cas9 system with increased mutagenesis frequency using the translational enhancer dMac3 and multiple guide RNAs in potato. *Sci. Rep.*, 8: 13753.

Li, X., Wang, Y., Liu, Y., Yang, B., Wang, X. et al. (2018). Base editing with a Cpf1–cytidine deaminase fusion. *Nat. Biotechnol.*, 36: 324–327.

Liu, W., Rudis, M.R., Cheplick, M.H., Millwood, R.J., Yang, J.P. et al. (2020). Lipofection-mediated genome editing using DNA-free delivery of the Cas9/gRNA ribonucleoprotein into plant cells. *Plant Cell Rep.*, 39: 245–257.

Lu, Y. and Zhu, J.K. (2017). Precise editing of a target base in the rice genome using a modified CRISPR/Cas9 system. *Mol. Plants*, 10: 523–525.

Ma, X., Zhang, X., Liu, H. and Li, Z. (2020). Highly efficient DNA-free plant genome editing using virally delivered CRISPR-Cas9. *Nat. Plants*, 6: 773–779.

Makhotenko, A.V., Khromov, A.V., Snigir, E.A., Makarova, S.S. et al. (2019). Functional analysis of coilin in virus resistance and stress tolerance of potato *Solanum tuberosum* using CRISPR-Cas9 editing. *Dokl. Biochem. Biophys.*, 484: 88–91.

Metje-Sprink, J., Menz, J., Modrzejewski, D. and Sprink, T. (2019). DNA-free genome editing: Past, present and future. *Front. Plant Sci.*, 9: 1957.

Mishra, R., Joshi, R.K. and Zhao, K. (2020). Base editing in crops: Current advances, limitations and future implications. *Plant Biotechnol. J.*, 18: 20–31.

Nadakuduti, S.S., Buell, C.R., Voytas, D.F., Starker, C.G. and Douches, D.S. (2018). Genome editing for crop improvement-applications in clonally propagated polyploids with a focus on potato (*Solanum tuberosum* L.). *Front. Plant Sci.*, 9: 1607.

Najera, V.A., Twyman, R.M., Christou, P. and Zhu, C. (2019). Applications of multiplex genome editing in higher plants. *Curr. Opin. Biotechnol.*, 59: 93–102.

Nakayasu, M., Akiyama, R., Lee, H.J., Osakabe, K., Osakabe, Y. et al. (2018). Generation of α-solanine-free hairy roots of potato by CRISPR/Cas9-mediated genome editing of the St16DOX gene. *Plant Physiol. Biochem.*, 131: 70–77.

Park, J. and Choe, S. (2019). DNA-free genome editing with preassembled CRISPR/Cas9 ribonucleoproteins in plants. *Transgenic Res.*, 28(Suppl 2): 61–64.

Potato Genome Sequencing Consortium. (2011). Genome sequence and analysis of the tuber crop potato. *Nature*, 475: 189–195.

Pyott, D.E., Sheehan, E. and Molnar, A. (2016). Engineering of CRISPR/Cas9-mediated potyvirus resistance in transgene-free *Arabidopsis* plants. *Mol. Plant Pathol.*, 17: 1276–1288.

Sandhya, D., Jogam, P., Allini, V.R., Abbagani, S. and Alok, A. (2020). The present and potential future methods for delivering CRISPR/Cas9 components in plants. *J. Genetic Eng. Biotechnol.*, 18: 25.

Sevestre, F., Facon, M., Wattebled, F. and Szydlowski, N. (2020). Facilitating gene editing in potato: a single-nucleotide polymorphism (SNP) map of the *Solanum tuberosum* L. cv. Desiree genome. *Sci. Rep.*, 10(1): 2045.

Tiwari, J.K., Buckseth, T., Singh, R.K., Kumar, M. and Kant, S. (2020). Prospects of improving nitrogen use efficiency in potato: lessons from transgenics to genome editing strategies in plants. *Front. Plant Sci.*, 11: 597481.

Tuncel, A., Corbin, K.R., Ahn-Jarvis, J., Harris, S., Hawkins, E. et al. (2019). Cas9-mediated mutagenesis of potato starch-branching enzymes generates a range of tuber starch phenotypes. *Plant Biotechnol. J.*, 17(12): 2259–2271.

United Nations. (2019). United Nations, Department of Economic and Social Affairs, Population Division, *World Population Prospects 2019: Highlights* (ST/ESA/SER.A/423).

Uranga, M., Aragonés, V., Selma, S., Vázquez-Vilar, M., Orzaez, D. and Daros J.-A. (2021a). Efficient Cas9 multiplex editing using unspaced sgRNA arrays engineering in a potato virus X vector. *Plant J.*, 106: 555–565.

Uranga, M., Vazquez-Vilar, M., Orzáez, D. and Daròs, J.A. (2021b). CRISPR-Cas12a genome editing at the whole-plant level using two compatible RNA virus vectors. *The CRISPR J.*, 4(5): 761–769.

Veillet, F., Chauvin, L., Kermarrec, M.P., Sevestre, F., Merrer, M. et al. (2019). The *Solanum tuberosum GBSSI* gene: a target for assessing gene and base editing in tetraploid potato. *Plant Cell Rep.*, 38(9): 1065–1080.

Veillet, F., Perrot, L., Chauvin, L., Kermarrec, M.P., Guyon-Debast, A. et al. (2019). Transgene-free genome editing in tomato and potato plants using *Agrobacterium*-mediated delivery of a CRISPR/Cas9 cytidine base editor. *Inter. J. Mol. Sci.*, 20(2): 402.

Veillet, F., Kermarrec, M.P., Chauvin, L., Chauvin, J.E. and Nogué, F. (2020a). CRISPR-induced indels and base editing using the *Staphylococcus aureus* Cas9 in potato. *PLoS ONE*, 15: e0235942.

Veillet, F., Kermarrec, M.-P., Chauvin, L., Guyon-Debast, A., Chauvin, J.E. et al. (2020b). Prime editing is achievable in the tetraploid potato, but needs improvement, *bioRxiv.* https://doi.org/10.1101/2020.06.18.159111.

Veillet, F., Perrot, L., Guyon-Debast, A., Kermarrec, M.P., Chauvin, L. et al. (2020c). Expanding the CRISPR toolbox in *P. patens* using SpCas9-NG variant and application for gene and base editing in Solanaceae crops. *Inter. J. Mol. Sci*., 21(3): 1024.

Wang, S., Zhang, S., Wang, W., Xiong, X., Meng, F. et al. (2015). Efficient targeted mutagenesis in potato by the CRISPR/Cas9 system. *Plant Cell Rep*., 34: 1473–1476.

Ye, M., Peng, Z., Tang, D., Yang, Z., Li, D., Xu, Y. et al. (2018). Generation of self-compatible diploid potato by knockout of *S-RNase*. *Nat. Plants*, 4(9): 651–654.

Zhan, X., Zhang, F., Zhong, Z., Chen, R. et al. (2019). Generation of virus-resistant potato plants by RNA genome targeting. *Plant Biotechnol. J*., 17: 1814–1822.

Zhang, T., Zheng, Q., Yi, X., An, H., Zhao, Y. et al. (2018). Establishing RNA virus resistance in plants by harnessing CRISPR immune system. *Plant Biotechnol. J*., 16: 1415–1423.

Zhao, X., Jayarathna, S., Turesson, H., Fält, A.S., Nestor, G. et al. (2021). Amylose starch with no detectable branching developed through DNA-free CRISPR-Cas9-mediated mutagenesis of two starch branching enzymes in potato. *Sci. Rep*., 11(1): 4311.

Zheng, Z., Ye, G., Zhou, Y., Pu, X., Su, W. and Wang, J. (2021). Editing sterol side chain reductase 2 gene (*StSSR2*) via CRISPR/Cas9 reduces the total steroidal glycoalkaloids in potato. *All Life*, 14: 401–413.

Zhou, X., Zha, M., Huang, J., Li, L., Imran, M. and Zhang, C. (2017). StMYB44 negatively regulates phosphate transport by suppressing expression of PHOSPHATE1 in potato. *J. Exp Bot*., 68: 1265–1281.

Chapter 9

Conventional True Potato Seed (TPS) to Diploid Hybrid Potato Technologies

1. Introduction

The cultivated potato is tetraploid (*Solanum tuberosum* L.; $2n = 4x = 48$). Potato is a vegetatively propagated crop through tubers. Seed is the most important input, which constitutes nearly half of the total production cost. Although seed replacement rate is low in potato, particularly in developing countries and, for example, nearly 20–25% of total seed requirements are supplied every year through the public and private sectors in India, while remaining growers/farmers use the previous season produce. This results in build up of viral diseases and degeneration of seed stocks in the subsequent generations, over the years. Therefore, healthy seed production is done through 'Seed Plot Technique' (SPT), which involves crop cultivation under low aphid pressure period to minimize virus transmission and adopt cultural practices to minimize disease spread. The production of healthy seed tubers through SPT is expensive and the low rate of tuber multiplication (normally six to eight times) provides only a limited quantity of quality tubers. Moreover, the low aphid areas suitable for producing healthy seed in the country lie in the northern plains and high hills (> 1800 m above mean sea level). Changing climate scenario is likely to shrink the seed-growing window for potato cultivation and increase pathogen/vector pressure, thereby seriously limiting quantity and quality of seed production in the country (Kumar et al., 2021).

To address the above issues, an alternative technology of True Potato Seed (TPS), i.e., botanical seeds, has shown a great potential for producing both disease-free healthy planting material in potato. The concept of exploring TPS technology was realized in early 1950s in the world. However, due to its certain disadvantages, this technology did not become popular. Potato is an auto-tetraploid, highly heterozygous and suffers from acute inbreeding depression. Selfing results in inbreeding depression and loss of vigor due to accumulation of deleterious or lethal alleles. Therefore, homozygous lines could be successful in tetraploid potatoes and F_1 hybrid development has not been successful so far and as a result, true-to-type genetic identity of clone is maintained either through tubers or tissue culture. Generally, varietal development programs take more than 10–12 years to develop new varieties. Hence, diploid potato hybrid has been demonstrated recently, using TPS as propagating material in potato breeding (Jansky et al., 2016). Thus, there is a great potential for new TPS technology based on self-compatible diploid inbred lines combined with apomixis and genome editing technology. This chapter describes briefly the classical TPS technology and the prospect of a new diploid F_1 hybrid potato technology in near future.

2. True Potato Seed (TPS) Technology

TPS is a botanical seed, which is a product of sexual reproduction. TPS is a small seed inside the fruit, called 'berry'. Depending upon genotype and environment, a single potato plant may have 50–100 berries. A berry contains 200–250 botanical seeds and 1 g TPS may contain 1,500–2,000 seeds. Hybrid TPS population is produced through bi-parental crosses and performs better than open-pollinated families. For example, three hybrid TPS populations are available for commercial cultivation in India (Table 1) (Kumar et al., 2021).

There are several advantages associated with TPS, such as (i) small quantity of TPS is required as initial planting material, (ii) TPS are free from viral diseases, except Potato Spindle Tuber Viroid (PSTVd) and potato virus T, (iii) TPS are free from storage insect-pest and diseases, (iv) no post-harvest losses, (v) low cost of production, (vi) easy storage and transportation and (vii) fast multiplication of planting material. On the contrary, major disadvantages of TPS technology are (i) longer dormancy period (about one year), (ii) longer crop maturity period

Table 1. TPS population used for potato production in India.

TPS population	Parents	Year	Salient features
92-PT-27	83-P-47 × D-150	2001	• Yield: 40–45 t/ha. • Maturity: Medium (via tuberlets: 90–110 days, via seedling tubers: 110–120 days). • Tubers: White-cream skin, round-ovoid shape and shallow to medium deep eyes. • Special traits: Early and fast bulking habit, vigorous, resistant to phoma leaf spot, early blight and late blight diseases.
TPS-C3	JT/C-107 × EX/A-680-16	1991	• Yield: 35–40 t/ha. • Maturity: Medium (via seedling tubers: 90–100 days). • Tubers: White-yellow skin, round-ovoid shape and fleet to medium eyes. • Special traits: Resistant to phoma leaf spot, early blight and late blight diseases.
HPS-1/13	MF-1 × TPS-13	1991	• Yield: 35–40 t/ha. • Maturity: Medium (via seedling tubers: 90–100 days). • Tubers: yellow skin, ovoid-round shape and fleet to medium eyes. • Special traits: Resistant to phoma leaf spot, early blight and late blight diseases.

(15–20 days longer than tubers), (iii) high heterogeneity of crop morphology and non-uniformity of tubers, and (iv) labour intensive due to seedling raising, transplanting and intensive irrigation.

3. Potato Production from TPS

The cultivated tetraploid potato is a complex genome and exhibits tetrasomic inheritance. Potato cultivars are highly heterozygous but show great uniformity due to vegetative propagation. The segregating tetraploid populations are highly heterogeneous and heterozygous. Therefore, parents producing relatively homogenous progenies are desirable for TPS development, which can result in acceptable levels of variation in plant morphology and tuber characters. Generally, TPS is used for commercial potato cultivation by two methods: (a) seedling transplant and (b) seedling tuberlets. In the first method, a commercial crop can be raised in the first year itself, while in the second method, tuberlets are produced from seedlings in the first year and then tuberlets are used as planting material in the next year for commercial cultivation (Fig. 1). The yield is higher when the crop is raised from tuberlets. Tuberlets method may be adopted in the northern Indo-Gangetic Plains, where winter is harsh, crop duration is short and healthy seed tubers can be produced following the seed plot technique. However, in areas where production and storage of healthy tubers is costly and difficult, temperatures are mild during the crop season (15–25°C ± 5°C) and crop duration is long (more than 120 days) for seedling transplants to be successful (Kumar et al., 2021).

Fig. 1. Conventional system of true potato seed (TPS) production system.

3.1 Seedling Transplant Method

- ***Nursery bed preparation***: Use 150 g TPS for a nursery bed (75 m^2) to raise one hectare crop. Mix soil and FYM (1:1), and add fertilizers (4–5 g N, 6–8 g P_2O_5 and 10 g K_2O per m^2 nursery bed area).
- ***Seedling raising***: Treat TPS with 0.01% Benlate or Bavistin, sowing 5 cm apart in a row during March (summer) and July–August (autumn) in north-eastern Indian hills.
- ***Field preparation and transplanting***: Prepare field as per usual practice and add half dose of N and full dose of P and K fertilizers. Remaining half N is applied in two split doses at the time of earthing up. Transplant the seedlings at 45–50 cm spacing.
- ***Harvesting and storage***: Cut haulms after 90 days of transplanting and harvest tubers after 15 days of dehaulming. Seed size tuber (10–40 g) are treated with 3% boric acid for 30 min, dried in shade and stored in cold storage for use as planting materials in the next season. Tubers with > 40 g may be used for fresh consumption.

3.2 Seedling Tuberlets Method

- ***Field preparation and planting***: Field preparation and recommended doses of fertilizers (NPK) are applied as normal crop. Sprouted tubers are used for planting. Optimum row to row spacing is 60 cm, whereas plant to plant spacing varies, such as 20, 15, 10 and 5 cm for 30–35, 25–30, 20–25 and 15–20 mm size tubers, respectively.
- ***Seed rate and tuber production***: For planting one hectare crop, 1.6 tons of seedling tubers (15–35 mm size) are required as compared to 2.5–3.0 ton of seed tubers (25–50 mm size). Normal crop produce is obtained from TPS crop that can be used for three successive clonal generations without any significant reduction in yield. Rest of the other cultural practices remain same as in normal crop.

4. Adoption of TPS Technology: A Case Study in India

TPS technology is highly useful in areas where quality seed tubers are costly or easily not available or difficult to produce due to high vector (aphids) pressure and shortage of cold storages. In India, such states are Karnataka, Maharashtra, Orissa, Madhya Pradesh, Bihar and north-eastern hill (NEH) states. Besides, TPS technology has also been used by farmers of various countries, like China, India, Nepal, Bangladesh, Vietnam, Peru, Nicaragua and Venezuela. The early period of TPS research at International Potato Centre, Lima, Peru triggered TPS breeding in India (Upadhyay et al., 1996), but, later TPS-related research activity has progressively declined the world over (Almekinders et al., 2009). Despite that, a strong TPS program has been established in Tripura state (NEH) in India. This technology could further be expanded in areas where good quality tubers can be produced but where shortage of breeder seeds is found. It can provide a cheap source of planting material for the small and marginal farmers and can play a significant role in bringing down the cost of potato cultivation. Though congenial climatic condition and assured irrigation are necessary for growing crop from seedling transplant method, seedling tuberlets can be used as seed material. If this technology picks up, the yields in low-productivity regions, like NEH, will certainly increase sharply, giving a boost to average national productivity. States like Meghalaya, Manipur, Nagaland and Arunachal Pradesh grow potatoes in two seasons. Hence, seedling tubers produced in one season can be stored easily under ambient conditions in diffused light, and later be utilized for commercial crop production in the next crop season (Kumar et al., 2021).

5. Diploid Hybrid (F_1) Potato Technology

Conventional tetraploid potato breeding has made a great contribution in developing varieties with a very high yield potential. This applies random segregation and fixation of heterosis in tetraploid clones in the first generation itself and then maintained clonally through tubers. Tetraploid potato is highly heterozygous and suffers from acute inbreeding depression upon selfing; therefore, inbred lines and F_1 hybrid is not possible yet. Haploid induction in potato has been preferably obtained via pollination with haploid inducer *S. tuberosum* Phureja Group or some through anther culture. Recently, Amundson et al. (2020) elucidated genomic outcomes of haploid induction crosses in potato and suggested that most potato dihaploids will be free of residual pollinator DNA. Further, they suggested that after tuberosum and phureja crosses, plants derived from seeds that did not express the purple spot marker and that display 2*x* genome content by flow cytometry, are likely to be clean dihaploids (free of pollinator genome). Zhou et al. (2020) provide a holistic view of the haplotype-resolved genome analysis of diploid potato, particularly on deleting deleterious mutations through efficient molecular selection and/or genome editing technologies. The inbred lines with complementary haplotypes are the core of diploid hybrid potato breeding. Until now, it is believed that higher ploidy level is necessary for vigor and heterosis in potato. But genomics research shows that vigor and yield are unrelated to ploidy and even diploid species can perform equivalent or better than tetraploid species. Interestingly, the cultivated potato originated from diploid species (*S. stenotomum* and *S. sparsipilum*). Selfing in autotetraploid species is not possible because it creates acute inbreeding depression due to expression of deleterious alleles, resulting in poor fertility, loss of vigor and low yield potential, which further failed the concept of developing F_1 hybrid in potato (Jansky et al., 2016). However, the recent research advancements have demonstrated that diploid F_1 hybrid potatoes have similar yield potential like tetraploid clones (Lindhout et al., 2011; Li et al., 2013). This work is a milestone in diploid potato breeding. Since then, researchers have now focused on development of diploid F_1 hybrid potato, using homozygous and vigorous inbred lines. Bachem et al. (2019) also elucidated knowledge on understanding of genetic load in potato and future potential of diploid hybrid potato production, and may be designed as the new potato (Stokstad, 2019).

Most potato species are diploid (70%) and self-incompatible but they are a rich source of desirable traits, like resistance to biotic and abiotic stresses and quality parameters. But wild species are sexually compatible with dihaploid of cultivated potato (Spooner et al., 2014). In diploid breeding, it is important to produce hybrids of different cross-combinations between wild and wild, wild and cultivated and cultivated and cultivated potato hybrids, where interaction and its genomic prediction is not possible (Jansky and Spooner, 2018). Now attention is drawn on diploid breeding through gene transfer from diploid wild species to dihaploid of cultivated *S. tuberosum* background. A recent study in diploid species indicates that deleterious alleles are line specific and there would be a high degree of hererosis, when two inbred lines are crossed (Zhang et al., 2019). Some of the major deleterious recessive mutations can be removed through the recombination process over the generation. Several factors need to be considered while resorting to diploid breeding, such as desirable agronomic traits, yield and tuber traits (starch quality, sugar content, texture, processing parameters, etc.). Although, the study shows that diploid lines grow faster and mature earlier than diploid, it is a desirable attribute for future potato breeding than the tetraploids. This technology allows stacking of genes and fixation of heterosis in the inbred lines in less time than the conventional system. Diploid F_1 hybrid breeding is an excellent opportunity to revolutionize potato breeding. Research advancement in diploid potato breeding is summarized in Table 2.

6. Strategies of Diploid Hybrid

6.1 Selection of Recipient Parent (Dihaploid/Diploid Clone)

Dihaploid of cultivated (*S. tuberosum*) or diploid wild potato species is the first and foremost requirement of diploid F_1 hybrid breeding. First, dihaploid of the cultivated tetraploid potato is produced through anther culture or wide hybridization by diploid inducer (*S. phureja*) through crossing with tetraploid lines or develop the self-compatible diploid line through genome editing technology. Although diploids are mostly self-incompatible and less vigorous, only vigorous plants having less deleterious alleles and less heterozygous are used as parent in diploid breeding. Dihaploids/diploids of potato show poor fertility, loss of vigor and lack of desirable agronomic traits, but hybridization between dihaploids and diploid wild/cultivated species can result in improvement of agronomics traits in F_1 hybrids. Second, diploid wild or cultivated species are used directly for the development of inbred lines for F_1 hybrid production. A schematic layout of diploid F_1 hybrid potato production is depicted in Figs. 2 and 3.

6.2 Selection of the Sli Gene Donor Parent

Diploid wild potato species show gametophytic self-incompatibility. Identification of the self-compatibility gene *Sli* (S-locus inhibitor) from the sexually compatible diploid wild species *Solanum chacoense* clones 'chc 525-3' (Hosaka and Hanneman, 1998; Phumichai et al., 2005) opened new avenues for diploid potato hybrid (F_1) breeding program (Ma et al., 2021). Later self-compatibility was also observed in other wild species. Maintenance of male fertility is an important issue in cross of cultivated and wild species. Therefore, it has been suggested to follow backcrossing to introduce small chromosomal regions from wild species into cultivated type (Jansky et al., 2016). Though anther culture is an effective method of inbred lines of development, its response to genotype is limited. Before inbred lines development,

Table 2. Advancement in diploid F_1 hybrid potato technology.

Sr. No.	Study	Key findings	References
1.	Genetics of *Sli* gene and homozygous inbred line development	Investigated genetics of self-compatibility in a self-incompatible wild diploid potato species *Solanum chacoense* and identified a dominant S locus inhibitor (*Sli*) gene. Mapped *Sli* gene at the distal end of chromosome 12 and used it to generate potato inbred lines. They suggest that *Sli* is a pollen-expressed gene with sporophytic action and homozygosity for *Sli* is lethal since homozygous *Sli* genotypes were absent in the F_8 population of *S. chacoense*. Developed inbred lines of self-incompatible cultivated diploid potatoes through incorporation of *Sli* gene.	Hosaka and Hanneman (1998), Birhman and Hosaka (2000)
2.	Homozygous inbred lines	Developed highly homozygous diploid potato lines using the self-compatibility controlling *Sli* gene obtained by repeated selfing and using the *Sli* gene.	Phumichai et al. (2005)
3.	Homozygous inbred lines and F_1 hybrid potato	Developed homozygous, fertile and vigorous diploid F_1 hybrid potato with desirable agronomic traits, based on the field tests. They suggested that homozygous fixation of donor alleles is possible, with simultaneous improvement of tuber shape and tuber size of the recipient inbred line.	Lindhout et al. (2011)
4.	M6, a highly homozygous inbred line	Developed a diploid potato inbred line, M6, for use in breeding and genetics research. M6 is a diploid self-compatible inbred line of *S. chacoense*, vigorous, fertile (both male and female), homozygous (90% of SolCAP SNP) line derived by self-pollinating up to 7th generations. M6 produces tubers under both short and long photoperiods. M6 has high dry matter content, good chip processing quality and resistance to soft rot and Verticillium wilt. They reinvented potato as a diploid inbred line-based crop with diploid hybrid potato in future.	Jansky et al. (2014, 2016)
5.	Genome sequence of diploid line Solyntus	Deciphered whole genome sequence of Solyntus, a highly homozygous, diploid, vigorous and self-compatible line. The 116 contigs provide a more direct and contiguous reference constructed by sequencing with state-of-the-art long- and short-read technology. This assembly contains 93.7% of the single-copy gene orthologs from the Solanaceae. This will increase knowledge on diploid potato breeding research.	van Lieshout et al. (2020)
6.	Diploid hybrid breeding	Enhanced understanding of the genetic load in potato for hybrid diploid breeding. Envisioned hybrid potato through the production of two elite lines which are achieved by self-fertilizing diploid potato clone of different genetic background. This will eliminate detrimental alleles (lethal alleles) naturally or by breeder selection. After several generations of selfing, the two elite lines become largely homozygous and retain their advantageous alleles. Then cross is performed between two homozygous elite inbred lines. The resulting hybrid progeny is identical and uniformly heterozygous. The combination of advantageous alleles results in hybrid vigor/heterosis.	Bachem et al. (2019)
7.	Genetics of inbreeding depression	Elucidated the genetic basis of inbreeding depression in potato for genome design of inbred lines. As a clonally propagated crop, potato suffers from severe inbreeding depression. They evaluated mutation in 151 diploid potatoes and obtained 3,44,831 predicted deleterious substitutions enriched in the pericentromeric regions and line specific. Most of the deleterious recessive alleles affect survival and growth vigor. One of these deleterious alleles disrupts a gene required for embryo development.	Zhang et al. (2019)

Table 2 contd. ...

...*Table 2 contd.*

Sr. No.	Study	Key findings	References
8.	Genome editing (CRISPR-Cas9) for self-compatible line development	Re-domestication of potato into inbred line-based diploid crop propagated by seed (TPS) represents a promising alternative over traditional clonally propagation by tubers of tetraploid potato. However, self-incompatibility has hindered the development of inbred lines. They developed self-compatible diploid potato lines by knocking out *S-RNase* (S-locus RNase) gene, using the CRISPR-Cas9 system. This strategy opens new avenues for diploid potato breeding.	Ye et al. (2018), Enciso-Rodriguez et al. (2019)
9.	Haplotype-based diploid genome analysis	Investigated haplotypes-based genome analyses in heterozygous diploid potato. They presented a holistic view of the genome organization of a clonally-propagated diploid species and provides insights into technological evolution in resolving complex genomes.	Zhou et al. (2020)
10.	Genome analysis of dihaploids and haploid inducer	Deciphered genomes of dihaploids and haploid inducer (*S. phureja*) and their genomic outcomes in potato crossing. Previous reports have described aneuploid and euploid progeny in haploid induction. Analyzed a population of 167 dihaploids for large-scale structural variations that underlie chromosomal addition from the haploid inducer, and for small-scale introgression of genetic markers. Deep sequencing indicated that occasional, short-tract signals appear to be of haploid inducer origin. Collectively, 52% of the assayed chromosomal loci were classified as dosage variable. This study elucidates the genomic consequences of potato haploid induction and suggests that most potato dihaploids will be free of residual pollinator DNA.	Amundson et al. (2020)
11.	*Sli* gene origin and occurrence	In fact, self-compatible diploid potatoes allow innovative potato breeding. Therefore, the *Sli* gene, originally described in *S. chacoense*, has received much attention. In fact, in elite *S. tuberosum* diploids, spontaneous berry set is occasionally observed. Hence, they studied the origin and widespread occurrence of *Sli*-based self-compatibility in potato of *S. tuberosum* origin. They showed that *Sli* is surprisingly widespread and indigenous to the cultivated gene pool of potato.	Clot et al. (2020)
12.	Fine mapping of the *Sli* gene	Mapped the *Sli* gene to a 12.6 kb interval on chromosome 12. They described an expression vector that converts self-incompatible genotypes into self-compatible and a CRISPR-Cas9 vector that converts self-compatible genotypes into self-incompatible ones. The *Sli* gene encodes an F-box protein that is specifically expressed in pollen from self-compatible plants. Thus, function of the *Sli* gene leads to self-compatibility and facilitates precision breeding in potato.	Eggers et al. (2021)
13.	Non S-locus F-box gene for self-compatibility	Demonstrated that a non S-locus F-box gene breaks self-incompatibility in diploid potatoes. They showed that a self-compatible diploid potato, RH89-039-16 (RH), can efficiently induce from self-incompatibility to self-compatibility, when crossed to self-incompatible lines. They identified *Sli* gene in RH, capable of interacting with multiple allelic variants of the pistil-specific S-ribonucleases (S-RNases). The *Sli* gene functions like a general S-RNase inhibitor to impart self-compatibility to RH and other self-incompatible potatoes.	Ma et al. (2021)
14.	Genomic design of hybrid potato	Demonstrated genomic design of hybrid potato through development of the first generation of potato inbred lines with high homozygosity to exploit heterosis. This will transform conventional tetraploid to fast iterative diploid breeding through seeds. Discussed four steps of diploid hybrid potato breeding: (i) selection of starting materials with self-compatible vigorous lines with diverse genetic background, (ii) genetic analysis of S_1 population with more beneficial alleles, (iii) development of homozygous inbred lines, and (iv) F_1 hybrid development by crossing two inbred lines.	Zhang et al. (2021)

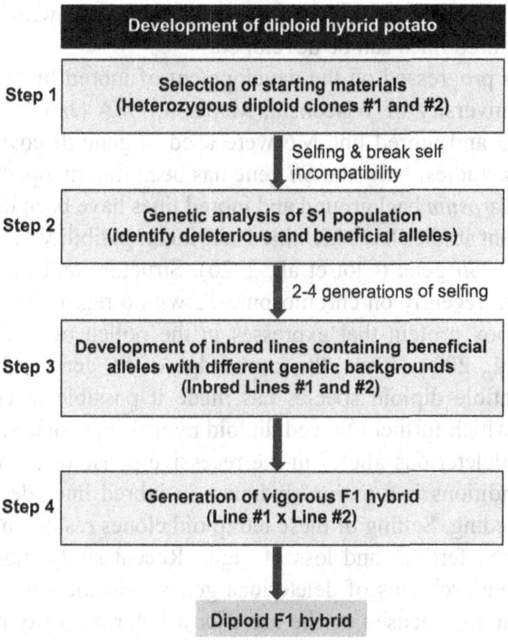

Fig. 2. Pipeline of development of diploid F_1 potato hybrid (*Source*: Modified and adapted from Zhang et al., 2021).

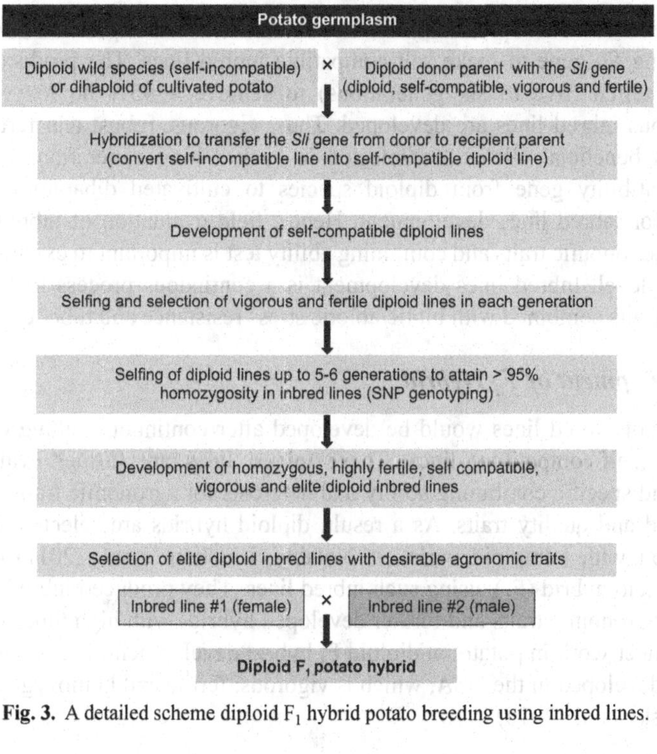

Fig. 3. A detailed scheme diploid F_1 hybrid potato breeding using inbred lines.

there must be available self-compatibility in donor parent with *Sli* gene, so that self-compatible inbred lines can be developed.

Research has progressed on the development of inbred lines using six diploid populations at University of Wisconsin, Madison, USA (Jansky et al., 2016). To illustrate, DM 1-3 and inbred line M6 were used in gene discovery, mapping and various genomics studies. Now the *Sli* gene has been fine-mapped and introgressed into diploid *S. tuberosum* background and inbred lines have been developed (Eggers et al., 2021). Recent studies have identified self-incompatibility in other wild species to be governed by *Sli* gene (Clot et al., 2020). Structure and function of *Sli* gene came to be known recently on chromosome 12 with a region of 12 kb and this *Sli* gene encodes F-box protein that expresses in the pollen of self-compatible lines only (Eggers et al., 2021). Thus, the knowledge of *Sli* gene and its introgression into self-incompatible diploid species has made it possible to convert them into self-compatible, which further allowed diploid hybrid (F_1) potato breeding. But, the accumulation of deleterious alleles in the recessive phase of tetraploid and highly heterozygous conditions is a major hindrance in inbred lines development and of hybrid potato breeding. Selfing of these tetraploid clones results in acute inbreeding depression, reduced fertility and loss of vigor. Recent study suggested a pipeline for overcoming the problems of deleterious genes over the generations for inbred lines development and focused on the use of low heterozygosity material with less deleterious alleles Zhang et al. (2021).

6.3 Development of Homozygous Diploid Inbred Lines

The above selected dihaploid/diploid lines are crossed with diploid self-compatible lines having *Sli* gene to make self-compatible inbred lines. The produced diploid lines are selfed (five to six generations) to achieve > 95% homozygosity and homozygous inbred lines are developed. Thus, vigorous, robust and fertile inbred lines with beneficial alleles are selected in each selfing generation. Transfer of self-compatibility gene from diploid species to cultivated dihaploid is equally essential for inbred lines development. Hence, field evaluation of inbred lines for desirable agronomic traits and combining ability test is important to exploit heterosis at diploid level. Inbred lines development is a continuous process to accumulate desirable traits combined with biotic/abiotic stress resistance and tuber quality traits.

6.4 Development of F_1 Hybrid

A number of inbred lines would be developed after continuous selfing of diploid/dihaploid (self-compatible) lines. These inbred lines are further evaluated for general and specific combining ability and heterosis for agronomic traits, including tuber yield and quality traits. As a result, diploid hybrids are selected for further comparison with tetraploid varieties. Notably, Lindhout et al. (2011) developed diploid potato hybrid (F_1), using such inbred lines. They produced inbred lines with desirable agronomic traits and further developed hybrids with high tuber yield. This was a pioneer work in potato on diploid F_1 hybrid development. Later, another line, M6, was developed in the USA, which is vigorous, fertile and homozygous with *Sli* gene (Jansky et al., 2016).

The idea of developing F_1 hybrid is not new but was proven realistic by Lindhout et al. (2011). Acute inbreeding depression and self-incompatibility of diploid germplasm lines blocked the development of inbred lines. However, back-crossing with line possessing *Sli* gene inhibiting gametophytic self-incompatibility produced self-compatible progenies. They demonstrated that fixation of homozygous donor allele is now possible with improvement of tuber traits. To achieve the F_1 hybrid seeds, development of diploid self-compatible inbred line (e.g., M6) is very crucial. M6 is vigorous and fertile (both male and female). This homozygous breeding line was derived by selfing the wild potato species *Solanun chacoense* for many generations as it possesses dominant self-incompatibility inhibitor gene *Sli* and has several desirable traits. M6 can be used to develop recombinant inbred lines for production of hybrid potato (Jansky et al., 2014). This elaborates production of diploid F_1 hybrid potato by crossing homozygous diploid parents as expected for hybrid vigor. Parent 1 (diploid, fertile and self-compatible like M6) is crossed with Parent 2 (diploid parents: D1, D2 and D3) and advance breeding lines were developed for desirable traits by crossing designs (selfing, back crossing, F3, etc.). Then F_1 hybrid potato is produced by crossing F_3 materials generated from the cross between Parent 1 × Parent 2. Yield attributes were claimed at about 200 g/plant (Patent US 2014/0115736A1). Applying the above concepts, Solynta, a leading potato seed breeding company based in Wageningen (NL), has developed the hybrid potato technology (http://solynta.com/).

7. Genomics in TPS Research

7.1 Apomixis

In flowering plants, there are two common modes of propagation, i.e., sexual and asexual or apomixis. The sexual method of propagation is commonly practiced in most seed crops and conventional TPS in potato, following hybridization of two parents and selection thereafter. The asexual method is commonly practiced in a vegetatively-propagated crop, like potato. Apomixis is asexual reproduction through seeds that bypass meiosis and fertilization and produce progenies genetically identical to the mother plant (Spillane et al., 2004). This offers an excellent opportunity to develop TPS in potato, which can be propagated for commercial tuber production. Apomictic TPS will follow mitosis cell division without any meiosis process to maintain mother plant identity. Heterosis is fixed in tetraploid clones which can be multiplied true-to-type through clone-specific TPS without any gamete formation and no meiotic segregation and recombination. Such TPS follows mitosis cell division; thereby it produces unreduced gametes which possess same ploidy, i.e., tetraploid (4*x*), like somatic mother cell (4*x*) instead of normal reduced gametes through meiosis, i.e., dihaploid (2*x*). The recent advancement in meiosis (recombination, cell cycle and chromosome distribution) has been reviewed by Crismani et al. (2013) and that can be applied to create apomixis and propagate new crop species. Apomixis is a common feature in some forage grasses and fruit trees, but not common in crops, including potato. With the recent advancements in apomixis technology in *Arabidopsis*, potato holds a great promise to develop TPS through apomixis for commercial cultivation via botanical seeds, instead of tubers. Potato hybrid TPS will play a key role in sustainable crop production through massive multiplication of disease-free quality planting materials. In clonally propagated potato, apomixis allows benefits of facing no phytosanitary threats or incompatibility barriers (Barcaccia and Albertini, 2013). Thus, potato hybrid TPS development through apomixis technology promises a potential to break the yield barriers in potato.

7.2 *Arabidopsis Apomictic Seeds: A Lesson for TPS*

With the research advancement in *Arabidopsis*, apomixis can be created in potato through targeted modifications during the sexual reproduction phase (Marimuthu et al., 2011). First, apomeiosis is followed for converting meiosis into mitosis cell division, which leads to formation of unreduced clonal gametes. This has been demonstrated in model plants that produce apomeiotic gametes, though most of these mutants are male/female sterile. Notable examples are available in *Arabidopsis*, where *MiMe* (mitosis instead of meiosis) mutants were created that follow mitosis division instead of meiosis and in turn, fertile gametes with high frequency were regenerated (D'Erfurth et al., 2009). In fact, *MiMe* mutants are an integration of three genes combinations (*SPO11-1*, *REC8* and *OSD1*) and affect homologous recombination, monopolar orientation of sister chromatids and second cell division. First, the *SPO11-1* gene eliminates meiosis recombination, second *REC8* gene results in separation of sister chromatids in first meiosis cell division, instead of distribution of homologous chromosomes, and finally, *OSD1* gene skips second meiosis cell division (D'Erfurth et al., 2009). Thus, *MiMe* mutant shows meiosis without recombination and separates sister chromatids in the first round of meiosis, thus mimicking the mitosis process. Haploid *Arabidopsis* plants were produced through seeds by manipulating the centromere-specific histone protein CENH3 (Ravi and Chan, 2010). As CENH3 is universal in eukaryotes, this method may be extended to produce haploids in any plant species. Then, hybridization of *MiMe* mutant with genome eliminated plant (i.e., CENH3 mutant) results in production of clonal offspring (Ravi and Chan, 2010; Marimuthu et al., 2011). Earlier, Ravi et al. (2008) demonstrated that mutation of the *Arabidopsis* gene *dyad*, a regulator of meiotic chromosome organization, leads to apomeiosis, a major component of apomixis. Most fertile ovules in *dyad* plants form seeds that are triploid and arise from fertilization of an unreduced female gamete by a haploid male gamete.

There are certain limitations in the application of *MiMe* technology. Although *MiMe* has been found efficient in model plants, its application to other crops species is still unclear. Literature evidences that the genes *SPO11-1* and *REC8* are conserved among the eukaryotes, and therefore, are homologous possibly across the plant species. However, gene *OSD1* seems to be a plant-specific gene with complex phylogeny. Secondly, *MiMe* mutants produce 2n gametes that are identical to their parents. But normal fertilization still occurs, leading to doubling of ploidy in the next generation. Therefore, it is a big challenge to develop *MiMe* mutant that produces gametes, causing direct formation of embryo without any other gametes. This can be achieved through genome elimination in which chromosomes of one parent is removed after fertilization in the zygote. Thus, *MiMe* combined with genome

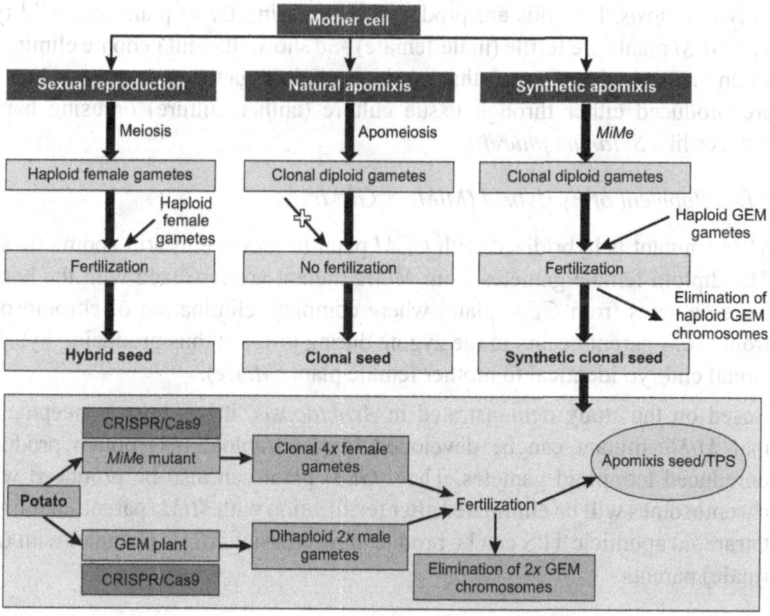

Fig. 4. Sexual, natural apomixis and synthetic apomixis seed reproduction systems in plants.

elimination though CENH3 techniques leads to normal ploidy without doubling in each sexual cycle (Marimuthu et al., 2011). The study demonstrates CENH3 gene manipulation in *Arabidopsis* but has not been transferred to other crops, except maize and that too with a very low frequency (Kelliher et al., 2016). A schematic layout of apomixis-based seed production is presented in Fig. 4.

7.2.1 Development of MiMe (Mitosis Instead of Meiosis) Mutant

- In *Arabidopsis*, first step is to develop *MiMe* mutant or *dyad* mutant that can produce diploid gametes, genetically identical to parents. During the cell cycle, mitosis produces diploid gametes that are identical to parents, unlike the meiosis process having haploid gametes from diploid organism.
- *MiMe* mutant is a combination of three genes, *OSD1*, *SPO11-1* and *REC8*, in which meiosis is totally replaced by mitosis. The *SPO11-1* and *REC8* genes mutant result in mitosis-like first meiosis, whereas *OSD1* eliminates second meiotic division. As a result, *MiMe* mutant produces diploid male and female gametes that are genetically identical to their diploid parent.

7.2.2 Development of GEM (Genome Elimination Mutant)

- Second step is to develop *GEM* where chromosomes are eliminated after fertilization. In *Arabidopsis*, *GEM* was produced through chromosome containing altered centromere-specific histone protein CENH3, whose chromosomes are eliminated after fertilization in the progeny. As CENH3 is universal in eukaryotes, this can be extended to produce haploids in any plant species.

- In *Arabidopsis*, haploids are produced by crossing *GEM* plant and wild type. The *GEM* plants are fertile (male/female) and show efficient genome elimination when crossed with a parent that produces diploid gametes. In potato, haploids are produced either through tissue culture (anther culture) or using haploid inducer like *Solanum phureja*.

7.2.3 Development of F_1 Hybrid (MiMe × GEM)

- *MiMe* mutant is hybridized with *GEM* plant to produce hybrid apomictic seed. The diploid female gametes from *MiMe* mutant are fertilized with the haploid male gametes from *GEM* plant, where complete elimination of chromosomes from *GEM* parent occurs in the zygote during mitosis; thus producing hybrid or clonal embryo identical to mother female plant (*MiMe*).
- Based on the study demonstrated in *Arabidopsis*, it has been conceptualized that *MiMe* mutant can be developed from tetraploid (4*x*) potato producing unreduced tetraploid gametes. Then, *GEM* plant can also be produced where chromosomes will be eliminated after fertilization with *MiMe* parent. In this way, tetraploid apomicic TPS can be produced by crossing *MiMe* (female) with *GEM* (male) parents.

7.3 Genome Editing in Diploid Hybrid Potato

Genome editing (CRISPR/Cas), an innovative technology has the potential to accelerate plant breeding research through easy manipulation of target genes of interest. Earlier, the *Arabidopsis MiMe* mutants creation with three gene combinations (*SPO11-1*, *REC8* and *OSD1*) was highly time consuming and also genotype- and species-specific. Further, development of *GEM* is also difficult through chromosomes containing altered centromere-specific histone protein CENH3, whose chromosomes are eliminated after fertilization. This facilitates genome editing at meiotic cell cycle genes to produce apomeiotic and CENH3 mutants for haploid production. Then both the mutants can be crossed to develop hybrid potato.

CRISPR/Cas9 offers an excellent opportunity to develop mutants with desirable gene loci. Programmable sequence-specific endonucleases facilitate precise editing of genomic loci for target traits. Recent advances in the field of genome editing are the key to enabling site-directed genome modifications and are sequence-specific nucleases that generate targeted double-stranded DNA breaks in genes of interest, including targeted mutations, gene insertions and gene replacements. The CRISPR-Cas9 technology has been demonstrated to overcome self-incompatibility in diploid potato by targeting *S-RNase* (S-locus RNase) and multiple *SLFs* (S-locus F-box proteins) which are expressed in style and pollen, respectively (Enciso-Rodriguez et al., 2019). They applied Cas9 technology to design a dual single-guided (sgRNA) and generated self-compatible diploid potato lines by targeting conserved exonic regions of the S-RNase gene and generate targeted knockouts (KOs) mutants. Self-compatibility was achieved in nine S-RNase KO T0 lines which contained bi-allelic and homozygous deletions/insertions in both genotypes, transmitting self-compatibility to T1 progeny. This study demonstrates an efficient approach to achieve stable, consistent self-compatibility through S-RNase KO for use in diploid potato breeding approaches (Enciso-Rodriguez et al., 2019). This new technology can be used to knockout the multiple gene of interest rather than introduction of foreign genes in plant chromosomes and elucidate gene function and develop new and valuable traits. One of the potential advantages of genome editing is that one can create a plant that has novel genetic variations and does not have a transgene. Genome editing in hybrid potato would help agriculture by fixing hybrid vigor and allowing perpetual multiplication of elite heterozygous genotype. In addition to above advantages of TPS, diploid potato hybrids allow application of simple genetics at diploid level to map and discover novel genes, gene stacking and higher genetic gain due to rapid accumulation of favorable alleles in inbred lines in sexual breeding cycles.

8. Challenges in Diploid Hybrid Production

Though diploid hybrid (F_1) TPS technology has several advantages, as mentioned above, it suffers from the several following disadvantages as well:

- The inbred lines produced to date in diploid potatoes are not completely homozygous for all the loci in the genome. This means that F_1 hybrids may lack uniformity for some traits governed by these loci. This could be a major deterrent to the acceptance of technology.
- The approach involves a complete transformation of potato breeding methodology; hence, it will require rigorous evaluation in various agro-ecologies to convince the stakeholders on the adoption of this technology. The trials in African countries have shown at par or less tuber yield of F_1 TPS in comparison to tuber-raised tetraploid varieties. At par tuber yield and quality to tetraploid varieties is a must for the adoption of diploid F_1 hybrid TPS technology.
- TPS is the starting material for growing potatoes. This will require additional time and resources to raise the crop in the nursery and transplanting seedlings in the field. The crop duration of TPS-based approach needs to match with tuber-raised crop to fit in the cropping system in India.
- Both scientists and farmers are accustomed to growing potatoes from tubers, i.e., conducting trials or raising crops. It will be important to standardize the agronomy of TPS-based crop as a complete package and practice the diploid F_1 hybrid TPS.

9. Conclusion

TPS technology is scientifically sound, technically feasible, economically viable and eco-friendly. TPS provides an opportunity to small farmers even to generate high quality planting material and assures high yields with low inputs as compared to bulky and costly tubers. The conventional TPS technology has encountered major problems, such as poor germination rates, non-uniform tubers, long dormancy and increased irrigation requirement. Therefore, diploid F_1 hybrid potato has proven successful in potato through development of self-compatible diploid inbred lines with *Sli* gene. The dihaploid of cultivated potato with desirable agronomic traits and diploid donor parent possessing *Sli* gene causing self-compatibility are the two essential requirements of hybrid development. The apomixis research has demonstrated, in model plant *Arabidopsis*, development of clonal seeds identical to mother plant, using *MiMe* mutant and *GEM* plants for chromosome elimination, which can be translated in potato breeding, though this is a challenging task and time consuming to develop various mutants. Hence, genome editing technology offers targeted mutation in the multiple genes with programmable nucleases to produce diploid hybrid potato. The genome editing (CRISPR-Cas9) has been demonstrated to overcome self-incompatibility in diploid potato by targeting *S-RNase* (S-locus RNase) and multiple *SLFs* (S-locus F-box proteins) to develop self-compatible potato lines. The genome-edited hybrid technology could make this a more attractive option for potato breeders and growers to ensure sustainable food production in the near future.

Acknowledgement

I am thankful to the Director, ICAR-Central Potato Research Institute, Shimla, and scientists/technicians/research fellows and other colleagues of the institute for their support under the institute research projects on biotechnology, germplasm, breeding, and seed research on aeroponics. I am also grateful to the funding agencies for support under the externally funded projects (CABin, ICAR-IASRI, New Delhi; ICAR-LBS Young Scientist Award Project, and DBT, Government of India).

References

Almekinders, C.J.M, Chujoy, E. and Thiele, G. (2009). The use of true potato seed as pro-poor technology: the efforts of an international agricultural research institute to innovating potato production. *Potato Res.*, 52: 275–293.

Amundson, K.R., Ordoñez, B., Santayana, M., Tan, E.H., Henry, I.M. et al. (2020). Genomic outcomes of haploid induction crosses in potato (*Solanum tuberosum* L.). *Genetics*, 214(2): 369–380.

Bachem, C., van Eck, H.J. and de Vries, M.E. (2019). Understanding genetic load in potato for hybrid diploid breeding. *Mol. Plant*, 12(7): 896–898.

Barcaccia, G. and Albertini, E. (2013). Apomixis in plant reproduction: a novel perspective on an old dilemma. *Plant Reprod.*, 26: 159–179.

Birhman, R.K. and Hosaka, K. (2000). Production of inbred progenies of diploid potatoes using an S-locus inhibitor (*Sli*) gene, and their characterization. *Genome*, 502: 495–502.

Clot, C.R., Polzer, C., Prodhomme, C., Schuit, C., Engelen, C.J. et al. (2020). The origin and widespread occurrence of *Sli*-based self-compatibility in potato. *Theor. Appl. Genet.*, 133(9): 2713–2728.

Crismani, W., Girard, C. and Mercier, R. (2013). Tinkering with meiosis. *J. Exp. Bot.*, 64: 55–65.

D'Erfurth, I., Jolivet, S., Froger, N., Catrice, O., Novatchkova, M. and Mercier, R. (2009). Turning meiosis into mitosis. *PLoS Biol.*, 7: e1000124.

Eggers, E.J., van der Burgt, A., van Heusden, S.A., de Vries, M.E., Visser, R.G. et al. (2021). Neofunctionalisation of the Sli gene leads to self-compatibility and facilitates precision breeding in potato. *Nat. Commun.*, 12(1): 1–9.

Enciso-Rodriguez, F., Manrique-Carpintero, N.C., Nadakuduti, S.S., Buell, C.R., Zarka, D. and Douches, D. (2019). Overcoming self-incompatibility in diploid potato using CRISPR-Cas9. *Front. Plant Sci.*, 10: 376.

Hosaka, K. and Hanneman, R.E. Jr. (1998). Genetics of self-compatibility in a self-incompatible wild diploid potato species *Solanum chacoense*. I. detection of an S locus inhibitor (*Sli*) gene. *Euphytica*, 99: 191–197.

Jansky, S.H., Chung, Y.S. and Kittipadukal, P. (2014). M6: A diploid potato inbred line for use in breeding and genetics research. *J. Plant Regist.*, 8: 195–199.

Jansky, S.H., Charkowski, A.O., Douches, D.S., Gusmini, G., Richael, C. et al. (2016). Reinventing potato as a diploid inbred line-based crop. *Crop Sci.*, 56: 1–11.

Jansky, S.H. and Spooner, D.M. (2018). The evolution of potato breeding. *Plant Breed. Rev.*, 41: 169–214.

Kelliher, T., Starr, D., Wang, W., McCuiston, J., Zhong, H. et al. (2016). Maternal haploids are preferentially induced by CENH3-tailswap transgenic complementation in maize. *Front. Plant Sci.*, 7: 1–11.

Kumar, S., Thakur, K.C., Sood, S., Tiwari, J.K., Kardile, H.B. et al. (2021). True potato seed for potato production. pp. 181–191. *In*: Chakrabarti, S.K. and Bhardwaj, V. (eds.). *Potato in Sub-tropics: A Saga of Success*. International Books & Periodical Supply Service, Delhi, India.

Li, Y., Li, G., Li, C., Qu, D. and Huang, S. (2013). Prospects of diploid hybrid breeding in potato. *Chinese Potato*, 27: 96–99.

Lindhout, P., Meijer, D., Schotte, T., Hutten, R.C.B., Visser, R.G.F. and VanEck, H.J. (2011). Towards F_1 hybrid seed potato breeding. *Potato Res.*, 54: 301–312.

Ma, L., Zhang, C., Zhang, B., Tang, F., Li, F. et al. (2021). A non S-locus F-box gene breaks self-incompatibility in diploid potatoes. *Nat. Commun.*, 12(1): 1–8.

Marimuthu, M.P., Jolivet, S., Ravi, M., Pereira, L., Davda, J.N. et al. (2011). Synthetic clonal reproduction through seeds. *Science*, 331: 876.

Phumichai, C., Mori, M., Kobayashi, A., Kamijima, O. and Hosaka, K. (2005). Toward the development of highly homozygous diploid potato lines using the self-compatibility controlling *Sli* gene. *Genome*, 48: 977–984.

Ravi, M., Marimuthu, M.P.A. and Siddiqi, I. (2008). Gamete formation without meiosis in *Arabidopsis*. *Nature*, 451: 1121–1124.

Ravi, M. and Chan, S.W.L. (2010). Haploid plants produced by centromere-mediated genome elimination. *Nature*, 464: 615–618.

Spillane, C., Curtis, M.D. and Grossniklaus, U. (2004). Apomixis technology development-virgin births in farmers' fields? *Nat. Biotechnol.*, 22: 687–691.

Spooner, D.M., Ghislain, M., Simon, R., Jansky, S.H. and Gavrilenko, T. (2014). Systematics, diversity, genetics and evolution of wild and cultivated potatoes. *The Botanical Rev.*, 80: 283–383.

Stokstad, E. (2019). The new potato. *Science*, 363: 574–577.

Upadhyay, M.D., Hardy, B., Gaur, P.C. and Ilangantilenke, S. (1996). Production and utilization of true potato seed in Asia, *CIP*, Lima, Peru and ICAR, New Delhi, India, pp. 231.

van Lieshout, N., van der Burgt, A., de Vries, M.E., Ter Maat, M., Eickholt, D. et al. (2020). Solyntus, the new highly contiguous reference genome for potato (*Solanum tuberosum*). *G3 (Bethesda)*, 10: 3489–3495.

Ye, M., Peng, Z., Tang, D., Yang, Z., Li, D. et al. (2018). Generation of self-compatible diploid potato by knockout of S-RNase. *Nat. Plants*, 4: 651–654.

Zhang, C., Wang, P., Tang, D., Yang, Z., Lu, F. et al. (2019). The genetic basis of inbreeding depression in potato. *Nat. Genet.*, 51: 374–378.

Zhang, C., Yang, Z., Tang, D., Zhu, Y., Wang, P. et al. (2021). Genome design of hybrid potato. *Cell*, 184: 3873–3883.

Zhou, Q., Tang, D., Huang, W., Yang, Z., Zhang, Y. et al. (2020). Haplotype-resolved genome analyses of a heterozygous diploid potato. *Nat. Genet.*, 52: 1018–1023.

Chapter 10

Pre-breeding Genomics in Somatic Hybridization

1. Plant Tissue Culture

Potato (*Solanum tuberosum* L.), a member of the family *Solanaceae,* is a successful example where tissue culture, particularly *in vitro* multiplication, has been extensively applied in germplasm management and genetic improvement. During the last few decades, problem-driven use of *in vitro* technology in potato had been instrumental in addressing many inherent problems associated with this vegetatively propagated, heterozygous and tetraploid crop. The propagation method and genetic nature of this crop imposes several limitations on seed multiplication, conservation of genetic resources and genetic improvement. Micropropagation of disease-free potato clones combined with conventional multiplication methods has become an integral part of seed production in many countries, including India. Several *in vitro* methods, including the cryo-conservations, have been applied in India for conservation of valuable potato genetic resources which is the largest in Asia. The developments in the fields of cellular selections, somaclonal variations, somatic hybridization and genetic transformation have not only improved present-day potato, but also generated novel genetic variability for selection of future potato cultivars.

Potato is a vegetatively propagated crop which is a good host for a large number of viruses besides other pathogens. Infection of planting material by pathogens, mainly potato viruses, cause severe reduction in yield. Therefore, disease-free/healthy *in vitro* multiplied plantlets are used in germplasm conservation and conventional seed production. Micropropagation allows large-scale multiplication of virus-free potato microplants. Nodal segments of virus-free potato microplants are cultured on semi-solid or liquid medium under aseptic conditions for obtaining new microplants. Murashige and Skoog's (MS) medium supplemented with 2.0 mg/L D-calcium pantothenate, 0.1 mg/L GA_3, 0.01 mg/L NAA and 30 g/L sucrose is best suited for propagation of potato microplants. Cultures are usually incubated under a 16-h photoperiod (50–60 μ mol $m^{-2}s^{-1}$ light intensity) at 20–22°C. Usually, two-three nodal cuttings (1.0–1.5 cm) are inoculated per culture tube (25 × 150 mm) and the tubes are closed with cotton plugs. Within three weeks, the axillary/apical buds of these cuttings grow into full plants. These plants can be further sub-cultured on fresh medium. At an interval of every 25 days of subculturing, theoretically 3^{15} (14.3 million) microplants can be obtained from a single virus-free microplant in a year. In the countries where disease-free seed potatoes cannot be produced due to lack of vector-free seed production areas, micropropagation can play an important role to maintain and multiply commercial cultivars under disease-free conditions.

2. Somatic Hybridization

Gene transfer is the basis of crop improvement, including that of potato. Conventionally, this is achieved through sexual hybridization, which limits a range of species from which gene flow can occur into a crop species. Wild species have contributed remarkably to the success of crop improvement and they allow crops to retain their commercial status. As a result, plant breeders have sought to utilize an increasing number of wild species as a source of valuable genes, ranging from disease resistance to grain yield, and produce quality. But many sources of useful genes cannot be included in crop improvement programme primarily because of sexual incompatibilities. Genetic transformation, a focused and direct gene transfer approach, requires identification, isolation and cloning of the concerned genes. Further, it is expensive and technically most exacting, although it may represent the ultimate strategy. However, some characters of interest may be governed by two or more and yet unknown genes; transfer of such characters through genetic transformation may pose many difficulties. Finally, transfer of cytoplasmic organells, viz. chloroplast and mitochondria, may often be the desired objectives; this, however is not possible through genetic transformation, while it can readily be achieved by somatic hybridization.

A huge genetic diversity is available in *Solanum* species for various desirable traits. Wild *Solanum* species have been used in potato breeding but they represent only a small fraction of the total *Solanum* diversity (Bradshaw et al., 2006). Utilization of the wild tuber-bearing diploid species has remained an untapped potential source for transferring resistance trait into the common potato. Wild tuber-bearing *Solanum* species are widely distributed from south-western USA to central Argentina and Chile. This extensive geographical range has resulted in types adapted to a broad range of climatic and soil conditions. In the course of evolution, these plants have also developed resistance/tolerance to different pathogens and pests. Much of this effort has involved the examination of wild species for various resistance traits related to potato. This trait is particularly attractive to breeders to widen the potato genetic base, but the barrier between the cultivated potato and wild species has proved a difficult task.

Many useful genes derived from wild sources cannot be transferred through the conventional technique because of sexual incompatibilities and primarily due to differences in ploidy and Endosperm Balance Number (EBN). It is extremely difficult to cross 1EBN wild species directly with common cultivated 4EBN potato. Limited success has been obtained by utilizing bridging species but the incompatibility of 1EBN wild species has generally prevented the use of this particularly valuable trait (Spooner and Salas, 2006). However, modern research and new techniques have made it possible to expand considerably the genetic resources available for use in breeding programs. A few methods have now become available to overcome this problem. These methods include manipulation of ploidy and EBN, bridge crosses, mentor pollination and embryo rescue, hormone treatment and reciprocal crosses (Jansky, 2006). Somatic hybridization removes pre-zygotic and some post-zygotic barriers between cultivated and wild species. Somatic hybridization can provide a means of bypassing sexual incompatibility between *Solanum* species, leading to fertile plants that can be used directly in breeding programs. Potato wild species, ploidy number and endosperm balance number are given in Table 1, and a few selected were assessed for various traits (Tiwari et al., 2019).

Somatic hybridization generates functional combinations of large sets of genetic material, which makes it similar to sexual hybridization. This method can also be used to overcome limitations of genetic transformation (Helgeson et al., 1993). Many of the important traits are predominantly polygenic, such as late blight resistance and thus unavailable as isolated and characterized sequences that are ready for genetic transformation. Therefore, efficient methods of transformation are yet to be made available for multiple genes that are expressed in a coordinated manner (Orczyk et al., 2003). On the other hand, somatic hybrids obtained directly after fusion contain all organelles from the cytoplasms of both the parents. Somatic hybridization via protoplast isolation, electrofusion and regeneration is a useful tool to transfer polygenic traits, such as late blight resistance, in a single step. It enables development of tetraploid somatic hybrid between diploid wild species and dihaploid of common potato. As a result, tetraploid somatic hybrids may be utilized in conventional breeding for late blight resistance and improvement of other traits. Thus, production of somatic hybrids between tetraploid 4EBN *S. tuberosum* and diploid 1EBN wild species has been envisaged for imparting durable resistance to late blight. As a consequence, the aim of this somatic fusion technology is creditable to enrich the cultivated potato gene pool by incorporating genes from new exotic wild species.

Hence, somatic hybridization is the technique that enables transfer of agronomically important traits by bypassing such sexual barriers, besides the conventional and recombinant-DNA technologies approaches. Despite these crossing-barriers, many researchers have used this technique and subsequently produced somatic hybrids with cultivated potato. Production of hybrid plants through the fusion of protoplasts of two different plant species/varieties is called 'somatic hybridization' and such hybrids are called 'somatic hybrids'. Thus, somatic hybridization aims to strengthen the potato gene pool by introducing genes from wild species. This technique allows several advantages over conventional breeding and transgenic methods, such as: (i) produces fertile somatic hybrids with target traits of wild *Solanum* species, (ii) provides access to basic pre-breeding material for effective utilization in breeding, (iii) enables easy transfer of monogenic and polygenic traits in one step, (iv) results in recombination of nuclear and cytoplasmic genomes, and (v) avoids biosafety regulatory issues associated with transgenics. During the past 40 years, over hundreds of symmetric somatic hybrids were produced, but only a limited number of asymmetric somatic hybrids (Orczyk et al., 2003; Tiwari et al., 2018).

Table 1. Potato wild species, ploidy number and endosperm balance number.

S. No.	Potato species	Origin country	Ploidy/EBN
Wild Species			
1.	*Solanum acaule* Bitter	ARG, BOL, PER	4x (2EBN)
2.	*S. achacachense* Cardenas	BOL	2x
3.	*S. acroglossum* Juz.	PER	2x (2EBN)
4.	*S. acroscopicum* Ochoa	PER	2x
5.	*S. agrimonifoloium* Rydb.	GUA, HON, MEX	4x (2EBN)
6.	*S. alandiae* Cardenas	BOL	2x
7.	*S. albicans* (Ochoa) Ochoa	ECU, PER	6x (4EBN)
8.	*S. albornozii* Correll	ECU	2x (2EBN)
9.	*S. amayanum* Ochoa	PER	2x (2EBN)
10	*S. ambiosinum* Ochoa	PER	2x (2EBN)
11.	*S. anamatophilum* Ochoa	PER	2x (2EBN)
12.	*S. ancophilum* (Correl) Ochoa	PER	2x (2EBN)
13.	*S. ancoripae* Ochoa	PER	2n
14.	*S. andreanum* Baker	COL, ECU	2x (2EBN)
15.	*S.* × *arahuayum* Ochoa	PER	2x
16.	*S. ariduphilum* Ochoa	PER	2x (2EBN)
17.	*S. arnezii* Cardenas	BOL	-
18.	*S. augustii* Ochoa	PER	2x (1EBN)
19.	*S. avilesii* Hawkes and Hjert	BOL	2x
20.	*S.* ayacuchense Ochoa	PER	2x (2EBN)
21.	*S. aymaraesense* Ochoa	PER	2x
22.	*S. berthaultii* Hawkes	BOL	2x (2EBN)
23.	*S. bill-hookeri* Ochoa	PER	2x
24.	*S.* × *blanco-galdosii* Ochoa	PER	2x (2EBN)
25.	*S. boliviense* Dunal	BOL	2x (2EBN)
26.	*S. bombycinum* Ochoa	BOL	4x
27.	*S. brachistotrichum* (Bitter) Rydb.	MEX	2x (1EBN)
28.	*S. brachycarpum* Correll	MEX	6x (4EBN)
29.	*S. brevicaule* Bitter	BOL	2x (2EBN)
30.	*S.* × *bruecheri* Correll	ARG	-
31.	*S. buesii* Vargag	PER	2x (2EBN)
32.	*S. bukasovii* Juz.	PER	2x (2EBN)
33.	*S. bulbocastanum* Dunal	MEX	2x (1EBN)
34.	*S. burkartii* Ochoa	PER	2x
35.	*S. burtonii* Ochoa	ECU	3x
36.	*S. cajamarquense* Ochoa	PER	2x (1EBN)
37.	*S. calacalinum* Ochoa	ECU	2x
38.	*S. calvescens* Bitter	BRA	3x
39.	*S. candolleanum* P. Berthault	BOL, PER	2x (2EBN)
40.	*S. cantense* Ochoa	PER	2x (2EBN)
41.	*S. cardiophyllum* Lindl.	MEX	2x (1EBN), 3x
42.	*S. chacoense* Bitter	ARG, BOL, PAR, URU	2x (2EBN)
43.	*S. chancayense* Ochoa	PER	2x (1EBN)

Table 1 contd. ...

...Table 1 contd.

S. No.	Potato species	Origin country	Ploidy/EBN
44.	*S. chilliasense* Ochoa	ECU	2x (2EBN)
45.	*S. chillonanum* Ochoa	PER	2x
46.	*S. chiquidenum* Ochoa	PER	2x (2EBN)
47.	*S. chomatophilum* Bitter	ECU, PER	2x (2EBN)
48.	*S. circaeifolium* Bitter	BOL	2x (1EBN)
49.	*S. clarum* Correll	MEX, GUA	2x
50.	*S. coelestipetalum* Vargas	PER	2x (2EBN)
51.	*S. colombianum* Bitter	COL, ECU, YEN	4x (2EBN)
52.	*S. commersonii* Dunal	ARG, BRA, URU	2x (1EBN)
53.	*S. contumazaense* Ochoa	-	2x (2EBN)
54.	*S. demissum* Lindl.	GUA, MEX	6x (4EBN)
55.	*S.* × *doddsii* Correll	BOL	2x(2EBN)
56.	*S. dolichocremastrum* Bitter	PER	2x(IEBN)
57.	*S. donachui* (Ochoa) Ochoa	COL	-
58.	*S.* × *edinense* P. Berthault	MEX	5x
59.	*S. etuberosum* Lindl.	CHL	2x (1EBN)
60.	*S. fendleri* A. Gray	MEX, USA	4x (2EBN)
61.	*S. fernandezianum* Phil.	CHL	2x (1EBN)
62.	*S. flahaultii* Bitter	COL	4x
63.	*S. flavoviridens* Ochoa	BOL	-
64.	*S. gandarillasii* Cardenas	BOL	2x (2EBN)
65.	*S. garcia-barrigae* Ochoa	COL	-
66.	*S. gracilifrons* Bitter	PER	2x
67.	*S. guerreroense* Correll	MEX	6x (4EBN)
68.	*S. guzmanguense* Whalen and Sagast	PER	2x (1EBN)
69.	*S. hastiforme* Correll	PER	2x (2EBN)
70.	*S. hintonii* Correl	MEX	-
71.	*S. hjertingii* Hawkes	MEX	4x (2EBN)
72.	*S. hoopesii* Hawkes and K.A. Okada	BOL	4x
73.	*S. hougasii* Correl	MEX	6x (4EBN)
74.	*S. huancabambense* Ochoa	PER	2x (2EBN)
75.	*S. huancavelicae* Ochoa	PER	2x (2EBN)
76.	*S. huarochiriense* Ochoa	PER	2x (2EBN)
77.	*S. humectophilum* Ochoa	PER	2x (1EBN)
78.	*S. hypacrarthrum* Bitter	PER	2x (1EBN)
79.	*S. immite* Dunal	PER	2x (1EBN), 3x
80.	*S. incahuasinum* Ochoa	PER	2x (1EBN)
81.	*S. incamayoense* K.A. Okada and A. M. Clausen	ARG	2x
82.	*S. incasicum* Ochoa	PER	2x (2EBN)
83.	*S.* × *indunii* K.A. Okada and A.M. Clausen	ARG	3x
84.	*S. injundibuliforme* Phil.	ARG,BOL	2x (2EBN)
85.	*S. ingijolium* Ochoa	PER	2x (1EBN)
86.	*S. iopetalum* (Bitter) Hawkes	MEX	6x (4EBN)

Table 1 contd. ...

...Table 1 contd.

S. No.	Potato species	Origin country	Ploidy/EBN
87.	*S. irosinum* Ochoa	PER	2x (2EBN)
88.	*S. jaenense* Ochoa	PER	6x (4EBN)
89.	*S. jalcae* Ochoa	PER	2x (2EBN)
90.	*S. jamesii* Torr.	MEX, USA	2x (1EBN)
91.	*S. kurtzianum* Bitter and Wittm.	ARG	2x (2EBN)
92.	*S. laxissimum* Bitter	PER	2x (2EBN)
93.	*S. leptophyes* Bitter	BOL,PER	2x (2EBN)/ 4x (4EBN)
94.	*S. leptosepalum* Correl	MEX,USA	-
95.	*S. lesteri* Hawkes and Hjert.	MEX	2x
96.	*S. lignicaule* Vargas	PER	2x (1EBN)
97.	*S. limbaniense* Ochoa	PER	2x (2EBN)
98.	*S.* × *litusinum* Ochoa	BOL	2x (2EBN)
99.	*S. lobbianum* Bitter	COL	4x (2EBN)
100.	*S. longiconicum* Bitter	CRI, PAN	4x
101.	*S. longiusculus* Ochoa	PER	2x
102.	*S. lopez-camarenae* Ochoa	PER	2x (1EBN)
103.	*S. macropilosum* Correl	MEX	-
104.	*S. maglia* Schltdl.	CHL	2x, 3x
105.	*S. marinasense* Vargas	PER	2x (2EBN)
106.	*S. matehualae* Hjert. and T.R. Tarn	MEX	4x
107.	*S. medians* Bitter	PER	2x (2EBN), 3x
108.	*S. megistacrolobum* Bitter	AGR, PER, BOL	2x (2EBN)
109.	*S.* × *michoacanum* (Bitter) Rydb.	MEX	2x
110.	*S. microdontum* Bitter	ARG, BOL	2x (2EBN), 3x
111.	*S. minutifoliolum* Correl	ECU	2x (1EBN)
112.	*S. mochiquense* Ochoa	PER	2x (1EBN)
113.	*S. morellijorme* Bitter and G. Muench	MEX, GUA, HON	2x
114.	*S. moscopanum* Hawkes	COL	6x (4EBN)
115.	*S. multiinterruptum* Bitter	PER	2x (2EBN)
116.	*S. nayaritense* (Bitter) Rydb.	MEX	-
117.	*S. nemorosum* Ochoa	PER	6x (4EBN)
118.	*S. neocardenasii* Hawkes and Hjert.	BOL	2x
119.	*S. neorosii* Hawkes and Hjert.	ARG	2x
120.	*S. neovalenzuelae* L. Lopez	COL	4x
121.	*S. neovargasii* Ochoa	PER	2x
122.	*S. neovavilovii* Ochoa	PER	2x (2EBN)
123.	*S.* × *neoweberbaueri* Wittm	PER	3x
124.	*S. nubicola* Ochoa	PER	4x (2EBN)
125.	*S. okadae* Hawkes and Hjert.	ARG, BOL	2x
126.	*S. olmosense* Ochoa	ECU, PER	2x (2EBN)
127.	*S. oplocense* Hawkes	ARG, BOL	2x (2EBN), 4x (4EBN), 6x (4EBN)

Table 1 contd. ...

...Table 1 contd.

S. No.	Potato species	Origin country	Ploidy/EBN
128.	*S. orocense* Ochoa	COL	-
129.	*S. orophilum* Correll	PER	2x (2EBN)
130.	*S. ortegae* Ochoa	PER	2x
131.	*S. otites* Dunal	COL, VEN	-
132.	*S. oxycarpum* Schiede	MEX	4x (2EBN)
133.	*S. palustre* Poepp.	ARG, CHL	2x (1EBN)
134.	*S. pampasense* Hawkes	PER	2x (2EBN)
135.	*S. pamplonense* L. Lopez	COL	4x
136.	*S. papita* Rydb.	MEX	4x (2EBN)
137.	*S. paucijugum* Bitter	ECU	4x (2EBN)
138.	*S. paucissectum* Ochoa	PER	2x (2EBN)
139.	*S. peloquinianum* Ochoa	PER	2x (2EBN)
140.	*S. pillahuatense* Vargas	PER	2x (2EBN)
141.	*S. pinnatisectum* Dunal	MEX	2x (1EBN)
142.	*S. piurae* Bitter	PER	2x (2EBN)
143.	*S. polyadenium* Greenm.	MEX	2x
144.	*S. polytrichon* Rydb.	MEX	4x (2EBN)
145.	*S. puchupuchense* Ochoa	BOL, PER	2x
146.	*S. raphanifolium* Cardenas and Hawkes	PER	2x (2EBN)
147.	*S. raquialatum* Ochoa	PER	2x (1EBN)
148.	*S.* × *rechei* Hawkes and Hjert.	ARG	2x,3x
149.	*S. regularijolium* Correll	ECU	2x
150.	*S. rhomboideilanceolatum* Ochoa	PER	2x (2EBN)
151.	*S.* × *ruiz-lealii* Briicher	ARG	-
152.	*S. salasianum* Ochoa	PER	2x
153.	*S. x sambucinum* Rydb.	MEX	2x
154.	*S. sanctae-rosae* Hawkes	ARG	2x (2EBN)
155.	*S. sandemanii* Hawkes	PER	2x (2EBN)
156.	*S. santolallae* Vargas	PER	2x (2EBN)
157.	*S. sarasarae* Ochoa	PER	2x (2EBN)
158.	*S. sawyeri* Ochoa	PER	2x (2EBN)
159.	*S. saxatilis* Ochoa	PER	2x (2EBN)
160.	*S. scabrijolium* Ochoa	PER	2x
161.	*S. schenckii* Bitter	MEX	6x (4EBN)
162.	*S.* × *semidemissum* Juz.	MEX	6x
163.	*S.* × *setulosistylum* Bitter	ARG	2x
164.	*S. simplicissimum* Ochoa	PER	2x (1EBN)
165.	*S. soestii* Hawkes and Hjert.	BOL	2x
166.	*S. sogarandinum* Ochoa	PER	2x (2EBN), $2n = 36$
167.	*S. solisii* Hawkes	ECU	-
168.	*S. sparsipilum* (Bitter) Juz. and Bukasov	BOL, PER	2x (2EBN)
169.	*S. spegazzinii* Bitter	ARG	2x (2EBN)
170.	*S. stenophyllidium* Bitter	MEX	2x (1EBN)

Table 1 contd. ...

...Table 1 contd.

S. No.	Potato species	Origin country	Ploidy/EBN
171.	*S. stoloniferum* Schltdl. and Bouchet	MEX	4x (2EBN)
172.	*S. subpanduratum* Ochoa	VEN	4x
173.	*S.* × *sucrense* Hawkes	BOL	4x (4EBN)
174.	*S. sucubunense* Ochoa	COL	-
175.	*S. tacnaense* Ochoa	PER	2x (2EBN)
176.	*S. tapojense* Ochoa	PER	2x (2EBN)
177.	*S. tarapatanum* Ochoa	PER	2x
178.	*S. tarijense* Hawkes	AGR, BOL	2x (2EBN)
179.	*S. tarnii* Hawkes and Hjert.	MEX	2x
180.	*S. taulisense* Ochoa	PER	2x (2EBN)
181.	*S. trifidum* Correll	MEX	2x (1EBN)
182.	*S. trinitense* Ochoa	PER	2x (1EBN)
183.	*S. tundalomense* Ochoa	ECU	6x (4EBN)
184.	*S. tuquerrense* Hawkes	COL, ECU	4x (2EBN)
185.	*S. ugentii* Hawkes and K.A. Okada	BOL	4x
186.	*S. urubambae* Juz.	PER	2x (2EBN)
187.	*S.* × *vallis-mexici* Juz.	MEX	3x
188.	*S. velardei* Ochoa	PER	2x
189.	*S. venturii* Hawkes and Hjert.	AGR	2x (2EBN)
190.	*S. vernei* Bitter and Wittm.	AGR	2x (2EBN)
191.	*S. verrucosum* Schltdl.	-	2x (2EBN)
192.	*S. vidaurrei* Cardenas	AGR, BOL	2x (2EBN)
193.	*S.* × *viirsoir:* K.A. Okada and A.M. Clausen	AGR	3x
194.	*S. violaceimarmoratum* Bitter	BOL	2x (2EBN)
195.	*S. virgultornm* (Bitter) Cardenas and Hawkes	BOL	2x
196.	*S. wittmackii* Bitter	PER	2x (1EBN)
197.	*S. woodsonii* Correll	PAN	-
198.	*S. yamobambense* Ochoa	PER	2x
199.	*S. yungasense* Hawkes	BOL, PER	2x (1EBN)
Cultivated Species			
200.	*S. ajanhuiri* Juz. and Bukasov	BOL, PER	2x
201.	*S. chaucha* Juz. and Bukasov	BOL, PER	3x
202.	*S. curtilobum* Juz. and Bukasov	BOL, PER	4x (4 EBN), most 5x
203.	*S. juzepczukii* Bukasov	BOL, PER	3x
204.	*S. phureja* Juz. and Bukasov subsp. *Phureja*	Widespread Andes	2x (2 EBN)
205.	*S. stenotomum* Juz. and Bukasov subsp. *Stenotomum*	BOL, PER	2x (2 EBN)
206.	*S. stenotomum* subsp. *goniocalyx* (Juz. and Bukasov) Hawkes	PER	2x (2 EBN)

Table 1 contd. ...

...Table 1 contd.

S. No.	Potato species	Origin country	Ploidy/EBN
207.	*S. tuberosum* L. subsp. *andigena* (Juz. and Bukasov) Hawkes	Widespread Andes	4x (4 EBN)
208.	*S. tuberosum* L. subsp. *tuberosum*	CHL (native), introduced worldwide and/or evolved from *andigena* in Europe	4x (4 EBN)

Origin country abbreviations: ARG - Argentina, BOL - Bolivia, BRA - Brazil, CHL - Chile, COL - Colombia, CRI - Costa Rica, ECU - Ecuador, GTM - Guatemala, HON - Honduras, MEX- Mexico, PAN - Panama, PER - Peru, USA - United States of America, and VEN - Venezuela.
Source: Modified and adapted from Spooner and Hijmans (2001)

Somatic hybridization technology offers the following advantages:

- Somatic hybrids can be produced between species, which cannot be hybridized sexually.
- Somatic hybrids can be readily used in the breeding programme for transfer of resistance genes.
- Hybrids can be produced even between such clones, which are completely sterile.
- Cytoplasm transfer can be done in one year, while back crossing may take five to six years. Even where backcrossing is not applicable, cytoplasm transfer can be made by using this approach.
- Mitochondria of one species can be combined with chloroplast of another species. This may be very important in some cases and is not achievable by sexual means, even between easily crossable species.
- Recombinant organelle genomes, especially of mitochondria, are generated in somatic hybrids. Some of these recombinant genomes may possess useful features.

However, somatic hybridization technology has the following limitations:

- Techniques for protoplast isolation, culture and fusion are very complicated.
- In many cases, chromosome elimination occurs from somatic hybrids leading to asymmetric hybrids. Such hybrids may be useful but there is no control on chromosome elimination.
- Many somatic hybrids show genetic instability, which may be an inherent feature of some species combinations.
- Many somatic hybrids either do not regenerate or give rise to sterile regenerants. Such hybrids are useful in crop improvement. All inter-family somatic hybrids are genetically unstable and/or morphologically abnormal, while intergeneric and intertribal hybrids are genetically stable, but produce abnormal and/or sterile plants.

3. Somatic Hybridization Strategies

3.1 Protoplast Isolation and Fusion

Cocking (1960) first reported protoplast isolation and Takebe et al. (1971) successfully regenerated plants from tobacco protoplasts. In the 1980s, the development of regenerants was more prevalent with the discovery of electrofusion method of protoplasts (Bates et al., 1987; Fish et al., 1988) or might be due to easy control of fusion parameters than the chemical method using polyethylene glycol (Wallin et al., 1974). The electrofusion method was first reported in potato between *Solanum tuberosum* and semi-cultivated (*S. phureja*) (Puite et al., 1986). Leaf-mesophyll cells of *in vitro*-grown plants were used largely to isolate protoplasts in protoplast-digestion medium (Binding et al., 1978). Then culture of fused products in the VKM medium (Binding and Nehls, 1977) was followed by shoot development on the $MS_{13}K$ medium (Behnke, 1975). Selection of somatic hybrids (*S. tuberosum* + *S. chacoense*), based on callus growth tagged with green fluorescent protein, was also observed (Rakosy-Tican and Aurori, 2015a).

3.2 Characterization of Somatic Hybrids

A large number of potato somatic hybrids have been produced via protoplast fusion between the common potato and wild species, such as *Solanum acaule*, *S. berthaultii*, *S. brevidens*, *S. bulbocastanum*, *S. cardiophyllum*, *S. chacoense*, *S. circaefolium*, *S. commersonii*, *S. etuberosum*, *S.* × *michoacanum*, *S. melongena*, *S. nigrum*, *S. phureja*, *S. pinnatisectum*, *S. tuberosum*, *S. sanctae-rosae*, *S. spegazzinii*, *S. stenotomum*, *S. tarnii*, *S. torvum*, *S. vernei*, *S. verrucosum*, and *S. villosum* (Table 2). Among them, *S. brevidens*, *S. commersonii*, *S. etuberosum* and *S. villosum* are the non-tuberous wild species. Putative somatic hybrid regenerants are characterized by using many methods, like cytological (ploidy analysis by flow cytometry, chromosome count, guard cell count, FISH-fluorescence *in situ* hybridization and GISH-genomic *in situ* hybridization), isozyme, molecular markers, like RAPD (random amplified polymorphic DNA), RFLP (restriction fragment length polymorphism), ISSR (inter simple sequence repeat), SSR (simple sequence repeat), AFLP (amplified fragment length polymorphism) and DArT (diversity array technology), morphological and pollen fertility. Somatic hybrids were analyzed for cytoplasm types (W/α, T/β, W/γ, W/δ and S/ε: Lössl et al., 1999), based on organelle (chloroplast and mitochondria) genome-specific markers as described by Lössl et al. (2000). Finally, somatic hybrids are evaluated for their fertility behavior and development of potato hybrids with desirable traits, along with the target traits from donor wild species. In the past six decades, numerous fusion experiments were attempted and over hundreds of potato somatic hybrids were produced and characterized (Table 2).

Table 2. Summary of production and characterization of potato somatic hybrids developed using *Solanum* species.

Sr. No.	Fusion parents (*Solanum* species)		Target trait	Characterization	References
1.	*S. acaule* dihaploid (2*x*)	*S. tuberosum* dihaploid (2*x*)	Potato virus X (PVX) resistance	Chromosome counting, RFLP, segregation pattern and BC_1 family	Yamada et al. (1997, 1998)
	S. acaule (4*x*) and dihaploid (2*x*)	*S. tuberosum* (4*x*) and *S. t.* dihaploid (2*x*)	Resistance to bacterial ring rot (*Clavibacter* spp.) and potato leaf roll virus (PLRV)	RAPD, flow cytometry, phenotypes, glycoalkaloid aglycone content, anther culture and BC_1 family	Rokka et al. (1998, 2005)
	S. acaule dihaploid (2*x*)	*S. tuberosum* (4*x*)	Glycolkaloids	Phenotypes	Kozukue et al. (1999)
2.	*S. berthaultii* (2*x*)	*S. tuberosum* dihaploid (2*x*)	Resistance to insects, PVY, soil-borne pathogens (*Fusarium*, *Pythium* and *Rhizoctonia* spp.) and salinity tolerance	Isozyme, cytology, ISSR, cytoplasmic DNA analysis, flow cytometry, phenotypes, nutrients content (Na, Cl and K) and tuber traits	Serraf et al. (1991), Bidani et al. (2007), Nouri-Ellouz et al. (2016)
3.	*S. brevidens* (2*x*)	*S. tuberosum* (4*x*) and *S. t.* dihaploid (2*x*)	PVY, PLRV, late blight, early blight, bacterial soft rot (*Erwini* spp.) and frost resistance	Chromosome counting, isozyme analysis, RFLP, phenotypes, RAPD analysis, back cross progenies (BC_1, BC_2 and BC_3), RFLP and FISH analyses and field trials	Austin et al. (1985a, 1985b, 1986, 1988), Gibson et al. (1988), Fish et al. (1988), Rokka et al. (1994), Polgar et al. (1999), Chen et al. (1999b), Tek et al. (2004)
	S. brevidens (2*x*)	*S. tuberosum* (4*x*)	PLRV and cold-stress resistance	Isozyme analysis, chromosome staining, phenotypes, multivariate analysis, X-ray irradiation, southern hybridization and phenotypes	Preiszner et al. (1991), Fehér et al. (1992), Xu et al. (1993)
	S. brevidens (2*x*)	*S. tuberosum* (4*x*)	Bacterial soft rot (*Erwini* spp.) resistance	Segregation analysis, back-cross progenies (BC_1, BC_2 and BC_3), and fluorescence *in situ* hybridization (FISH)	McGrath et al. (1996, 2002), McGrath and Helgeson (1998)
	S. brevidens (2*x*)	*S. tuberosum* (4*x*)	Common scab (*Streptomyces* spp.) resistance	Chromosome counting, BC_1 and BC_2 progenies, and phenotypes	Ahn and Park (2013)
	S. brevidens (2*x*)	*S. tuberosum* (4*x*)	Glycoalkaloids	GISH and Gas chromatography-mass spectrometry (GC-MS) analyses and genomic composition of somatic hybrids	Laurila et al. (2001)

4.	*S. bulbocastanum* (2*x*)	*S. tuberosum* (4*x*)	Nematode (*Meloidogyne* spp.) resistance	Chromosome counting, isozyme analysis, phenotypes, crossability and tuber traits	Austin et al. (1993), Mojtahedi et al. (1995)
	S. bulbocastanum (2*x*)	*S. tuberosum* haploid (2*x*)	Late blight resistance	X-ray irradiation, flow cytometry and RFLP	Oberwalder et al. (1997, 1998)
	S. bulbocastanum (2*x*)	*S. tuberosum* (4*x*)	Late blight resistance	RFLP, RAPD, and back cross progenies (BC_1 and BC_2) and gene mapping	Helgeson et al. (1998), Naess et al. (2000, 2001)
	S. bulbocastanum (2*x*)	*S. tuberosum* (4*x*)	Late blight resistance	Cytology (chromosome counting and GISH), flow cytometry, AFLP and SSR analyses, fertility, phenotypes and hybridization	Rakosy-Tican et al. (2015b)
	S. bulbocastanum (2*x*)	*S. tuberosum* dihaploid (2*x*)	Late blight resistance	Plant regeneration, RAPD, flow cytometry and phenotypes	Szczerbakowa et al. (2000, 2001, 2003a), Bołtowicz et al. (2005), Greplová et al. (2008)
	S. bulbocastanum (2*x*)	*S. tuberosum* dihaploid (2*x*)	Late blight resistance	ISSR, cytoplasmic DNA, genomic *in situ* hybridization (GISH), ploidy analysis and anther culture	Iovene et al. (2007, 2012), Aversano et al. (2009)
5.	*S. cardiophyllum* (2*x*)	*S. tuberosum* (4*x*)	Late blight resistance	RAPD, chromosome counting and phenotypes	Shi et al. (2006)
	S. cardiophyllum (2*x*)	*S. tuberosum* (4*x*)	Late blight, PVY and Colorado potato beetle resistance	SSR, AFLP, flow cytometry, phenotypes and fertility of hybrids	Thieme et al. (2004, 2010)
	S. cardiophyllum (2*x*)	*S. tuberosum* dihaploid (2*x*)	Late blight resistance	RAPD, ISSR, SSR, AFLP, cytoplasm type (chloroplast and mitochondrial genomes), flow cytometry, phenotypes, fertility of hybrids and field evaluation	Chandel et al. (2015), Luthra et al. (2018)
6.	*S. chcoense* (2*x*)	*S. tuberosum* dihaploid (2*x*)	Colorado potato beetle resistance	Morphological, biochemical (anthocyanin pigment) and isozyme markers	Cheng et al. (1995)
	S. chcoense (2*x*)	*S. tuberosum* (4*x*)	Bacterial wilt (*Ralstonia* spp.) resistance	RAPD, flow cytometry, nuclear and cytoplasmic genomes, genomic stability, meiotic behaviour of pollen mother cells and SSR	Cai et al. (2004), Guo et al. (2010), Chen et al. (2013)

Table 2 contd. ...

...Table 2 contd.

Sr. No.	Fusion parents (*Solanum* species)		Target trait	Characterization	References
7.	*S. circaefolium* (2*x*)	*S. tuberosum* dihaploid (2*x*)	Late blight resistance	Flow cytometer, RFLP analyses and fertility of hybrids	Oberwalder et al. (1997, 2000)
8.	*S. commersonii* (2*x*)	*S. tuberosum* dihaploid (2*x*)	Verticillium wilt, tuber soft rot (*Erwinia* spp.) and frost resistance	RAPD, ploidy (chloroplast count in stomata guard cell, chromosome counts and flow cytometer), phenotypes, fertility, cell sorting, southern analysis and organelle DNA, BC_1 and F_2 progenies, multivariate analysis, AFLP and organelles DNA	Cardi et al. (1993, 1999, 2002), Cardi (1998), Nyman and Waara (1997), Waara et al. (1998), Bastia et al. (2000, 2001, 2002), Carputo et al. (2000), Barone et al. (2002), Scotti et al. (2003, 2004)
	S. commersonii (2*x*)	*S. tuberosum* dihaploid (2*x*)	Bacterial wilt resistance and cold-stress/freezing tolerance	RAPD, isozyme, chromosome counting, selfing/crossing, back cross progenies (BC_1 and BC_2) and phenotypes	Kim et al. (1993), Laferriere et al. (1999), Kim-Lee et al. (2005), Chen et al. (1999a, 1999c)
9.	*S. etuberosum* (2*x*)	*S. tuberosum* dihaploid (2*x*)	PVY, PLRV and green peach aphid resistance	Genomic and chloroplast DNA, RFLP, GISH, cytology, phenotypes, tuber characteristics, back cross progenies (BC_1, BC_2 and BC_3) and field trials	Novy and Helgeson (1994a, 1994b), Dong et al. (1999), Novy et al. (2002, 2007), Gillen and Novy (2007)
	S. etuberosum (2*x*)	*S. tuberosum* dihaploid (2*x*)	PVY resistance	Isozyme, SSR markers, GISH and back-cross progenies (BC_1 and BC_2) and flow cytometry	Thieme et al. (1999, 2004), Gavrilenko et al. (2003)
	S. etuberosum (2*x*)	*S. tuberosum* dihaploid (2*x*)	PVY resistance	RAPD, SSR, flow cytometry, cytoplasm type and phenotypes	Tiwari et al. (2010, 2015c)
10.	*S. malmeanum* (2*x*)	*S. tuberosum* dihaploid (2*x*)	Freezing tolerance	Ploidy, plant morphology, SSR and freezing tolerance	Tu et al. (2021)
11.	*S. melongena* (2*x*)	*S. tuberosum* dihaploid (2*x*)	Bacterial wilt resistance	Ploidy level (chromosome count and flow cytometry), SSR, GISH karyotypes and cytoplasmic DNA	Yu et al. (2013)
12.	*S.* × *michoacanum* (2*x*)	*S. tuberosum* dihaploid (2*x*)	Late blight resistance	RAPD, SSR, CAPS, ploidy level (guard cell count and flow cytometry), phenotypes, fertility, and F_1 and BC_1 progenies	Szczerbakowa et al. (2010), Smyda et al. (2013), Smyda-Dajmund et al. (2017)

13.	*S. nigrum* (6*x*)	*S. tuberosum* dihaploid (2*x*)	Late blight resistance	Plant regeneration, RAPD, flow cytometry, phenotypes and crossing	Szczerbakowa et al. (2000, 2001, 2003b, 2011), Zimnoch-Guzowska et al. (2003)
	S. nigrum (6*x*)	*S. tuberosum* (4*x*)	Atrazine resistance	Chromosome counting and phenotypes	Binding et al. (1982)
	S. nigrum (6*x*) complex species	*S. tuberosum* (4*x*) and dihaploid (2*x*)	-	Isozyme, flow cytometry, selectable markers and phenotypes	Horsman et al. (1997)
14.	*S. phureja* (2*x*)	*S. tuberosum* dihaploid (2*x*)	Late blight resistance	Fluorescein diacetate staining, cytology, chromosome number and phenotypes	Puite et al. (1986), Mattheij and Puite (1992)
	S. phureja (2*x*)	*S. tuberosum* dihaploid (2*x*)	Bacterial wilt resistance	Isozyme, RAPD, SSR, flow cytometry, chloroplast genome and phenotypes	Fock et al. (2000)
	S. phureja monoploids (1*x*)	*S. phureja* monoploids (1*x*)	Long photoperiods	SSR, flow cytometry, field evaluations, Sequence-Specific Amplification Polymorphism (S-SAP) and retrotransposon-based markers	Lightbourn and Veilleux (2007), Lightbourn et al. (2007)
15.	*S. pinnatisectum* (2*x*)	*S. tuberosum* dihaploid (2*x*)	Late blight resistance	Flow cytometry, RFLP, phenotypes and tuber traits	Menke et al. (1996)
	S. pinnatisectum (2*x*)	*S. tuberosum* dihaploid (2*x*) and *S. phureja* (2*x*)	Late blight resistance	γ-irradiation, isozyme analysis, chromosome counting and chloroplast DNA	Sidorov et al. (1987)
	S. pinnatisectum (2*x*) & hybrid clone (*S. pinnatisectum* × *S. bulbocastanum*)	*S. tuberosum* dihaploid (2*x*)	Late blight resistance	Isozyme, RAPD, flow cytometry, cytological, phenotypes, crossability and fertility	Thieme et al. (1997)
	S. pinnatisectum (2*x*)	*S. tuberosum* dihaploid (2*x*)	Late blight resistance	RAPD, cytological and phenotypes	Szczerbakowa et al. (2005)
	S. pinnatisectum (2*x*)	*S. tuberosum* dihaploid (2*x*)	Late blight resistance	RAPD, flow cytometry and phenotypes	Greplová et al. (2008)

Table 2 contd. ...

...Table 2 contd.

Sr. No.	Fusion parents (*Solanum* species)		Target trait	Characterization	References
	S. pinnatisectum (2*x*)	*S. tuberosum* (4*x*)	Late blight resistance	SSR, organelle DNA analysis and phenotypes	Polzerová et al. (2011)
	S. pinnatisectum (2*x*)	*S. tuberosum* dihaploid (2*x*)	Late blight resistance	Efficient electrofusion system, RAPD, SSR, cytoplasm type, flow cytometry, phenotypes, field trial, breeding potential and crossing (BC_1 progeny)	Sharma et al. (2011), Sarkar et al. (2011), Tiwari et al. (2013a), Luthra et al. (2016)
16.	*S. tuberosum* dihaploid (2*x*)	*S. tuberosum* dihaploid (2*x*)	PVX and PVY resistance	Chromosome counting, isozyme and phenotypes	Austin et al. (1985b), Thach et al. (1993)
	S. tuberosum (4*x*)	*S. tuberosum* (4*x*)	Cytoplasmic male sterility	Phenotypes, flower morphology and true potato seed	Perl et al. (1990)
	S. tuberosum dihaploid (2*x*)	*S. tuberosum* dihaploid (2*x*)	Resistance to *Globodera* spp. and metribuzin herbicide	Isozyme and phenotypic markers	Möllers and Wenzel (1992)
	S. tuberosum dihaploid (2*x*)	*S. tuberosum* dihaploid (2*x*)	Resistance to potato cyst nematode and late blight	Phenotypic markers, chromosome counting, flower morphology and RAPD analyses	Cooper-Bland et al. (1994), Rasmussen et al. (1996, 1998)
	S. tuberosum dihaploid (2*x*)	*S. tuberosum* dihaploid (2*x*)	-	Isozyme, chromosome counting, RAPD analysis and phenotypes	Gavrilenko et al. (1999), Przetakiewicz et al. (2007)
	S. tuberosum dihaploid (2*x*)	*S. tuberosum* dihaploid (2*x*)	Resistance to PLRV, PVY, late blight and soft rot (*Erwiniai* spp.)	RAPD analysis, chromosome counting, flow cytometry, phenotypes, fertility and back crossing	Rokka et al. (1994, 2000), Valkonen and Rokka (1998)
	S. tuberosum dihaploid (2*x*)	*S. tuberosum* dihaploid (2*x*)	Resistance to PVY and storage rot (*Pythium aphanidermatum*)	Isoenzyme analysis, chromosome counting, microtuberization, SSR, ISSR and organelle DNA	Nouri-Ellouz et al. (2006)
17.	*S. sanctae-rosae* (2*x*)	*S. tuberosum* (4*x*)	Potato cyst nematode resistance	SSR, RFLP mitochondrial profiles and organelle genomes	Harding and Millam (2000)
18.	*S. spegazzinii* (2*x*) & hybrid clone (2*x*) (*S. microdontum* x *S. vernei*)	*S. tuberosum* (4*x*) and dihaploid (2*x*)		X-ray irradiation, micromanipulation, fluorescence-activated cell sorting (FACS), cytoplasmic genomes and RAPD	Rasmussen et al. (1997, 2000)

19.	*S. stenotomum* (2*x*)	*S. tuberosum* dihaploid (2*x*)	Bacterial wilt resistance	Ploidy level, pollen viability, isozyme, SSR, chloroplast DNA, protein, carboxylase assay, Western blot, microtuberization, sucrose and glucose	Fock et al. (2001)
20.	*S. tarnii* (2*x*)	*S. tuberosum* (4*x*)	Late blight, Colorado potato beetle and PVY resistance	Flow cytometric, SSR, AFLP and BC_1 progenies	Thieme et al. (2004, 2008)
21.	*S. torvum* (2*x*)	*S. tuberosum* dihaploid (2*x*)	*Verticillium* wilt resistance	Isozyme, morphological markers and phenotypes	Jadari et al. (1992)
22.	*S. vernei* (2*x*)	*S. tuberosum* dihaploid (2*x*)	Salt tolerance	Isoenzyme, RAPD, ISSR, organelle DNA, flow cytometry and phenotypes	Trabelsi et al. (2005)
23.	*S. verrucosum* (2*x*)	*S. tuberosum* dihaploid (2*x*)	Resistance to PLRV	RAPD, chloroplast counting, phenotypes and field trials	Carrasco et al. (2000)
24.	*S. villosum* (4*x*)	*S. tuberosum* dihaploid (2*x*)	Late blight resistance	RAPD, GISH and reactive oxygen species (ROS) production by pathogen	Tarwacka et al. (2013)

4. Application of Somatic Hybrids

4.1 Establishment of In Vitro Conservation Protocol

Genetic fidelity of *in vitro* propagated potato microtubers has been assessed using molecular markers (RAPD, ISSR, SSR and AFLP). We have demonstrated that genetic stability of *in vitro* conserved microplants for three years on MS medium supplemented with 20 g/L sucrose, 40 g/L sorbitol and 7 g/L agar at low temperature (7 ± 1°C) under light and temperature-controlled conditions is a safe protocol for regeneration of true-to-type potato genotypes. In another study, genetic and epigenetic changes (DNA methylation) have been examined by Amplified Fragment Length Polymorphism (AFLP) and Methylation-Sensitive Amplified Polymorphism (MSAP) in the tissue culture-propagated somatic hybrids. Study showed that *in vitro* propagated somatic hybrids could be multiplied 'true-to-type' up to 30th cycles of sub-culturing from micropropagated plants for their future use in potato breeding (Tiwari et al., 2013c, 2015f).

4.2 Genetics

Genetics of potato somatic hybrids and their segregating progenies have been studied in nuclear genome to dissect the recombination patterns. Chromosomal segregation pattern was analyzed in hexaploid somatic hybrid (*S. brevidens* + *S. tuberosum*) and their progenies with *S. tuberosum*, using RFLP (Williams et al., 1993) and RAPD markers (McGrath et al., 1996). The study suggests that *S. brevidens* ribosomal (r) DNA loci are primarily contributed to isochromosome formation in the hybrids and progenies (McGrath and Helgeson, 1998). Further, a chromosome substitution line was developed in *S. breviden*-somatic hybrids and progenies (BC_1, BC_2 and BC_3) with *S. tuberosum*. The study further demonstrates that a single copy of chromosome 8 from *S. brevidens* replaced the same in the BC_3 clone and has significant impact on transferring resistance to tuber soft rot and early blight (Tek et al., 2004). The importance of combining GISH and DNA markers was also suggested to study chromosomal behavior in potato (Dong et al., 1999). Poor chromosomal pairing was observed in somatic hybrid (*S. etuberosum* + *S. tuberosum*) and progenies (BC_1 and BC_2), and further suggests that genome dosage affects tuber formation but has less effect on potato virus Y (PVY) resistance (Gavrilenko et al., 2003). Evidence of tetrasomic inheritance was investigated in a tetraploid somatic hybrid (*S. commersonii* + *S. tuberosum*) and F_2 progeny (90 individuals), using RAPD and AFLP markers. Segregation pattern was investigated by RFLP markers in a progeny of hexaploid somatic hybrids (*S. acaule* + *S. tuberosum*) crossed with *S. tuberosum*. The study concludes that somatic hybridization allows an effective use of *S. acaule* genes into cultivated potato (Yamada et al., 1998). Recently, somatic hybrid (*S. chacoense* + *S. tuberosum*) exhibited tetrasomic or disomic segregation ratio using SSR markers and suggested that pentaploid hybrid exhibits tetraploid inheritance pattern (Chen et al., 2016).

Interaction between nuclear and cytoplasmic genes can affect fertility and agronomic traits of somatic hybrids and progenies (Lössl et al., 1994). Segregation and recombination patterns of organelle genomes were investigated in potato (Frei et al., 1998) and analyzed for variations (Tiwari et al., 2014, 2016). Majority of somatic hybrids follow recombination of mitochondrial genome from both parents while the chloroplast pattern from only one parent, such as *S. bulbocastanum* (Iovene et al., 2007), *S. pinnaticetum* (Sarkar et al., 2011) and *S. chacoense* (Chen et al., 2013) except that recombination of chloroplast genome was observed only once in *S. vernei*-somatic hybrid (Trabelsi et al., 2005). Loss of male sterility was observed in nuclear-mitochondrial genomes re-arrangement in *S. commersonii*-somatic hybrids and BC_1 progeny (Cardi et al., 1999). The study suggests the possibility of exploitation of novel cytoplasm in potato breeding (Scotti, 2003), especially variation in a hot spot mitochondrial region (*rpl5–rps14*) (Scotti et al., 2004). Recently, random and non-random segregations of organelle genomes were observed in somatic hybrid (*S.* × *michoacanum* + *S. tuberosum*), using DArT markers (Smyda-Dajmund et al., 2016).

In search of new genes, a few potato somatic hybrids and their progenies were exploited in linkage mapping studies. The *RB* gene (*Rpi-blb1*) originates from diploid wild species, *S. bulbocastanum* of somatic hybrid (*S. bulbocastanum* + *S. tuberosum*) and confers durable resistance to late blight (Helgeson et al., 1998; Song et al., 2003). The *RB* gene was mapped to potato chromosome 8 through analysis of somatic hybrid progenies (BC_1 and BC_2) (Naess et al., 2000, 2001). Besides, genetic stability of *in vitro* plants of somatic hybrids (*S. tuberosum* dihaploid 'C-13' + *S. pinnatisectum*, and 'C-13' + *S. etuberosum*) was confirmed, using methylation-sensitive amplified polymorphism (MSAP) and AFLP markers (Tiwari et al., 2013b, 2015d, 2015e).

4.3 Molecular Markers and Breeding

Improvement of somatic hybrids is essential for desirable agronomic traits through breeding methods to decrease the undesirable effects of wild species. In addition, transfer of disease/pest resistance traits from somatic hybrids to progenies is also important. Assessment of genetic and phenotypic variations among somatic hybrids (Gavrilenko et al., 1999) and development of their advanced progenies, such as F_2, BC_1, BC_2 and BC_3, led to effective utilization somatic hybrids in potato breeding (Table 1). SSR alleles were identified for bacterial wilt-resistance breeding in *S. chacoense*-somatic hybrids and backcross progenies (Chen et al., 2016). Diverse cytoplasm (chloroplast and mitochondrial genomes) types were observed in inter-specific potato somatic hybrids, using organelles (chloroplast and mitochondria) genome-specific molecular markers. The study showed that our somatic hybrids possess W/α, W/γ and T/β types of diverse cytoplasm. Thus, somatic hybridization has the unique potential to widen the cytoplasm types of the cultivated gene pools from wild species through introgression by breeding methods. SSR alleles linked to late blight resistant somatic hybrid have been identified for molecular breeding, using somatic hybrids as parental material. In another study, we profiled somatic hybrids, using SSR markers and identified markers linked to potato somatic hybrid-derived progenies (Tiwari et al., 2020). A total of 50 breeding lines (parents and progenies) were analyzed, using SSR markers and distinguished the parents, i.e., somatic hybrid P8 (*Solanum tuberosum* dihaploid C-13 + wild species *S. pinnatisectum*) and potato cv. Kufri Jyoti, and their progenies (MSH-14/112, MSH-14/113, MSH-14/114, MSH-14/115, MSH-14/116, MSH-14/122 and MSH-14/123). STM0003 showed three distinct alleles (103, 132 and 144 bp), where both P8 and progenies contained 103 and 144 bp, while Kufri Jyoti had 132 and 144 bp alleles. On the other hand, STI0001 distinguished progenies, namely MSH/17-16 (Kufri Garima × Crd10), MSH/17-25 (Kufri Garima × P10), and MSH/17-27 (Kufri Jyoti × Crd 16) with respect to their parents and STI0001 contained six alleles (169, 172, 175, 178, 184 and 188 bp). The study suggests that STM0003 and STI0001 are diagnostic markers to identify these somatic hybrid derived progenies and parents (Tiwari et al., 2022).

Somatic hybrids performed better in field trials in terms of tuber traits and phenotypes (Carrasco et al., 2000) and a few hybrids produced higher yield, tuber number and weight than parents (Möllers et al., 1994). Tuber yield per plant of back-crossed progenies of somatic hybrid improved considerably than that of parents (Carputo et al., 2000). Recently, potential *S. pinnatisectum*-somatic hybrids (P4, P8 and P10) were selected for adaptability, tuber traits, late blight resistance and keeping quality traits in the sub-tropical plains of India, where nearly 90% of potato is grown (Luthra et al., 2016). Further, hybrid progeny (BC_1) was also generated that can be utilized in potato breeding (Luthra et al., 2016). Good tuber yield and quality were observed in *S. tarnii*-somatic hybrids derived BC_1 progenies in the field trials (Thieme et al., 2008). Further, the study confirmed that somatic hybrid had resistance to both PVY and late blight, of which only PVY resistance was transferred to BC_1 progeny. Multiple years of field evaluations of *S. etuberosum*-somatic hybrids and progenies showed stable transmission and expression of PLRV and PVY resistances in three (BC_1, BC_2 and BC_3) and two (BC_1 and BC_2) generations, respectively (Novy et al., 2007). Besides, resistances to PVY, Potato Leaf Roll Virus (PLRV), and green peach aphid in BC_2 progeny were also observed (Novy et al., 2002). Another study demonstrated that late blight resistance can be transferred successfully through breeding from tetraploid somatic hybrids (*S.* × *michoacanum* + *S. tuberosum* and autofused *S.* × *michoacanum*) to common varieties (Smyda-Dajmund et al., 2017). Bacterial wilt resistance was transferred to advanced progenies of *S. commersonii*-somatic hybrids and BC_1 and BC_2 were selected as breeding materials (Kim-Lee et al., 2005). The effect of genetic constitution of *S. tuberosum* was investigated in *S. bulbocastanum*-somatic hybrids and progenies (BC_1 and BC_2) (Rakosy-Tican et al., 2015b). Field performance for foliage maturity and tuber traits (tuber yield, tuber number, tuber weight and specific gravity) was observed in somatic hybrids (*S. brevidens*/*S. commersonii* + *S. tuberosum*) progenies and were implicated to breeding for freezing tolerance (Chen et al., 1999a, 1999b, 1999c). Somatic hybrids (*S. commersonii* + *S. tuberosum*) were observed to be more similar to cultivated potato for phenotypes in field evaluations and developed F_2 progeny, suggesting fast transfer of useful traits from *S. commersonii* into cultivated background (Cardi, 1998; Cardi et al., 2002). Late blight resistance genes (*Rpi-blb1*/RB, *Rpi-blb2*, *Rpi-blb3* and *Rpi-bt1*) have been identified in *S. bulbocastanum*-derived somatic hybrid and their progenies (BC_2) with common potato varieties and were found resistant in glasshouse and field trials (Rakosy-Tican et al., 2020). A large number of traits were monitored in somatic hybrids, such as recent physiological age and antioxidant after cold storage (Kammoun et al., 2020), profiling of phytochemicals (phenolic acids and anthocyanins) in tuber peels (Ben Jeddou et al., 2021). To our knowledge, there are investigations up to BC_3 generation and all attempts to exploit somatic hybrid material so far have not resulted in registration of a potato cultivar.

4.4 Genomics

Somatic hybrids and their parents have been exploited in genomics studies. A recent study indicates that new phenotypes of potato are induced by mismatch repair deficiency and somatic hybridization in potato (Rakosy-Tican et al., 2019). Whole genome sequences of chloroplast genome of wild potato species (*S. commersonii*), a commonly used fusion parent, was deciphered and identified two Indel markers for application in chloroplast genotyping (Cho et al., 2016). High-throughput genotyping of somatic hybrids (*S.* × *michoacanum* + *S. tuberosum*) showed presence of both parental chromosomes and loss of some markers (13.9–29.6%) in the hybrids using 5358 DArT markers analysis (Smyda-Dajmund et al., 2016). In functional genomics, genes controlling potato tuberization in tuber-bearing somatic hybrids (*S. tuberosum* + *S. etuberosum*) vs. control parent (*S. etuberosum*-non tuberous) were identified, using microarray. Findings suggest that candidate genes expression in leaf tissue of somatic hybrid are implicated in tuber growth and development process, such as transport, carbohydrate metabolism, phytohormones and transcription/translation/binding functions (Tiwari et al., 2015a, 2015b). In another study, late light resistance genes were identified in somatic hybrid (*S. tuberosum* + *S. pinnatisectum*) by microarrays and the study suggests a broad spectrum of candidate genes involved in late blight resistance in the hybrid (Singh et al., 2016a). Further, recent identification of eight miRNAs (miR395, 821, 1030, 1510, 2673, 3979, 5021 and 5213) in *S. pinnatisectum*-somatic hybrid for late blight resistance genes and their targets has led to a new insight in potato biology (Singh et al., 2016b). Most of the predicted target genes are associated with different biological processes, such as disease resistance proteins and transcription factors families. These miRNAs could be manipulated through RNAi technique for transgenic development (Singh et al., 2016b).

The *Solanum tuberosum* dihaploid 'C-13' ($2n = 2x = 24$) was developed from potato cv. Kufri Chipsona-2 ($2n = 2x = 48$) by anther culture at our institute. The dihaploid 'C-13' has been used as fusion parent for development of somatic hybrids. We sequenced the whole genome of 'C-13' inhouse, using Illumina approach. Overall, ~ 810 Mb genome size with 30,241 predicted genes were identified in the C-13 genome. A total of 11,22,388 SNPs and 48,145 Indels were also identified in C-13. In other studies, potato somatic hybrid 'P8' (C-13 + *S. pinnatisectum*) and fusion parents were genome sequenced, using Illumina technology. We identified genomic variation in P8, *S. pinnatisectum* and hybrid progeny (P8 × cv. Kufri Jyoti). More than 125 million reads per sample were processed, having 12–15X genome coverage and 65–75% mapping to the reference potato genome (Tiwari et al., 2021). On the other hand, *RB* gene homologous were identified in wild species used in protoplast fusion experiment, such as *S. chacoense*, *S. pinnatisectum*, *S. polyadenium*, *S. trifidum*, *S. cardiophyllum*, *S. lesteri*, *S. huancabambense*, *S. verrucosum*, *S. jamesii*, *S. polytrichon* and *S. stoloniferum*. These homologs may serve as an important genomic resource for the novel gene discovery in late blight-resistance breeding programs (Tiwari et al., 2015a). Further, we have also identified genes controlling potato tuberization in somatic hybrids of 'C-13' (+) *S. etuberosum* by microarray technology. A total of 468 genes (94 up-regulated and 374 down-regulated) was identified and this was statistically significant and differentially expressed in tuber-bearing potato somatic hybrid (E1-3) versus control non-tuberous wild species *S. etuberosum* (Etb). Overall, the findings showed that candidate genes induced in leaves of E1-3 were implicated in the tuberization process, such as transport, carbohydrate metabolism, phytohormones and transcription/translation/binding functions (Tiwari et al., 2015b). Thus, with the advancement in next-generation sequencing technologies, there is a huge scope to deploy these novel tools in somatic hybrids to study potato biology and improvement.

5. Development of Potato Somatic Hybrids

Application of this technology has been observed widely in the production of somatic hybrids for various agronomic traits and biotic/abiotic stress resistance/tolerance. Potato somatic hybrids developed worldwide are summarized in Table 2. We have produced potato somatic hybrids of three wild species, as briefed here. We have produced interspecific potato somatic hybrids (four genotypes) via protoplast fusion between potato dihaploid *Solanum tuberosum* dihaploid 'C-13' ($2n = 2x = 24$) and wild *Solanum cardiophyllum* (PI 341233) for late blight resistance with wider genetic base (Chandel et al., 2015). These hybrids are tertraploid, male fertile, resistant to late blight introgressed from wild *Solanum cardiophyllum* and have wider genetic base possessing W/α, W/γ and T/β diverse cytoplasm types. Further, promising somatic hybrid clones, Crd 6, Crd 10 and Crd 16, were identified, based on field test. They possess high tuber dry matter content (≥ 24%), excellent storage quality and high late blight resistance along with adaptability under sub-tropical plain conditions (Luthra et al., 2018). All somatic hybrids and advance hybrids along with ploidy and molecular characterization are depicted in Figs. 1, 2, 3, 4 and 5.

Similarly, interspecific potato somatic hybrids (11 genotypes) were produced via protoplast fusion between 'C-13' and wild *Solanum pinnatisectum* (CGN No.: 17745) for very high resistance to late blight disease (Sarkar et al., 2011). These somatic hybrids have immense potential in potato breeding to transfer durable resistance to late blight by breeding methods and to widen the gene pool of the cultivated potato. Somatic hybrids were analyzed for cytoplasm types, using organelles (chloroplast and mitochondria) genome-specific markers. Study showed that the above somatic hybrids possess W/α, W/γ and T/β diverse cytoplasm types. Besides, protocols have been standardized for an efficient cell system in potato for somatic cell genetic manipulations from stoloniferous shoot protoplast. For symmetric somatic hybridization (electrofusion) between 'C-13' and diploid wild species *S. pinnatisectum*, protoplasts isolated from 0.1 M sucrose-induced stoloniferous shoots were also found to be most responsive (Sharma et al., 2011). Somatic hybridization has the unique potential to widen the cytoplasm types of the cultivated gene pools from wild species through introgression by breeding methods. Further, genetic improvement of these somatic hybrids (C-13 + *S. pinnatisectum*) was done through breeding. The promising somatic hybrids, namely P-4, P-8 and P-10, were identified and have been utilized in breeding important characters, such as high tuber

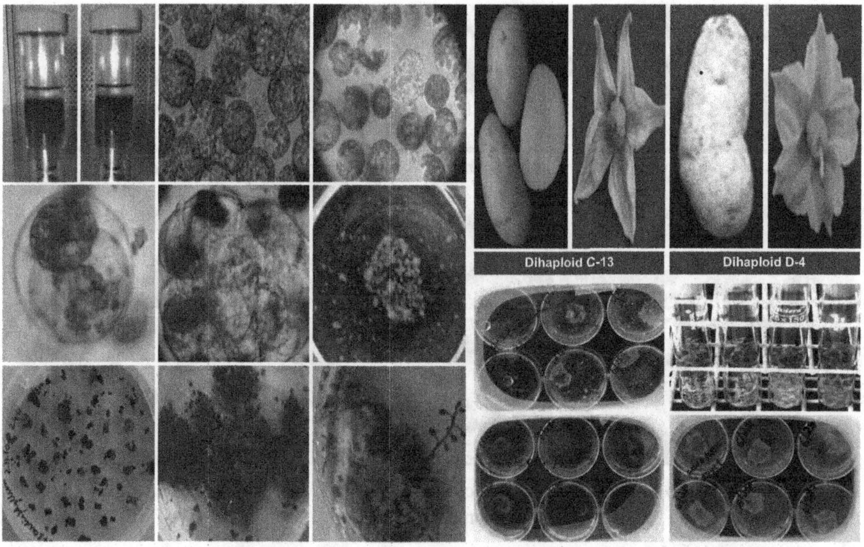

Fig. 1. Protoplast fusion and development of somatic hybrids in potato.

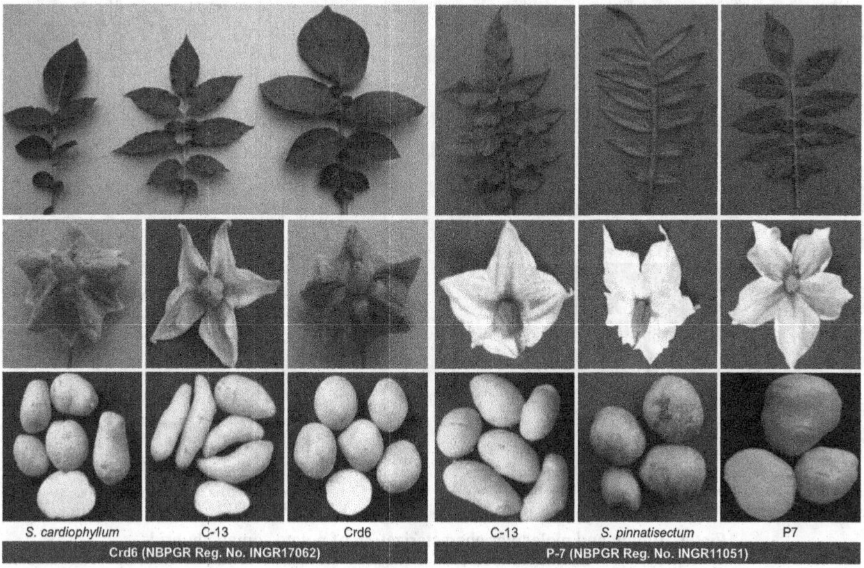

Fig. 2. Late blight resistant interspecific potato somatic hybrids registered as elite genetic stock with ICAR-NBPGR, New Delhi, India. *Right*: Crd6 (*S. tuberosum* dihaploid C-13 (+) *S. cardiophyllum*; *Left*: C-13 (+) *S. pinnatisectum*.

dry matter, resistance to late blight and excellent storing quality in the cultivated potato (Luthra et al., 2016). One promising advance stage hybrid, MSH/14-7, was identified for All Indian Coordinated Research Project (AICRP)-Potato for multi-location testing, under Indian varietal development programs. Besides, five promising hybrids, namely MSH/14-112, -113, -115, -122 and -123 (P8 × Kufri

Fig. 3. Advance potato hybrids developed by crossing somatic hybrids (C-13 + *S. pinnatisectum*) with common potato varieties.

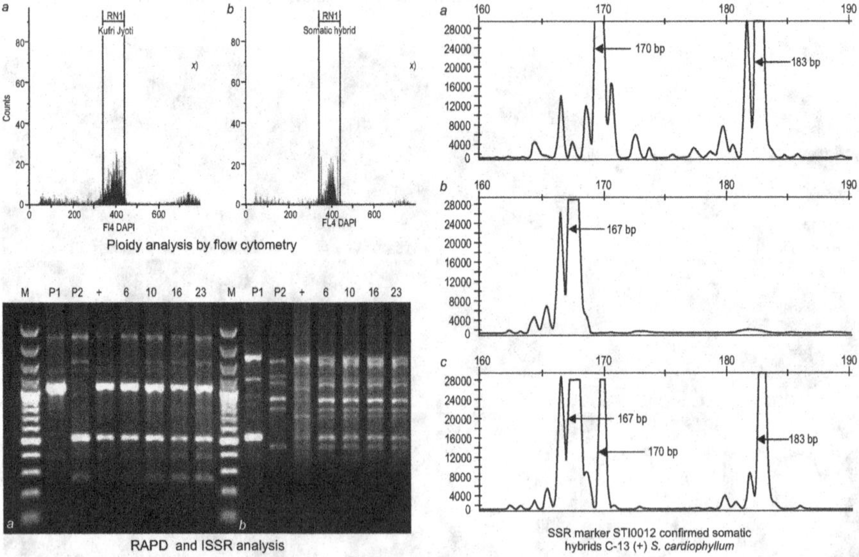

Fig. 4. Ploidy and molecular characterization of potato somatic hybrids.

Jyoti), were selected for very high resistance to late blight and high dry matter content (20.75–22.10%). SSR alleles (STM0003) linked to late blight resistant somatic hybrid (C-13 + *S. pinnatisectum*) parent P8 (103 and 144 bp) and Kufri Jyoti (132 and 144 bp), their progenies (103 and 144) were identified. Progenies generated by cross of P8 × Kufri Jyoti. SSR allele 103 bp of P8 were transmitted into the progenies and identified for genetic fidelity testing and breeding application.

We have also developed interspecific potato somatic hybrids (21 genotypes) via protoplast fusion between dihaploid *Solanum tuberosum* 'C-13' and wild *Solanum*

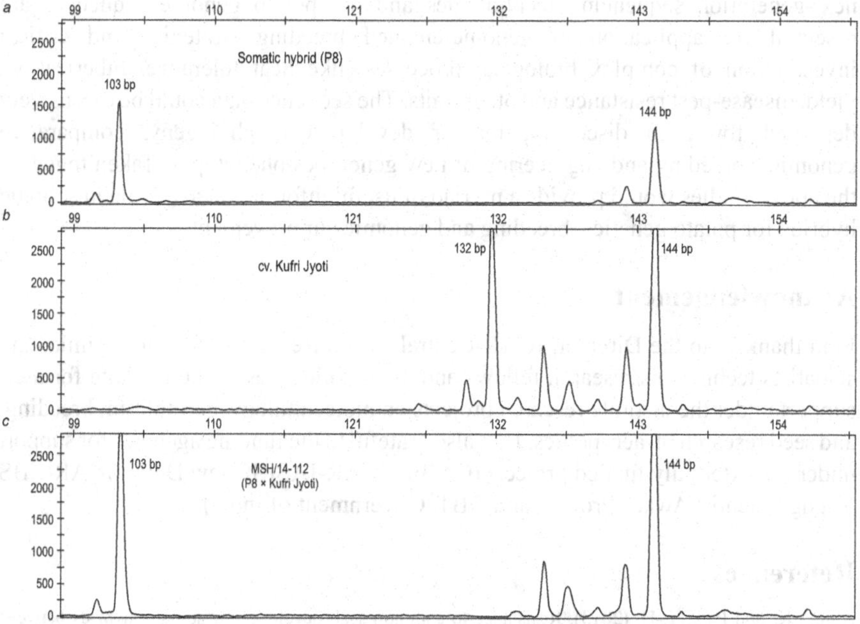

Fig. 5. Identification of SSR alleles (STM0003) in somatic hybrids parents: (a) P8, (b) Kufri Jyoti, and (c) progeny MSH/14-112.

etuberosum (CGN No.: 23066) for potato virus Y resistance. The wild species, *Solanum etuberosum,* is no-tuberous species, whereas somatic hybrids form tubers but not like tetraploid potato. Somatic hybrids were analyzed for cytoplasm types, using organelles (chloroplast and mitochondria) genome-specific markers and showed diverse cytoplasm types (W/α, W/γ and T/β). Potato cytoplasm types are described as based on primer pairs: (i) primers ALC1/ALC3: 381 bp (T/β) and 622 bp (W/α and W/γ); (ii) ALM1/ALM3: 1.2 kb (W/α and W/γ); (iii) ALM4/ALM5: 1.6 kb (T/β), 2.4 kb (W/α); and (iv) ALM6/ALM7: 2.4 kb (W/γ) (Lössl et al., 1999, 2000).

6. Conclusion

Limited utilization of *Solanum* species and therefore narrow genetic base of the cultivated potato is the cause of concern in yield stagnation. Development of potato somatic hybrids, using wild *Solanum* species with desirable attributes, could be helpful to address this issue. Analyzing breeding potential of somatic hybrids, development of advanced progenies by hybridization with common varieties and identification of linked molecular markers are important for their successful application in potato breeding. Further, information on nuclear-cytoplasmic interaction between cultivated and wild species could also be useful while conducting exploitation of somatic hybrids in breeding. Novel genomics tools, like whole genome sequencing, DArT markers-based genotyping, microRNAs, microarrays and many others could strengthen the somatic hybridization research in potato. Moreover, access to the next-generation sequencing technologies and the potato genome sequences are essential for application of genome-enabled breeding strategies and critical investigation of complex biological processes, like heat tolerance, tuberization, yield, disease-pest resistance and other traits. The sequence data could be extensively deployed for gene discovery, marker development, phylogeny, comparative genomics, breeding and engineering of new genotypes/phenotypes. Taken together, the above studies would provide a myriad of useful information available in somatic hybrids for potato genetics, breeding and genomics improvement.

Acknowledgement

I am thankful to the Director, ICAR-Central Potato Research Institute, Shimla, and scientists/technicians/research fellows and other colleagues of the institute for their support under the institute research projects on biotechnology, germplasm, breeding, and seed research on aeroponics. I am also grateful to the funding agencies for support under the externally funded projects (CABin, ICAR-IASRI, New Delhi; ICAR-LBS Young Scientist Award Project, and DBT, Government of India).

References

Ahn, Y.K. and Park, T.-H. (2013). Resistance to common scab developed by somatic hybrids between *Solanum brevidens* and *Solanum tuberosum*. Acta *Agri. Scand*., Sect. B, 63: 595–603.

Austin, S., Baer, M., Ehlenfeldt, M., Kazmierczak, P.J. and Helgeson, J.P. (1985a). Intra-specific fusions in *Solanum tuberosum*. *Theor. Appl. Genet*., 71: 172–175.

Austin, S., Baer, M.A. and Helgeson, J.P. (1985b). Transfer of resistance to potato leaf roll virus from *Solanum brevidens* into *Solanum tuberosum* by somatic fusion. *Plant Sci*., 9: 75–82.

Austin, S., Ehlenfeldt, M.K., Baer, M.A. and Helgeson, J.P. (1986). Somatic hybrids produced by protoplast fusion between *S. tuberosum* and *S. brevidens*: phenotypic variation under field conditions. *Theor. Appl. Genet*., 71: 682–90.

Austin, S., Lojkowska, E., Ehlenfeldt, M.K., Kelman, A. and Helgeson, J.P. (1988). Fertile interspecific hybrids of *Solanum*: A novel source of resistance to *Erwinia* soft rot. *Phytopathology*, 78: 1216–1220.

Austin, S., Pohlman, J.D., Brown, C.R., Mojtahedi, H., Santo, G.S. et al. (1993). Interspecific somatic hybridization between *Solanum tuberosum* L. and *S. bulbocastanum* Dun. as a means of transferring nematode resistance. *Am. Potato J*., 70: 485–495.

Aversano, R., Savarese, S., Nova, J.M.D., Frusciante, L., Punzo, M. and Carputo, D. (2009). Genetic stability at nuclear and plastid DNA level in regenerated plants of *Solanum* species and hybrids. *Euphytica*, 165: 353–361.

Barone, A., Li, J., Sebastiano, A., Cardi, T. and Frusciante, L. (2002). Evidence for tetrasomic inheritance in a tetraploid *Solanum commersonii* (+) *S. tuberosum* somatic hybrid through the use of molecular markers. *Theor. Appl. Genet*., 104: 539–546.

Bastia, A., Li, J., Sebastiano, A., Cardi, T. and Frusciante, L. (2002). Evidence for tetrasomic inheritance in a tetraploid *Solanum commersonii* (+) *S. tuberosum* somatic hybrid through the use of molecular markers. *Theor. Appl. Genet*., 104: 539–546.

Bastia, T., Carotenuto, N., Basile, B., Zoina, A. and Cardi, T. (2000). Induction of novel organelle DNA variation and transfer of resistance to frost and *Verticillium* wilt in *Solanum tuberosum* through somatic hybridization with 1EBN *S. commersonii*. *Euphytica*, 116: 1–10.

Bastia, T., Scotti, N. and Cardi, T. (2001). Organelle DNA analysis of *Solanum* and *Brassica* somatic hybrids by PCR with universal primers. *Theor. Appl. Genet*., 102: 1265–1272.

Bates, G.W., Nea, L.J. and Hasenkampf, C.A. (1987). Electrofusion and plant somatic hybridization. pp. 479–496. *In*: Sowers, A.E. (ed.). *Cell Fusion*. Plenum Press, New York.

Behnke, M. (1975). Regeneration in Gewebekulturen einiger dihaplider *Solanum tuberosum*-Klone. *Z. Pflanzenzücht*, 75: 262–265.

Ben Jeddou, K., Kammoun, M., Hellström, J., Gutiérrez-Quequezana, L., Rokka, V.M. et al. (2021). Profiling beneficial phytochemicals in a potato somatic hybrid for tuber peels processing: Phenolic acids and anthocyanins composition. *Food Sci. Nutr.*, 9(3): 1388–1398.

Bidani, A., Nouri-Ellouz, O., Lakhoua, L., Sihachakr, D., Cheniclet, C. et al. (2007). Interspecific potato somatic hybrids between *Solanum berthaultii* and *Solanum tuberosum* L. showed recombinant plastome and improved tolerance to salinity. *Plant Cell Tiss. Organ Cult.*, 91: 179–189.

Binding, H. and Nehls, R. (1977). Regeneration of isolated protoplasts to plant *Solanum dulcamara* L. *Z. Pflanzenphysiol.*, 85: 279–280.

Binding, H., Nehls, R., Schieder, O., Sopory, S.K. and Wenzel, G. (1978). Regeneration of mesophyll protoplasts isolated from dihaploid clones of *Solanum tuberosum* L. *Plant Physiol.*, 43: 52–54.

Binding, H., Jain, S.M., Finger, J., Mordhorst, G., Nehls, R. and Gressel, J. (1982). Somatic hybridization of an atrazine resistant biotype of *Solanum nigrum* with *Solanum tuberosum*, Part 1, Clonal variation of morphology and in atrazine sensitivity. *Theor. Appl. Genet.*, 63: 273–277.

Bołtowicz, B., Szczerbakowa, A. and Wielgat, B. (2005). RAPD analysis of the interspecific somatic hybrids *Solanum bulbocastanum* (+) *S. tuberosum*. *Cell Mol. Biol. Lett.*, 10: 151–162.

Bradshaw, J.E., Bryan, G.J. and Ramsay G (2006) Genetic resources (including wild and cultivated *Solanum* species) and progress in their utilisation in potato breeding. *Potato Res.*, 49: 49–65.

Cai, X.K., Liu, J. and Xie, C.H. (2004). Mesophyll protoplast fusion of *Solanum tuberosum* and *Solanum chacoense* and their somatic hybrid analysis. *Acta Hortic. Sin.*, 31: 623–626.

Cardi, T., Ambrosio, F.D., Consoli, D., Puite, K.J. and Ramulu, K.S. (1993). Production of somatic hybrids between frost-tolerant *Solanum commersonii* and *S. tuberosum*: characterization of hybrid plants. *Theor. Appl. Genet.*, 87: 193–200.

Cardi, T. (1998). Multivariate analysis of variation among *Solanum commersonii* (+) *S. tuberosum* somatic hybrids with different ploidy levels. *Euphytica*, 99: 35–41.

Cardi, T., Bastia, T., Monti, L. and Earle, E.D. (1999). Organelle DNA and male fertility variation in *Solanum* spp. and interspecific somatic hybrids. *Theor. Appl. Genet.*, 99: 819–828.

Cardi, T., Mazzei, M. and Frusciante, L. (2002). Field variation in a tetraploid progeny derived by selfing a *Solanum commersonii* (+) *S. tuberosum* somatic hybrid: A multivariate analysis. *Euphytica*, 124: 111–119.

Carputo, D., Basile, B., Cardi, T. and Frusciante, L. (2000). *Erwinia* resistance in backcross progenies of *Solanum tuberosum* × *S. tarijense* and *S. tuberosum* (+) *S. commersonii* hybrids. *Potato Res.*, 43: 135–142.

Carrasco, A., De Galarreta, J.I.R., Rico, A. and Ritter, E. (2000). Transfer of PLRV resistance from *Solanum verrucosum* Schlechdt to potato (*S. tuberosum* L.) by protoplast electrofusion. *Potato Res.*, 43: 31–42.

Chandel, P., Tiwari, J.K., Ali, N., Devi, S., Sharma, S. et al. (2015). Interspecific potato somatic hybrids between *Solanum tuberosum* and *S. cardiophyllum*, potential sources of late blight resistance breeding. *Plant Cell Tiss. Organ Cult.*, 123: 579–589.

Chen, L., Guo, X., Xie, C., He, L., Cai, X. et al. (2013). Nuclear and cytoplasmic genome components of *Solanum tuberosum* + *S. chacoense* somatic hybrids and three SSR alleles related to bacterial wilt resistance. *Theor. Appl. Genet.*, 126: 1861–1872.

Chen, L., Guo, X., Wang, H., Xie, C., Cai, X. et al. (2016). Tetrasomic inheritance pattern of the pentaploid *Solanum chacoense* (+) *S. tuberosum* somatic hybrid (resistant to bacterial wilt) revealed by SSR detected alleles. *Plant Cell Tiss. Organ Cult.*, 127: 315–323.

Chen, Y.-K.H., Bamberg, J.B. and Palta, J.P. (1999a). Expression of freezing tolerance in the interspecific F_1 and somatic hybrids of potatoes. *Theor. Appl. Genet.*, 98: 995–1004.

Chen, Y.-K.H., Palta, J.P., Bamberg, J.B., Kim, H., Haberlach, G.T. and Helgeson, J.P. (1999b). Expression of non-acclimated freezing tolerance and cold acclimation capacity in somatic hybrids between hardy wild *Solanum* species and cultivated potatoes. *Euphytica*, 107: 1–8.

Chen, Y.-K.H., Palta, J.P. and Bamberg, J.B. (1999c). Freezing tolerance and tuber production in selfed and backcross progenies derived from somatic hybrids between *Solanum tuberosum* L. and *S. commersonii* Dun. *Theor. Appl. Genet.*, 99: 100–107.

Cheng, J., Saunders, J.A. and Sinden, S.L. (1995). Colorado potato beetle resistant somatic hybrid potato plants produced via protoplast electrofusion. *In Vitro Cell Dev. Biol.*, 31: 90–95.

Cho, K.-S., Cheon, K.-S., Hong, S.-Y., Cho, J.-H., Im, J.-S. et al. (2016). Complete chloroplast genome sequences of *Solanum commersonii* and its application to chloroplast genotype in somatic hybrids with *Solanum tuberosum*. *Plant Cell Rep.*, 35: 2113–2123.

Cocking, E.C. (1960). A method for the isolation of plant protoplasts and vacuoles. *Nature*, 187: 962–963.

Cooper-Bland, S., Main, D.M.J., Fleming, M.H., Phillips, M.S., Powell, W. and Kumar, A. (1994). Synthesis of intraspecific somatic hybrids of *Solanum tuberosum*: assessments of morphological, biochemical and nematode (*Globodera pallida*) resistance characteristics. *J. Exp. Bot.*, 45: 1319–1325.

Dong, F., Novy, R.G., Helgeson, J.P. and Jiang, J. (1999). Cytological characterization of potato-*Solanum etuberosum* somatic hybrids and their backcross progenies by genomic *in situ* hybridization. *Genome*, 42: 987–992.

Fehér, A., Preizner, J., Litkey, Z., Casanádi, G. and Dudits, D. (1992). Characterization of chromosome instability in interspecific somatic hybrids obtained by X-ray fusion between potato (*Solanum tuberosum* L.) and *S. brevidens* Phil. *Theor. Appl. Genet.*, 84: 880–890.

Fish, N., Karp, A. and Jones, M.G.K. (1988). Production of somatic hybrids by electrofusion in *Solanum*. *Theor. Appl. Genet.*, 76: 260–266.

Fock, I., Collonier, C., Purwito, A., Luisetti, J., Sonvannavong, V. et al. (2000). Resistance to bacterial wilt in somatic hybrids between *Solanum tuberosum* and *S. phureja*. *Plant, Sci.*, 160: 165–176.

Fock, I., Collonier, C., Purwito, A., Luisetti, J., Sonvannavong, V. et al. (2001). Use of *S. stenotomum* for introduction of resistance to bacterial wilt in somatic hybrids of potato. *Plant Physiol. Biohem.*, 39: 899–908.

Frei, U., Stattmann, M., Lössl, A. and Wenzel, G. (1998). Aspects of fusion combining ability of dihaploid *Solanum tuberosum* L.: influence of the cytoplasm. *Potato Res.*, 41: 155–162.

Gavrilenko, T., Thieme, R. and Tiemann, H. (1999). Assessment of genetic and phenotypic variation among interspecific somatic hybrids of potato, *Solanum tuberosum* L. *Plant Breed.*, 118: 205–213.

Gavrilenko, T., Thieme, R., Heimbach, U. and Thieme, T. (2003). Fertile somatic hybrids of *Solanum etuberosum* + dihaploid *Solanum tuberosum* and their backcrossing progenies: Relationships of genome dosage with tuber development and resistance to potato virus Y. *Euphytica*, 131: 323–332.

Gibson, R.W., Jones, M.G.K. and Fish, N. (1988). Resistance to potato leaf roll virus and potato virus Y in somatic hybrids between dihaploid *Solanum tuberosum* and *S. brevidens*. *Theor. Appl. Genet.*, 76: 113–117.

Gillen, A.M. and Novy, R.G. (2007). Molecular characterization of the progeny of *Solanum etuberosum* identifies a genomic region associated with resistance to potato leafroll virus. *Euphytica*, 155: 403–415.

Greplová, M., Polzerová and Vlastníková, H. (2008). Electrofusion of protoplasts from *Solanum tuberosum*, *S. bulbocastanum* and *S. pinnatisectum*. *Acta Physiol. Plant.*, 30: 787–796.

Guo, X., Xie, C., Cai, X., Song, B., He, L. and Liu, J. (2010). Meiotic behavior of pollen mother cells in relation to ploidy level of somatic hybrids between *Solanum tuberosum* and *S. chacoense*. *Plant Cell Rep.*, 29: 1277–1285.

Harding, K. and Millam, S. (2000). Analysis of chromatin, nuclear DNA and organelle composition in somatic hybrids between *Solanum tuberosum* and *Solanum sanctae-rosae*. *Theor. Appl. Genet.*, 101: 939–947.

Helgeson, J.P., Haberlach, G.T., Ehlenfeldt, M.K., Hunt, G., Pohlman, J.D. and Austin, S. (1993). Fertile somatic hybrids of potato and wild *Solanum* species: Potential for use in breeding programs. *Am. Potato J.*, 70: 437–452.

Helgeson, J.P., Pohlman, J.D., Austin, S., Haberlach, G.T., Wielgus, S.M. et al. (1998). Somatic hybrids between *Solanum bulbocastanum* and potato: A new source of resistance to late blight. *Theor. Appl. Genet.*, 96: 738–742.

Horsman, K., Bergervoet, J.E.M. and Jacobsen, E. (1997). Somatic hybridization between *Solanum tuberosum* and species of the *S. nigrum* complex: Selection of vigorously growing and flowering plants. *Euphytica*, 96: 345–352.

Iovene, M., Savarese, M., Cardi, T., Frusciante, L., Scott, N. et al. (2007). Nuclear and cytoplasmic genome composition of *Solanum bulbocastanum* (+) *S. tuberosum* somatic hybrids. *Genome*, 50: 443–450.

Iovene, M., Aversano, R., Savarese, S., Caruso, I., Dimatteo, A. et al. (2012). Interspecific somatic hybrids between *Solanum bulbocastanum* and *S. tuberosum* and their haploidization for potato breeding. *Biol. Plant.*, 56: 1–8.

Jadari, R., Sihachakr, D., Rossignol, L. and Ducreux, G. (1992). Transfer of resistance to *Verticillium dahliae* Kleb. from *Solanum torvum* S.W. into potato (*Solanum tuberosum* L.) by protoplast electrofusion. *Euphytica*, 64: 39–47.

Jansky, S. (2006). Overcoming hybridization barriers in potato. *Plant Breed*, 125: 1–12.

Kammoun, M., Essid, M.F., Ksouri, F., Rokka, V.M., Charfeddine, M. et al. (2020). Assessment of physiological age and antioxidant status of new somatic hybrid potato seeds during extended cold storage. *J. Plant Physiol.*, 254: 153279.

Kim, H., Coi, S.U., Chae, M.S., Weilgus, S.M. and Helgeson, J.P. (1993). Identification of somatic hybrids produced by protoplast fusion between *Solanum commersonii* and *S. tuberosum* haploid. *Korean J. Plant Tissue Cult.*, 20: 337–344.

Kim-Lee, H., Moon, J.S., Hong, Y.J., Kim, M.S. and Cho, H.M. (2005). Bacterial wilt resistance in the progenies of the fusion hybrids between haploid of potato and *Solanum commersonii*. *Am. J. Potato Res.*, 82: 129–137.

Kozukue, N., Misoo, S., Yamada, T., Kamijima, O. and Friedman, M. (1999). Inheritance of morphological characters and glycoalkaloids in potatoes of somatic hybrids between dihaploid *Solanum acaule* and tetraploid *Solanum tuberosum*. *J. Agric. Food Chem.*, 47: 4478–4483.

Laferriere, L.T., Helgeson, J.P. and Allen, C. (1999). Fertile *Solanum tuberosum* + *S. commersonii* somatic hybrids as sources of resistance to bacterial wilt caused by *Ralstonia solanacearum*. *Theor. Appl. Genet.*, 98: 1272–1278.

Laurila, J., Laasko, I., Larkka, J., Gavrilenko, T., Rokka, V.-M. and Pehu, E. (2001). The proportions of glycoalkaloid aglycones are dependent on the genome constitutions of interspecific hybrids between two *Solanum* species (*S. brevidens* and *S. tuberosum*). *Plant Sci.*, 161: 677–683.

Lightbourn, G.J. and Veilleux, R.E. (2007). Production and evaluation of somatic hybrids derived from monoploid potato. *Am. J. Potato Res.*, 84: 425–435.

Lightbourn, G.J., Jelesko, J.G. and Veilleux, R.E. (2007). Retrotransposon-based markers from monoploids used in somatic hybridization. *Genome*, 50: 492–501.

Lössl, A., Frei, U. and Wenzel, G. (1994). Interaction between cytoplasmic composition and yield parameters in somatic hybrid of *S. tuberosum* L. *Theor. Appl. Genet.*, 89: 873–878.

Lössl, A., Adler, N., Horn, R., Frei, U. and Wenzel, G. (1999). Chondriome-type characterization of potato: mt α, β, γ, δ, ε and novel plastid-mitochondrial configurations in somatic hybrids. *Theor. Appl. Genet.*, 98: 1–10.

Lössl, A., Götz, M., Braun, A. and Wenzel, G. (2000). Molecular markers for cytoplasm in potato: male sterility and contribution of different plastid-mitochondrial configurations to starch production. *Euphytica*, 116: 221–230.

Luthra, S.K., Tiwari, J.K., Lal, M, Chandel, P. and Kumar, V. (2016). Breeding potential of potato somatic hybrids: evaluations for adaptability, tuber traits, late blight resistance, keeping quality and backcross (BC_1) progenies. *Potato Res.*, 59: 375–391.

Luthra, S.K., Tiwari, J.K., Kumar, V. and Lal, M. (2018). Evaluation of interspecific somatic hybrids of potato (*Solanum tuberosum*) and wild *S. cardiophyllum* for adaptability, tuber dry matter, keeping quality and late blight resistance. *Agric. Res.*, 8: 158–164.

Mattheij, W.M. and Puite, K.J. (1992). Tetraploid potato hybrids through protoplast fusions and analysis of their performance in the field. *Theor. Appl. Genet.*, 83: 807–812.

McGrath, J.M., Wielgus, S.M. and Helgeson, J.P. (1996). Segregation and recombination of *Solanum brevidens* synteny groups in progeny of somatic hybrids with *S. tuberosum*: intra-genomic equals or exceeds inter-genomic recombination. *Genetics*, 142: 1335–1348.

McGrath, J.M. and Helgeson, J.P. (1998). Differential behavior of *Solanum brevidens* ribosomal DNA loci in a somatic hybrid and its progeny with potato. *Genome*, 41: 435–439.

McGrath, J.M., Williams, C.E., Haberlach, G.T., Wielgus, S.M., Uchytil, T.F. and Helgeson, J.P. (2002). Introgression and stabilization of *Erwinia* tuber soft rot resistance into potato after somatic hybridization of *Solanum tuberosum* and *S. brevidens*. *Am. J. Potato Res.*, 79: 19–24.

Menke, U., Schilde-Rentschler, L., Ruoss, B., Zanke, C., Hemleben, V. and Ninnemann, H. (1996). Somatic hybrids between the cultivated potato *Solanum tuberosum* L. and the 1EBN wild species

Solanum pinnatisectum Dun.: Morphological and molecular characterization. *Theor. Appl. Genet.*, 92: 617–626.

Mojtahedi, H., Brown, C.R. and Santo, G.S. (1995). Characterization of resistance in a somatic hybrid of *Solanum bulbocastanum* and *S. tuberosum* to *Meloidogyne chitwoodi*. *J. Nematol.*, 27: 86–93.

Möllers, C. and Wenzel, G. (1992). Somatic hybridization of dihaploid potato protoplasts as a tool for potato breeding. *Botanica Acta*, 105: 133–39.

Möllers, C., Frei, U. and Wenzel, G. (1994). Field evaluation of tetraploid somatic potato hybrids. *Theor. Appl. Genet.*, 88: 147–52.

Naess, S.K., Bradeen, J.M., Wielgus, S.M., Haberlach, G.T., McGrath, J.M. and Helgeson, J.P. (2000). Resistance to late blight in *Solanum bulbocastanum* is mapped to chromosome 8. *Theor. Appl. Genet.*, 101: 697–704.

Naess, S.K., Bradeen, J.M., Wielgus, S.M., Haberlach, G.T., McGrath, J.M. and Helgeson, J.P. (2001). Analysis of the introgression of *Solanum bulbocastaum* DNA into potato breeding lines. *Mol. Genet. Genomics*, 265: 694–704.

Nouri-Ellouz, O., Gargouri-Bouzid, R., Sihachakr, D., Triki, M.A., Ducreux, G. et al. (2006). Production of potato intraspecific somatic hybrids with improved tolerance to PVY and *Pythium aphanidermatum*. *J. Plant Physiol.*, 163: 1321–1332.

Nouri-Ellouz, O., Triki, M.A., Jbir-Koubaa, R., Louhichi, A., Charfeddine, S. et al. (2016). Somatic hybrids between potato and *S. berthaultii* show partial resistance to soil-borne fungi and potato virus Y. *J. Phytopathol.*, 164: 485–496.

Novy, R.G. and Helgeson, J.E. (1994a). Somatic hybrids between *Solanum etuberosum* and diploid, tuber bearing *Solanum* clones. *Theor. Appl. Genet.*, 89: 775–782.

Novy, R.G. and Helgeson, J.P. (1994b). Resistance to potato virus Y in somatic hybrids between *Solanum etuberosum* and *S. tuberosum* × *S. berthaultii* hybrid. *Theor. Appl. Genet.*, 89: 783–786.

Novy, R.G., Nasruddin, A., Ragsdale, D.W. and Radcliffe, E.B. (2002). Genetic resistances to potato leafroll virus, Potato virus Y, and Green peach aphid in progeny of *Solanum etuberosum*. *Am. J. Potato Res.*, 79: 9–18.

Novy, R.G., Gillen, A.M. and Whitworth, J.L. (2007). Characterization of the expression and inheritance of potato leafroll virus (PLRV) and potato virus Y (PVY) resistance in three generations of germplasm derived from *Solanum etuberosum*. *Theor. Appl. Genet.*, 114: 1161–1172.

Nyman, M. and Waara, S. (1997). Characterisation of somatic hybrids between *Solanum tuberosum* and its frost-tolerant relative *Solanum commersonii*. *Theor. Appl. Genet.*, 95: 1127–1132.

Oberwalder, B., Ruoâ, B., Schilde-Rentschler, L., Hemleben, V. and Ninnemann, H. (1997). Asymmetric protoplast fusion between wild and cultivated species of potato (*Solanum* ssp.) detection of asymmetric hybrids and genome elimination. *Theor. Appl. Genet.*, 94: 1104–1112.

Oberwalder, B., Schilde-Rentschler, L., Ruoâ, B., Wittemann, S. and Ninnemann, H. (1998). Asymmetric protoplast fusion between wild species and breeding lines of potato: Effect of recipient and genome stability. *Theor. Appl. Genet.*, 97: 1347–1354.

Oberwalder, B., Schilde-Rentschler, L., Loffelhardt-Ruob, B. and Ninnemann, H. (2000). Differences between hybrids of *Solanum tuberosum* L. and *Solanum circaeifolium* Bitt. obtained from symmetric and asymmetric fusion experiments. *Potato Res.*, 43: 71–82.

Orczyk, W., Przetakiewicz, J. and Nadoloska-Orczyk, A. (2003). Somatic hybrids of *Solanum tuberosum*-application to genetics and breeding. *Plant Cell Tiss. Organ Cult.*, 74: 1–13.

Perl, A., Aviv, D. and Galun, E. (1990). Potato fusion derived CMS potato cybrids: potential seed parents for hybrid, true potato seeds. *J. Heredity*, 81: 438–442.

Polgar, Z., Wielgus, S.M., Horvath, S. and Helgeson, J.P. (1999). DNA analysis of potato + *Solanum brevidens* somatic hybrid lines. *Euphytica*, 105: 103–107.

Polzerová, H., Patzak, J. and Greplová, M. (2011). Early characterization of somatic hybrids from symmetric protoplast electrofusion of *Solanum pinnatisectum* Dun. and *Solanum tuberosum* L. *Plant Cell Tiss. Organ Cult.*, 104: 163–170.

Preiszner, J., Feher, A., Veisz, O., Sutka, J. and Dudits, D. (1991). Characterization of morphological variation and cold resistance in interspecific somatic hybrids between potato (*Solanum tuberosum* L.) and *S. brevidens* Phil. *Euphytica*, 57: 37–49.

Przetakiewicz, J., Nadolska-Orczyk, A., Kuc, D. and Orczyk, W. (2007). Tetraploid somatic hybrids of potato (*Solanum tuberosum* L.) obtained from diploid breeding lines. *Cell Mol. Biol. Lett.*, 12: 253–267.

Puite, K.J., Roest, S. and Pijnacker, L.P. (1986). Somatic hybrid potato plants after electrofusion of diploid *Solanum tuberosum* and *Solanum phureja*. *Plant Cell Rep.*, 5: 262–265.

Rakosy-Tican, E. and Aurori, A. (2015a). Green fluorescent protein (GFP) supports the selection based on callus vigorous growth in the somatic hybrids *Solanum tuberosum* L. + *S. chacoense* Bitt. *Acta Physiol. Plant*, 37: 201.

Rakosy-Tican, E., Thieme, R., Nachtigall, M., Molnar, I. and Denes, T.-E. (2015b). The recipient potato cultivar influences the genetic makeup of the somatic hybrids between five potato cultivars and one cloned accession of sexually incompatible species *Solanum bulbocastanum* Dun. *Plant Cell Tiss. Organ Cult.*, 122: 395–407.

Rakosy-Tican, E., Lörincz-Besenyei, E., Molnár, I., Thieme, R., Hartung, F. et al. (2019). New phenotypes of potato co-induced by mismatch repair deficiency and somatic hybridization. *Front. Plant Sci.*, 10: 3.

Rakosy-Tican, E., Thieme, R., König, J., Nachtigall, M., Hammann, T. et al. (2020). Introgression of two broad-spectrum late blight resistance genes, *Rpi-Blb1* and *Rpi-Blb3*, from *Solanum bulbocastanum* Dun plus race-specific R genes into potato pre-breeding lines. *Front. Plant Sci.*, 11: 699.

Rasmussen, J.O., Nepper, J.P. and Rasmussen, O.S. (1996). Analysis of somatic hybrids between two sterile dihaploid *Solanum tuberosum* L. breeding lines. Restoration of fertility and complementation of *G. pallida* Pa2 and Pa3 resistance. *Theor. Appl. Genet.*, 92: 403–410.

Rasmussen, J.O., Waara, S. and Rasmussen, O.S. (1997). Regeneration and analysis of interspecific asymmetric potato-*Solanum* ssp. hybrid plants selected by micromanipulation or fluorescence activated cell sorting (FACS). *Theor. Appl. Genet.*, 95: 41–49.

Rasmussen, J.O., Nepper, J.P., Kirk, H.-G., Tolstrup, K. and Rasmussen, O.S. (1998). Combination of resistance to potato late blight in foliage and tubers by interspecific dihaploid protoplast fusion. *Euphytica*, 102: 363–370.

Rasmussen, J.O., Lössl, A. and Rasmussen, O.S. (2000). Analysis of the plastome and chondriome origin in plants regenerated after asymmetric *Solanum* ssp. protoplast fusions. *Theor. Appl. Genet.*, 101: 336–343.

Rokka, V.-M., Xu, Y.-S., Kankila, J., Kuusela, A., Pulli, S. and Pehu, E. (1994). Identification of somatic hybrids of dihaploid *Solanum tuberosum* lines and *S. brevidens* by species specific RAPD patterns and assessment of disease resistance of the hybrids. *Euphytica*, 80: 207–217.

Rokka, V.-M., Tauriainen, A., Pietilä, L. and Pehu, E. (1998). Interspecific somatic hybrids between wild potato *Solanum acaule* Bitt. and anther-derived dihaploid potato (*Solanum tuberosum* L.). *Plant Cell Rep.*, 18: 82–88.

Rokka, V.-M., Valkonen, J.P.T., Tauriainen, A., Pietila, L., Lubecka, R. et al. (2000). Production and characterization of 'second generation' of somatic hybrids derived from protoplast fusion between interspecific somatohaploid and dihaploid *Solanum tuberosum* L. *Am. J. Potato Res.*, 77: 149–159.

Rokka, V.-M., Laurila, J., Tauriainen, A., Laakso, I., Larkka, J. et al. (2005). Glycoalkaloid aglycone accumulations associated with infection by *Clavibacter michiganensis* ssp. *sepedonicus* in potato species *Solanum acaule* and *Solanum tuberosum* and their interspecific somatic hybrids. *Plant Cell Rep.*, 23: 683–691.

Sarkar, D., Tiwari, J.K., Sharma, S.H., Poonam, Sharma, S.A. et al. (2011). Production and characterization of somatic hybrids between *Solanum tuberosum* L. and *S. pinnatisectum* Dun. *Plant Cell Tiss. Organ Cult.*, 107: 427–440.

Scotti, N., Monti, L. and Cardi, T. (2003). Organelle DNA variation in parental Solanum spp. genotypes and nuclear-cytoplasmic interactions in *Solanum tuberosum* (+) *S. commersonii* somatic hybrid-backcross progeny. *Theor. Appl. Genet.*, 108: 87–94.

Scotti, N., Marechal-Drouard, L. and Cardi, T. (2004). The *rpl5–rps14* mitochondrial region: A hot spot for DNA rearrangements in Solanum spp. somatic hybrids. *Curr. Genet.*, 45: 378–382.

Serraf, I., Sihachakr, D., Ducreux, G., Brown, S.C., Allot, M. et al. (1991). Interspecific somatic hybridization in potato by protoplast electrofusion. *Plant Sci.*, 76: 115–126.

Sharma, S., Sarkar, D., Pandey, S.K., Chandel, P. and Tiwari, J.K. (2011). Stoloniferous shoot protoplast, an efficient cell system in potato for somatic cell genetic manipulations. *Scientia Hort.*, 128: 84–91.

Shi, Y.Z., Chen, Q., Li, H.Y., Beasley, D. and Lynch, D.R. (2006). Somatic hybridization between *Solanum tuberosum* and *S. cardiophyllum*. *Can. J. Plant Sci.*, 86: 539–545.

Sidorov, V.A., Zubko, M.K., Kuchko, A.A., Komarnitsky, I.K. and Gleba, Y.Y. (1987). Somatic hybridization in potato: use of γ-irradiated protoplasts of *Solanum pinnatisectum* in genetic reconstruction. *Theor. Appl. Genet.*, 74: 364–368.

Singh, R., Tiwari, J.K., Rawatm S., Sharma, V. and Singh, B.P. (2016a). Monitoring gene expression pattern in somatic hybrid of *Solanum tuberosum* and *S. pinnatisectum* for late blight resistance using microarray analysis. *Plant Omics J.*, 9: 99–105.

Singh, R., Tiwari, J.K., Rawat, S., Sharma, V. and Singh, B.P. (2016b). *In silico* identification of candidate microRNAs and their targets in potato somatic hybrid *Solanum tuberosum* (+) *S. pinnatisectum* for late blight resistance. *Plant Omics J.*, 9: 159–164.

Smyda, P., Jakuczun, H., Debski, K., Śliwka, J., Thieme, R. et al. (2013). Development of somatic hybrids *Solanum* × *michoacanum* Bitter. (Rydb.) (+) *S. tuberosum* L. and autofused 4*x S.* × *michoacanum* plants as potential sources of late blight resistance for potato breeding. *Plant Cell Rep.*, 32: 1231–1241.

Smyda-Dajmund, P., Sliwka, J., Wasilewicz-Flis, I., Jakuczun, H. and Zimnoch-Guzowska, E. (2016). Genetic composition of interspecific potato somatic hybrids and autofused 4x plants evaluated by DArT and cytoplasmic DNA markers. *Plant Cell Rep.*, 35: 1345–1358.

Smyda-Dajmund, P., Śliwka, J., Wasilewicz-Flis, I., Jakuczun, H. and Zimnoch-Guzowska, E. (2017). BC_1 and F_1 progeny from *Solanum* × *michoacanum* (+) *S. tuberosum* somatic hybrids, autofused 4x *S. michoacanum* and cultivated potato. *Am. J. Potato Res.*, 94: 323–333.

Song, J., Bradeen, J.M., Naess, S.K., Raasch, J.A., Wielgus, S.M. et al. (2003). Gene *RB* cloned from *Solanum bulbocastanum* confers broad spectrum resistance to potato late blight. *Procd. Natl. Acad. Sci., U.S.A.*. 100: 9128–9133.

Spooner, D.M. and Hijmans, R.J. (2001). Potato systematics and germplasm collecting, 1989–2000. *Am. J. Potato Res.*, 78: 237–268.

Spooner, D.M. and Salas, A. (2006). Structure, biosystematics and genetic resources. pp. 1–40. *In*: Gopal, J. Khurana and Paul, S.M. (eds.). *Handbook of Potato Production, Improvement and Postharvest Management*. Food Product Press, New York.

Szczerbakowa, A., Maciejewska, U. and Wielgat, B. (2000). Plant regeneration from the protoplasts of *Solanum tuberosum, S. nigrum* and S. *bulbocastanum*. *Acta Physiol. Plant.*, 22: 3–10.

Szczerbakowa, A., Maciejewska, U., Pawlowski, P., Skierski, J.S. and Wielgat, B. (2001). Electrofusion of protoplasts from *Solanum tuberosum, S. nigrum* and *S. bulbocastanum*. *Acta Physiol. Plant.*, 23: 169–179.

Szczerbakowa, A., Boltowicz, D. and Wielgat, B. (2003a). Interspecific somatic hybrids *Solanum bulbocastanum* (+) *S. tuberosum* H-8105. *Acta Physiol. Plant.*, 25: 365–373.

Szczerbakowa, A., Maciejewska, U., Zimnoch-Guzowska, E. and Wielgat, B. (2003b). Somatic hybrids *Solanum nigrum* (+) *S. tuberosum*: morphological assessment and verification of hybridity. *Plant Cell Rep.*, 21: 577–584.

Szczerbakowa, A., Boltowicz, D., Lebecka, R., Radomski, P. and Wielgat, B. (2005). Characteristics of the interspecific somatic hybrids *Solanum pinnatisectum* (+) *S. tuberosum* H-8105. *Acta Physiol. Plant.*, 27: 265–273.

Szczerbakowa, A., Tarwacka, J., Oskiera, M., Jakuczun, H. and Wielgat, B. (2010). Somatic hybridization between the diploids of *S.* × *michoacanum* and *S. tuberosum*. *Acta Physiol. Plant.*, 32: 867–873.

Szczerbakowa, A., Tarwacka, J., Sliwinska, E. and Wielgat, B. (2011). Nuclear DNA content and chromosome number in somatic hybrid allopolyploids of *Solanum*. *Plant Cell Tiss. Organ Cult.*, 106: 373–380.

Takebe, I., Labib, G. and Melchers, G. (1971). Regeneration of whole plants from isolated mesophyll protoplasts of tobacco. *Naturwissenschaften*, 58: 318–320.

Tarwacka, J., Polkowska-Kowalczyk, L., Kolano, B., Śliwka, J. and Wielgat, B. (2013). Interspecific somatic hybrids *Solanum villosum* (+) *S. tuberosum*, resistant to *Phytophthora infestans*. *J. Plant Physiol.*, 170: 1541–1548.

Tek, A.L., Stevenson, W.R., Helgeson, J.P. and Jiang, J. (2004). Transfer of tuber soft rot and early blight resistances from *Solanum brevidens* into cultivated potato. *Theor. Appl. Genet.*, 109: 249–254.

Thach, N.Q., Frei, U. and Wenzel, G. (1993). Somatic fusion for combining virus resistances in *Solanum tuberosum* L. *Theor. Appl. Genet*., 85: 863–867.

Thieme, R., Darsow, U., Gavrilenko, T., Dorokhov, D. and Tiemann, H. (1997). Production of somatic hybrids between *S. tuberosum* L. and late blight resistant Mexican wild potato species. *Euphytica*, 97: 189–200.

Thieme, R., Gavrilenko, T., Thieme, T. and Heimbach, U. (1999). Production of potato genotypes with resistance to potato virus Y by biotechnological methods. pp. 557–560. *In*: Altmann, A. et al. (eds.). *Plant Biotechnology and In Vitro Biology in the 21st Century*. Kluwer Academic Publishers.

Thieme, R., Darsow, U., Rakosy-Tican, L., Kang, Z., Gavrilenko, T. et al. (2004). Use of somatic hybridization to transfer resistance to late blight and potato virus Y (PVY) into cultivated potato. *Plant Breed Seed Sci*., 50: 113–118.

Thieme, R., Rakosy-Tican, E., Gavrilenko, T., Antonova, O., Schubert, J. et al. (2008). Novel somatic hybrids (*Solanum tuberosum* L. + *Solanum tarnii*) and their fertile BC_1 progenies express extreme resistance to potato virus Y and late blight. *Theor. Appl. Genet*., 116: 691–700.

Thieme, R., Rakosy-Tican, E., Nachtigall, M., Schubert, J., Hammann, T. et al. (2010). Characterization of the multiple resistance traits of somatic hybrids between *Solanum cardiophyllum* Lindl. and two commercial potato cultivars. *Plant Cell Rep*., 29: 1187–1201.

Tiwari, J.K., Poonam, Sarkar, D., Pandey, S.K., Gopal, J. and Kumar, S.R. (2010). Molecular and morphological characterization of somatic hybrids between *Solanum tuberosum* L. and *S. etuberosum* Lindl. *Plant Cell Tiss. Organ Cult*., 103: 175–187.

Tiwari, J.K., Poonam, Kumar, V., Singh, B.P., Sharma, S. et al. (2013a). Evaluation of potato somatic hybrids of dihaploid *S. tuberosum* (+) *S. pinnatisectum* for late blight resistance. *Potato J*., 40: 176–179.

Tiwari, J.K., Saurabh, S., Chandel, P., Singh, B.P. and Bhardwaj, V. (2013b). Analysis of genetic and epigenetic variation in potato somatic hybrid by AFLP and MASP markers. *Electronic J. Biotechol*. Doi.: 10.2225/vol16-issue 6-full text-9.

Tiwari, J.K., Chandel, P., Gupta, S., Gopal, J., Singh, B.P. and Bhardwaj, V. (2013c). Analysis of genetic stability of *in vitro* propagated potato microtubers using DNA markers. *Physiol. Mol. Biol. Plants*, 19: 587–595.

Tiwari, J.K., Chandel, P., Singh, B.P. and Bhardwaj, V. (2014). Analysis of plastome and chondriome genome types in potato somatic hybrids from *Solanum tuberosum* × *Solanum etuberosum*. *Genome*, 57: 29–35.

Tiwari, J.K., Devi, S., Sharma, S., Chandel, P., Rawat, S. and Singh, B.P. (2015a). Allele mining in *Solanum* germplasm: cloning and characterization of *RB*-homologous gene fragments from late blight resistant wild potato species. *Plant Mol. Biol. Rep*., 33: 1584–1598.

Tiwari, J.K., Devi, S., Sundaresha, S., Chandel, P., Ali, N. et al. (2015b). Microarray analysis of gene expression patterns in the leaf during potato tuberization in the potato somatic hybrid *Solanum tuberosum* and *Solanum etuberosum*. *Genome*, 58: 305–313.

Tiwari, J.K., Poonam, Saurabh, S., Devi, S., Ali, N. et al. (2015c). Molecular characterization of potato somatic hybrids by inter simple sequence repeat (ISSR) markers. *Potato J*., 42: 1–7.

Tiwari, J.K., Saurabh, S., Chandel, P., Singh, B.P. and Bhardwaj, V. (2015d). Assessment of genetic and epigenetic variations in potato somatic hybrids by methylation-sensitive ISSR and RAPD markers. *Bangladesh J. Bot*., 44: 45–50.

Tiwari, J.K., Saurabh, S., Chandel, P., Devi, S., Ali, N. et al. (2015e). Analysis of genetic and epigenetic changes in potato somatic hybrids between *Solanum tuberosum* and *S. etuberosum* by AFLP and MSAP markers. *Agric. Res*., 4: 339–346.

Tiwari, J.K., Gupta, S., Gopal, J., Kumar, V., Bhardwaj, V. and Singh, B.P. (2015f). Molecular analysis of genetic stability of *in-vitro* conserved potato microplants. *Potato J*., 42: 137–145.

Tiwari, J.K., Devi, S., Chandel, P., Ali, N., Bhardwaj, V. and Singh, B.P. (2016). Organelle genome analysis in somatic hybrids between *Solanum tuberosum* and *S. pinnatisectum* revealed diverse cytoplasm type in potato. *Agric. Res*., 5: 22–28.

Tiwari, J.K., Devi, S., Ali, N., Luthra, S.K., Kumar, V. et al. (2018). Progress in somatic hybridization research in potato during the past 40 years. *Plant Cell Tiss. Organ Cult*, 132: 225–238.

Tiwari, J.K., Kumar, V., Zinta, R., Dalamu, Bhardwaj, V. et al. (2019). Characterization of wild potato species for molecular, morphological and late blight resistance traits. *Potato J*., 46(2): 107–114.

Tiwari, J.K., Luhra, S.K., Zinta, R., Dalamu, Buckseth, T. et al. (2020). Development of SSR fingerprints of interspecific potato somatic hybrids. *Potato J.*, 47(1): 71–79.

Tiwari, J.K., Rawat, S., Luthra, S.K., Zinta, R. and Sahu, S. (2021). Genome sequence analysis provides insights on genomic variation and late blight resistance genes in potato somatic hybrid (parents and progeny). *Mol. Biol. Rep.*, 48: 623–635.

Tiwari, J.K., Zinta, R., Luthra, S.K., Dalamu, Rawat, S. et al. (2022). Identification of simple sequence repeat (SSR) markers linked to interspecific potato somatic hybrids. *Indian J. Agric. Sci.*, 92(3): Art. No. 105337.

Trabelsi, S., Gargouri-Bouzid, R., Vedel, F., Nato, A., Lakhoua, L. and Drira, N. (2005). Somatic hybrids between potato *Solanum tuberosum* and wild species *Solanum verneï* exhibit a recombination in the plastome. *Plant Cell Tiss. Organ Cult.*, 83: 1–11.

Tu, W., Dong, J., Zou, Y., Zhao, Q., Wang, H. et al. (2021). Interspecific potato somatic hybrids between *Solanum malmeanum* and *S. tuberosum* provide valuable resources for freezing-tolerance breeding. *Plant Cell Tiss. Organ Cult.*, 147: 73–83.

Valkonen, J.P.T. and Rokka, V.-M. (1998). Combination and expression of two virus resistance mechanisms in interspecific somatic hybrids of potato. *Plant Sci.*, 131: 85–94.

Waara, S., Nyman, M. and Johannisson, A. (1998). Efficient selection of potato heterokaryons by flow cytometric sorting and the regeneration of hybrid plants. *Euphytica*, 101: 293–299.

Wallin, A., Glimelius, K. and Eriksson, T. (1974). The induction of aggregation and fusion of *Daucus carota* protoplasts by polyethylene glycol. *Z. Pflanzenphysiol.*, 74: 64–80.

Williams, C.E., Wielgus, S.M., Haberlach, G.T., Guenther, C., Kim-Lee, H. and Helgeson, J.P. (1993). RFLP analysis of chromosomal segregation in progeny from an interspecific hexaploid somatic hybrid between *Solanum brevidens* and *Solanum tuberosum*. *Genetics*, 135: 1167–1173.

Xu, Y.S., Murto, M., Dunckley, R., Jones, M.G.K. and Pehu, E. (1993). Production of asymmetric hybrids between *Solanum tuberosum* and irradiated *S. brevidens*. *Theor. Appl. Genet.*, 85: 729–734.

Yamada, T., Misoo, S., Ishii, T., Ito, Y., Takaoka, K. and Kamijima, O. (1997). Characterization of somatic hybrids between tetraploid *Solanum tuberosum* L. and dihaploid *S. acaule*. *Breed. Sci.*, 47: 229–236.

Yamada, T., Hosaka, K., Kaide, N., Nakagawa, K., Misoo, S. and Kamijima, O. (1998). Cytological and molecular characterization of BC_1 progeny from two somatic hybrids between dihaploid *Solanum acaule* and tetraploid *S. tuberosum*. *Genome*, 41: 743–750.

Yu, Y., Ye, W., He, L., Cai, X., Liu, T. and Liu, J. (2013). Introgression of bacterial wilt resistance from eggplant to potato via protoplast fusion and genome components of the hybrids. *Plant Cell Rep.*, 32: 1687–1701.

Zimnoch-Guzowska, E., Lebecka, R., Kryszczuk, A., Maciejewska, U., Szczerbakowa, A. and Wielgat, B. (2003). Resistance to *Phytophthora infestans* in somatic hybrids of *Solanum nigrum* L. and diploid potato. *Theor. Appl. Genet.*, 107: 43–48.

Chapter 11

Potato Transgenics

1. Introduction

Potato is the world's number one non-grain food crop and staple food of humanity that meets food demand of fast growing human population and therefore it has been regarded as food for future. The introduction of the first genetically modified biotech crop plants in the mid-1990s, made the agriculture industry has a steady increase in the acreage of those crops planted and harvested worldwide each year. In 2014, a record 18 million farmers in 28 countries planted 447 million acres of biotech soybean, maize, cotton, canola, zucchini squash, papaya, alfalfa, poplar, sugar beet, tomato, eggplant, and sweet pepper (James, 2015). This represents a more than 100-fold increase in usage between 1996 and 2014. This increase is largely due to the economic, environmental and productivity benefits derived from their use. The vast majority of biotech crops grown worldwide continue to be used primarily for animal feed (soybean, maize, alfalfa) or for fiber products that are not directly consumable (cotton), although many of the foods we eat contain ingredients derived from biotech crops (e.g., oils, starches, sugars).

The appearance of consumer for biotech crops on the market is increasing as biotech potato, including sweet corn, wheat, apple and papaya, have already completed or are currently awaiting the completion of the regulatory clearance process. The costs associated with the development of a new biotech crop variety make it difficult for research scientists to carry out the entire process without industry and market support (Miller and Bradford, 2010). However, in some cases where a devastating disease threatens crops, we have seen the relatively rapid release and acceptance of biotech crops containing resistance, such as with virus-resistant Rainbow papaya (Gonsalves, 1998). In the US, potato annually accounts for $4.2 billion in production value and the crop is grown on just over a million acres (James, 2015). Potato is an ideal crop for the introduction of traits using biotechnology. In fact, after virus-resistant tobacco in 1992 in China, and the FlavrSavr tomato in the USA in 1994, potato was one of the first crops to be genetically modified; it was grown commercially as New Leaf™ by Monsanto in 1995. This indicates use of transgenic technologies in crop improvement.

In potato, several transgenic plants have been developed for traits, like biotic stress, abiotic stress, tuber quality and plant phenotypes. A few transgenics have been commercialized worldwide by multinational companies (Table 1). Recently, Nahirñak et al. (2022) reviewed the genetic engineering technology in potato, including transgenics to next generation breeding technique, like genome editing. Potato improvement has shown promising results through conventional breeding and biotechnology. Conventional breeding has mostly focused on yield enhancement, quality and disease-pest resistance, but limited focus was given to abiotic stress tolerance and wild genes introgression (Kikuchi et al., 2015). Since conventional breeding is time consuming and challenging, genetic engineering provides the opportunity to overexpress or knockout the target gene of interest without altering allelic combinations of the genotype or modifying gene sequence through genome editing approach. Genetic modification techniques can be used to address the above problems in potato, particularly for introgression of genes derived from wild or related species into the cultivated potato species. Currently, genetic modification techniques have been applied in potato via gene overexpression or gene knockout mechanisms (RNAi) and recently, genome editing technology (Martínez-Prada et al., 2021). Genome editing has been discussed in a separate chapter. The traits which have been genetically modified in potato are resistance/tolerance biotic and abiotic stresses and tuber quality traits, such as late blight, viruses (PVX, PVY,

Table 1. Summary of potato transgenics commercialized worldwide.

Developer	**Commercial trade name**	**Trait**	**Genetic modificaiton**	**Country (year)**
Monsanto®	NewLeaf™	Colorado potato beetle resistance	*Cry3A* gene introduction	US, Canada (1995)
	NewLeaf™ Plus	Colorado potato beetle and potato leaf roll virus resistance	*Cry3A* and PLRV replicase and helicase genes introduction	US (1998)
	NewLeaf™ Y	Colorado potato beetle and potato virus Y resistance	*Cry3A* and PVY coat protein introduction	US (1998)
J.R. Simplot®	Innate® 1.0	Reduced acrylamide formation and black spot bruise	*Asn1* and *Ppo2* down-regulation	US (2015)
	Innate® 2.0	Reduced acrylamide formation, black spot bruise and CIS; *Phytophthora infestans* resistance	*Asn1, Ppo2* and *Vlnv* down-regulation; *Rpi-vnt1* introduction	US (2017)
BASF Plant Science	Amflora™	Reduced amylose formation	*GBSSI* down-regulation	EU (2010)
	Starch potato	Reduced amylose formation	*GBSSI* down-regulation	EU (2014)
Russian Academy of Sciences	Elizaveta Plus/ Lugovskoi Plus	Colorado potato beetle resistance	*Cry3A* introduction	RU (2005/2007)

Source: Adapted from Martínez-Prada et al. (2021)

PLRV), bacterial wilt, aphids, Colorado potato beetle, potato cyst nematode, potato tuber moth, salinity, heat, drought, cold, starch quality, storage and chip quality and plant phenotype.

2. Gene Cloning

Molecular cloning or recombinant DNA technology is an important practice in research labs to be used to create copies of a particular gene for downstream applications, such as sequencing, mutagenesis, genotyping or heterologous expression of a protein. Heterologous expression is used to produce large amounts of a protein of interest for functional and biochemical analyses. It is a traditional technique frequently used to learn about biochemical and functional aspects which involve the transfer of a DNA fragment of interest from one organism to another via a self-replicating genetic material plasmid. Cloned genes have been integrated randomly into the host genome for the production of transgenic organism called Genetically Modified Organism (GMO). Foreign DNA is introduced into the organism and then transmitted through the germ line so that every cell, including germ cells, of the organism contain the same modified genetic material. But in case of vegetatively propagated crops, stability of transgene needs to be checked in first two generations for inheritance of the gene and traits. This has been well proven from the transgenic potato events from various research groups, including India (Ottaviani et al., 1992; Sanju et al., 2015; Shandil et al., 2017; Tomar et al., 2018).

Gene cloning has become much easier with the availability of huge bioinformatics database. For example, the NCBI's (http://www.ncbi.nlm.nih.gov) GenBank database maintains more than 71 million sequences, out of which around 90,781 are from potato. Many of those nucleotide sequences represent identified genes. For potato, about 39,800 gene sequences are available from the PGSC database (http://solanaceae.plantbiology.msu.edu/pgsc_download.shtml). This information can be utilized for functional genomic research studies, including the development of transgenics. Availability of suitable regeneration protocol is a pre-requisite for undertaking genetic transformation work in any crop. A rapid and efficient *Agrobacterium tumefaciens*-mediated transformation protocol, based on direct organogenesis from internodal stem explants of *in vitro* potato plants, has been deployed the world over. It ensures development of a large number of transgenic plants in a short period of four to six weeks without formation of intermediary callus phase. The protocol gives about 70% regeneration efficiency and is being used routinely for genetic transformation of potato.

Before a cloning experiment is undertaken for such genes, scientists must understand how many genes are involved, how they are regulated, what other effects they might have on the plant and how it interacts with other genes active in the same biochemical pathway. Public and private research programs are investing heavily in new technologies, like transcriptome analysis of gene expression to investigate these issues and to rapidly sequence and determine functions of complex genes. These efforts should result in identification of a large number of genes, which are potentially useful for producing transgenic plants.

3. Genetic Transformation

3.1 Agrobacterium Tumefaciens-mediated Transformation

Availability of suitable regeneration protocol is a pre-requisite for undertaking genetic transformation and target-specific genome editing work in any crop. A rapid and efficient *Agrobacterium tumefaciens*-mediated transformation protocol based on direct organogenesis from inter-nodal stem explants of *in vitro* potato plants has been standardized. It ensures development of a large number of transgenic plants in a short period of four to six weeks without formation of intermediary callus phase. The protocol gives about 70% regeneration efficiency and is being used routinely for transformation work in various laboratories. Using these many transgenics have been developed and which possess resistance to various biotic/abiotic stresses and tuber quality traits.

3.2 Gene Gun-mediated Transformation

Biolistic (biological ballistics) plant transformation was initially developed in 1987 for transformation of monocots but later, the technology was used in most crops. Gene gun is very much required for the protoplast transformation in any crop, including potato. A gene gun-mediated transformation protocol has been successfully developed in potato. Though, potato is amicable to the *Agrobacterium*-mediated or indirect method of transformation, use of gene gun not only is a must for plastid transformation but also enhances the transformation efficiency in general. Here tungsten or gold particles coated with the DNA are accelerated to a high speed to bombard the target tissue. The technique has been successfully used to transfer the plastid-specific cassette for tuber-specific expression of *cry1Ab* and a fused *cry1Ab+cry1B* genes to develop transgenic potato resistant to potato tuber moth.

4. Development of Potato Transgenics

Genetic transformation, also called genetic modification, has many advantages for plant breeding and these advantages are even more striking in crops with polyploid complex inheritance, such as potato. While conventional breeding manipulates genomes in a largely uncontrolled fashion, requiring generations of selection to assemble and fix the maximum number of desirable traits, transformation offers a direct approach, allowing introgression of a single distinct gene without linkage drag. Thus, genetic modification allows rapid and often powerful improvement of crop plants and is not limited by compatibility barriers. In cases where genetic diversity among sexually compatible relatives of crop species is insufficient for a particular trait, genetic modification may represent the only possibility for improvement in that trait. Transformation offers a highly effective means of adding single gene to existing elite potato clones with no or very minimal disturbances. Potato, being highly amenable to genetic transformation, attracts attention of researchers to assess the impact of development of transgenic potatoes harbouring diverse traits. Genetic engineering in potatoes has a rather long history with the first transgenic potato developed about 30 years ago, and it is envisaged that many of the transgenic plant products to be commercialized in the present decade are likely to be potatoes with enhanced characteristics. In general, the policy of using genetic transformation technique is applied to improving those traits which cannot be manipulated by conventional breeding, since conventional breeding of potato is a complex and time-consuming exercise. Mutation breeding is also not feasible in a crop like potato that is autotetraploid, heterozygous and vegetative propagated. Transformation of potato with transgenes or cisgenes is, therefore, a new possibility to improve existing varieties without causing any major phenotypic side effect. Several gene sequences being available from the potato genome sequence consortium have been used for cloning and genetic transformation of potato cultivars. A few priority traits, namely durable disease resistance (late blight, viruses, bacterial wilt and soil and tuber borne diseases), pests (aphids, white fly, potato tuber moth) reduction of cold-induced sweetening, nutritional enhancement, abiotic stress tolerance (heat, drought and nutrient use efficiency) and tuberization under high temperature are important for improvement by genetic engineering. Table 2 summarizes different recent transgenics developed in potato for biotic stress, abiotic stress quality traits and plant phenotypes in potato. Below is the summary of important traits targeted for transgenics development in India.

- *Late blight resistance transgenics*: Since none of the varieties having durable resistance, lines have been developed expressing *RB* gene in popular Indian potato cultivar using cis gene approach followed by back cross breeding.
- *Virus resistance transgenics*: Due to unavailability of host resistance sources, transgenics have been developed in potato against tomato leaf curl New Delhi Virus (ToLCNDV), using *Replicase* gene of RNAi constructs and potato virus Y (PVY) using coat protein gene.
- *Transgenic with reduced cold-induced sweetening*: Indian potato varieties lack resistant against sweetening during cold storage, so there is transgenics with reduced cold-induced sweetening potatoes using RNAi technology.
- *Transgenics with dwarf plant architecture*: Dwarf potato lines have been developed for better harvest index, particularly hilly conditions.
- *Bacterial wilt resistance transgenics by RNAi*: None of the germplasm have resistance source, hence employed RNAi technology to suppress the susceptible factor to achieve resistance against bacterial wilt.
- *Transgenics for heat and drought-tolerant*: Potato transgenics have been developed for heat and drought stress tolerance to meet the changing climate scenario.

4.1 Late Blight Resistance

Late blight caused by the oomycetes fungus, *Phytophthora infestans*, is a major disease of potato. The disease has the history of causing catastrophic famine in Ireland, where people depended heavily on this crop. Since Irish famine in 1845–49 till date, potato production across the globe is severely affected by this notorious disease. Globally € 12 billion crop are lost annually and in India, average crop losses

Table 2. A summary of recent transgenics developed through gene overexpression and silencing approaches in potato.

Sr. No.	Trait	Gene and genotype	Transgenic approach	Key findings	References
1.	PVX, PVY and PLRV resistance	ORF2 (PVX), Helper Component Protease gene (PVY), and Coat Protein (CP) gene (PLRV) cv. Desiree and Kuroda	RNAi, *A. tumefaciens*	Simultaneous RNA silencing demonstrated that 20% of the transgenics plants were immune to all three viruses by accumulation of specific siRNAs.	Arif et al. (2012)
2.	Phenotype	cvs. Lady Olympia, Granola, Agria, Désirée and Innovator	Overexpression, *A. tu mefaciens*	Development of efficient, reproducible and stable *Agrobacterium*-mediated genetic transformation methods using leaf discs and intermodal explants, of which, intermodal explants showed best results in cv. Lady Olympia.	Bakhsh (2020)
3.	Periderm formation	*S. tuberosum RS2-INTERACTING KH PROTEIN* (*StRIK*) cv. Desiree	Overexpression, *A. tumefaciens*	*StRIK* gene plays a regulatory role in potato tuber periderm formation through stress signaling and RNA metabolism.	Boher et al. (2021)
4.	Bacterial wilt resistance	*Arabidopsis thaliana* elongation factor-Tu (EF-Tu) receptor (*AtEFR*) cv. INIA Iporá and 09509.6	Overexpression, *A. tumefaciens*	The combination of *AtEFR* expression combined with quantitative resistance introgressed from wild *S. commersonii* is a promising strategy to develop bacterial wilt-resistance potato plants.	Boschi et al. (2017)
5.	Salt and drought stress tolerance	Drought-responsive element binding protein gene (*StDREB1*) cv. Nicola, and Belle de Fontenay	Overexpression, *A. tumefaciens*	Overexpression of *StDREB1* transcription factor improved tolerance to salt and drought stress in transgenic potato plants.	Bouaziz et al. (2013)
6.	Starch quality	STARCH BRANCHING ENZYME II (*SBEII*) and granule-bound starch synthase (*GBSS*) cv. Karaka	*A. tumefaciens*	Overexpression of *SBEII* gene is an effective way to modify potato starch physicochemical properties for commercial applications. Study indicated that an increased ratio of short to long amylopectin branches produces commercially useful starch properties.	Brummell et al. (2015)
7.	PVY resistance	CP, untranslated region 3'-UTR) (UR), and CP-UR genes cv. Marfona	RNAi, *A. tumefaciens*	Study showed more than 67% of transgenic potato plants were PVY resistant and all three gene constructs showed similar degrees of resistance. The highest numbers of transgenic lines with high resistance levels were found in the CP-UR followed by CP and UR lines, respectively.	Byarugaba et al. (2021)

8.	Salt stress tolerance	Potato ethylene responsive factor (*StERF94*) cv. Spunta	*A. tumefaciens*	Overexpression of the *StERF94* transcription factor increased salinity tolerance by improving plant growth, osmoprotectant synthesis and antioxidant activity leading to low oxidative stress damage.	Charfeddine et al. (2019)
9.	Cold storage and processing traits	Vacuolar invertase gene (*VInv*) cv. Ranger Russet	RNAi (TALEN), Protoplast transformation	The targeted gene (*VInv*) knockout demonstrated improved cold storage and processing traits in potato. Tubers of full *VInv*-knockout plants showed an undetectable level of reducing sugars, acceptable chip colour and reduced acrylamide content.	Clasen et al. (2016)
10.	*M. chitwoodi* resistance	Effector gene *Mc16D10L* cv. Rutgers	RNAi, *A. tumefaciens*	RNA interference approach demonstrated that specific silencing of the putative effector gene *Mc16D10L* led to *M. chitwoodi* resistance in *Arabidopsis* and potato.	Dinh et al. (2014)
11.	Root-knot nematode resistance	*Meloidogyne* effector gene *16D10* cv. Russet Burbank	RNAi, *A. tumefaciens*	The plant-mediated RNAi silencing of the *16D10* effector gene resulted in significant resistance against different root-knot nematode species, viz. *M. arenaria, M. chitwoodi, M. hapla, M. incognita* and *M. javanica*.	Dinh et al. (2015)
12.	Potato–PVY interaction	Mitogen-activated protein kinase genes (*StWIPK* and St*MKK6*), *S. venturii* (VNT366-2)	Virus-induced gene silencing (VIGS), *A. tumefaciens*	Study revealed that *S. venturii* is a suitable model for potato-PVY interactions via VIGS of *StMKK6* gene. Silencing of *StMKK6* enabled a faster spread of the virus throughout the plant, while silencing of *WIPK* had no effect on virus spread.	Dobnik et al. (2016)
13.	Heat stress tolerance	*Arabidopsis thaliana* cold-regulatory C-repeat binding factor (*CBF*) gene *AtCBF3* cv. Luyin NO.1	*A. tumefaciens*	Results suggested that overexpression of the *AtCBF3* gene enhanced heat stress tolerance up to 40°C or even higher in potato plants. This might be due to increase in gene expression related to stress resistance, photosynthesis and antioxidant defence, but independent of HSP70 regulatory pathways.	Dou et al. (2014)
14.	PVY resistance	Host factor *eIF4E-1* variant *Eva1*, *S. chacoense, S. demissum, S. etuberosum*	Overexpression and RNAi, *A. tumefaciens*	Overexpression of the wild potato *eIF4E-1* variant *Eva1* conferred PVY resistance in transgenic potato silenced for the native *eIF4E-1* gene.	Duan et al. (2012)
15.	Starch content and tuber sprouting	*Isoamylase 1* (*ISA1*), *isoamylase 2* (*ISA1*), *isoamylase 3* (*ISA3*) cv. Solara	RNAi, *A. tumefaciens*	Simultaneous RNAi silencing of *ISA1*, *ISA2* and *ISA3* genes showed reduced content of starch and sucrose but more sucrose in potato tubers of transgenic plants. High sucrose levels might also explain the increased number of growing sprouts per tuber.	Ferreira et al. (2017)

Table 2 contd. ...

...Table 2 contd.

Sr. No.	Trait	Gene and genotype	Transgenic approach	Key findings	References
16.	Late blight resistance	*RB* and *Rpi-blb2* (*S. bulbocastanum*) and *Rpi-vnt1.1* (*S. venturii*) cvs. Desiree, Victoria	*A. tumefaciens*	Stacking of three late-blight resistance genes (*RB* and *Rpi-blb2*, and *Rpi-vnt1.1*) provided complete field resistance to local races in African highland. In the field, 13 resistant transgenic events with all three R genes did not show any infection and without any fungicide spray, whereas non-transgenic plants were fully infected.	Ghislain et al. (2019)
17.	Late blight and Andean potato Weevil resistance	Phenylalanine-derived benzylglucosinolate (BGLS) genes: *SUR1, UGT74B1, SOT16, CYP79A2, CYP83B1* and *GGP1* cv. Desiree	*A. tumefaciens*	Results showed that transgenic events produced as high as 5.18 pmol BGLS/mg fresh weight compared to undetectable in non-transgenics. Preliminary bioassays exhibited resistance to late blight and the Andean potato weevil (*Premnotrypes suturicallus*). However, transgenic lines had abnormal leaf morphology, thickness, curly leaves and less tuber yield.	González-Romero et al. (2021)
18.	PVX, PVY and PVS resistance	CP gene cv. Desiree	RNAi, *A. tumefaciens*	Nearly 100% resistance against PVX, PVY and PVS infection was observed in potato transgenics. The study demonstrated the efficacy of RNAi-mediate simultaneous resistance to three viruses, using the CP gene as a potential target for induction of stable resistance.	Hameed et al. (2017)
19.	Transgenic root & non-transgenic shoot	The *uidA* gene encoding b-glucuronidase (GUS) cvs. Albatros, Desiree, Sabina and Saturna	*Agrobacterium rhizogenes*	The study showed that antisense, inverted-repeat and hairpin constructs can be used to induce gene silencing in hairy roots of composite plants. This may be useful for study of systemic antiviral silencing or induced systemic resistance caused by infection of the roots by viruses or microbes, respectively.	Horn et al. (2014)
20.	Colorado potato beetle (CPB) resistance	Ecdysone receptor (*EcR*) gene of CPB cvs. Agria and Lady Olympia	RNAi, *A. tumefaciens*	The plant-mediated RNAi silencing approach showed enhanced resistance against CPB through the suppression of transcripts level of molting-associated Ecdysone receptor (EcR) gene of CPB (*Leptinotarsa decemlineata*, Say) and leaf bioassays of transgenic plants exhibited 20–80% of mortality of CPB.	Hussain et al. (2019)
21.	Late blight resistance	*P. infestans* genes: *PiGPB1*, *PiCESA2*, *PiPEC* and *PiGAPDH* cv. Desiree	*A. tumefaciens*	Study revealed that the *hp-PiGPB1* gene targeting the G protein β-subunit is important for pathogenicity and provides high resistance to late blight. They suggest that an RNAi strategy could be processed in host plant for targeting pathogen genes to provide late blight resistance.	Jahan et al. (2015)

22.	PVY resistance	Coat protein (CP) gene, untranslated region (UTR) (UR), and CP-UR cv. Marfona	RNAi, *A. tumefaciens*	RNAi-based transgenic potato plants exhibited resistance to PVY through targeting CP, UR and CP-UR genes. Bioassay analysis revealed that more than 67% of transgenic plants were resistant to PVY. CP-UR lines exhibited relatively high resistance followed by CP and UR expressing lines, respectively.	Jahromi et al. (2022)
23.	Late blight resistance	*Rpi-sto1* and *Rpi-vnt1.1* cvs. Atlantic, Desiree, Bintje, Potae9	*A. tumefaciens*	Developed a marker-free transformation pipeline to select potato plants functionally expressing R genes, *Rpi-sto1* and *Rpi-vnt1.1* originating from wild species, *S. stoloniferum* and *S. venturii*, respectively. This study provides an important cis-genic approach for the successful deployment of R genes in potato.	Jo et al. (2014)
24.	Drought and heat stress	*Arabidopsis thaliana* hexokinase 1 (*AtHXK1*), SELF-PRUNING 6A (*SP6A*) genes cvs. Désirée, Solara	*A. tumefaciens*	Demonstrated drought and heat stress tolerance in potato by co-overexpression of *AtHXK1* and *SP6A* genes. The guard cell-specific expression resulted in decreased stomatal conductance and improved water use efficiency. Further, co-expression of *AtHXK1* with the FT-homolog *SP6A* improved drought and heat stress tolerance.	Lehretz et al. (2021)
25.	Freezing tolerance	Stearoyl-acyl carrier protein desaturase (SAD) genes *ScoSAD, SaSAD, ScaSAD* and *StSAD* cv. Zhongshu 8	*A. tumefaciens*	Cloning and functional characterization of SAD genes, *ScoSAD* (*S. commersonii*), *SaSAD* (*S. acaule*), *ScaSAD* (*S. cardiophyllum*) and *StSAD* (*S. tuberosum*), showed that overexpression of the *ScoSAD* gene in transgenic plants significantly enhanced freeze tolerance. This led to increased level of linoleic acid content suggesting that linoleic acid possibly plays a key role in improving freeze tolerance.	Li et al. (2015)
26.	Aphid (*Myzus persicae*) resistance	*Myzus persicae* the gap *hunchback* gene (*Mphb*) Tobacco cv. Samsun NN)	RNAi (dsRNA)	Plant-mediated RNAi silencing of the *Mphb* gene enhanced aphid tolerance in tobacco. When aphid nymphs were fed on *Mphb* dsRNA-expressing plant, *Mphb* mRNA level was reduced and aphid fecundity was impaired.	Mao et al. (2014)
27.	Transgene stacking	10-stack T-DNA/genes cv. B5141-6 (Lenape)	*Agrobacterium rhizogenes* GAANTRY system	The study showed successful transgene stacking in potato using the GAANTRY (Gene Assembly in *Agrobacterium* by Nucleic acid Transfer using Recombinase technologY) system. The GAANTRY system efficiently generated high quality transgenics with stacked genes.	McCue et al. (2019)

Table 2 contd. ...

...Table 2 contd.

Sr. No.	Trait	Gene and genotype	Transgenic approach	Key findings	References
28.	Potato tuber moth resistance	*Phthorimaea operculella* chitin synthase A (*PhoCHSA*) gene	RNAi	RNAi-mediated knockdown of three different regions of gene *PhoCHSA* of *Phthorimaea operculella* showed different efficacy against larvae and dsRNA targeting the 5′ region has the highest efficacy. This indicates that *PhoCHSA* gene can be a suitable RNAi target for insect control.	Mohammed et al. (2017)
29.	PLRV resistance	Coat protein (*CP*) and recombinase (*CreloxP*) genes cv. Desiree	RNAi, *A. tumefaciens*	An inverted repeat construct of PLRV coat protein gene under the control of a constitutive promoter with a heat inducible *Cre-loxP* system was created to excise the *nptII* antibiotic resistance marker gene. Of 58 total transgenic events evaluated for PLRV resistance, seven were highly resistant and four were extremely resistant.	Orbegozo et al. (2016)
30.	Shoot branching, stolon and tuber growth	*CAROTENOID CLEAVAGE DIOXYGENASE8* (CCD8) gene cv. Desiree	RNAi	The *CCD8* gene play a key role in strigolactones (SLs) phytohormones biosynthesis pathway controlling shoot branching. The *CCD8*-RNAi potato plants showed significantly more lateral and main branches, reduced stolon formation and dwarfing phenotypes. The study demonstrates the role of *CCD8* gene in potato stolon and tuber development.	Pasare et al. (2013)
31.	PVY^{NTN} resistance	Viral *HC-Pro* gene cv. Agria	RNAi (dsRNA)	The study demonstrated a new way of controlling PVY by blocking replication and transmission through the plant by RNAi-based (dsRNA) control. Blocking the *HC-Pro* gene of PVY^{NTN} in newly grown plant was established. The old leaves remained infected but later defoliated, leaving the plant virus free.	Petrov et al. (2015)
32.	Drought stress tolerance	RING finger protein (RFP) gene *StRFP2* cv. Atlantic and Qingshu 9	*A. tumefaciens*	Developed transgenic plants overexpressing RING-finger protein gene *StRFP2* in potato, showing significantly higher plant phenotypes and drought tolerance under polyethylene glycol (PEG) osmotic stress conditions.	Qi et al. (2020)
33.	Late blight resistance	RXLR effector *Avr3a* gene cvs. Kufri Khyati, Kufri Pukhraj	RNAi	The study showed that host-mediated gene silencing of the RXLR effector *Avr3a* gene of the pathogen imparts partial resistance in potato.	Sanju et al. (2015)

34.	Abiotic stress tolerance (drought, salt, low temperature, ABA)	*StGA2ox1* gene cv. Qingshu 9 (Q9) and Mingshu 1	*A. tumefaciens*	Overexpression of *StGA2ox1* gene increases tolerance to different abiotic stress conditions, such as dehydration, low temperature, abscisic acid treatment, salt, drought, exogenous hormones. Results suggested that *StGA2ox1* is involved in the regulation of plant growth and provides abiotic stress tolerance by regulating the gibberellin synthesis pathway.	Shi et al. (2019)
35.	Late blight and powdery mildew resistance	*Arabidopsis* multiple susceptibility (S) gene: *Defense No Death 1 (DND1)* cv. Desiree, and diploid SH83-92-488 (SH)	RNAi	RNAi silencing of the *DND1* ortholog in potato and tomato resulted in resistance to late blight (*P. infestans*) and two powdery mildew species (*Oidium neolycopersici* and *Golovinomyces orontii*). Late blight resistance in potato was found effective to four different isolates from complete or partial resistance.	Sun et al. (2016)
36.	Late blight resistance	*A. thaliana* susceptibility (S) genes (*CESA3, DMR1, DMR6, DND1, SR1, PMR4, BIK1, CPR5, DND2, PMR5,* PMR6) orthologs in potato cv. Desiree	RNAi	RNAi silencing of five S-genes orthologos in genes showed complete resistance to late blight but sixth S-gene resulted in reduced susceptibility.	Sun et al. (2016)
37.	Drought stress tolerance	Potato E3 ubiquitin ligase *PUB27* cv. Atlantic, and Qingshu 9	Overexpression and RNAi *A. tumefaciens*	Overexpression of the *StPUB27* gene accelerated dehydration of detached leaves and greater stomatal conductance, while the RNAi silencing of *StPUB27* showed smaller stomatal conductance and thus higher tolerance to osmotic stress. The study demonstrates that potato *PUB27* negatively regulates drought tolerance by mediating stomatal conductance.	Tang et al. (2020)
38.	Vitamin E and abiotic stress tolerance (salt and heavy metal)	*Arabidopsis* homogentisate-phytyltransferase (*At-HPT*) and γ-tocopherol-methyltransferase (*At-γ-TMT*) cv. Kufri jyoti	Overexpression, *A. tumefaciens*	The transgenic plants resulted in significant increase in vitamin E content (173-258 %) and also had increased cellular antioxidant enzymes, proline, osmolyte and glutathione contents that are directly correlated with abiotic stress tolerance to salt (NaCl) and heavy metal ($CdCl_2$).	Upadhyaya et al. (2021)
39.	Drought stress tolerance	Plasma membrane intrinsic proteins (PIPs) aquaporins gene *StPIP1* cv. Shepody	*A. tumefaciens*	The overexpression of *StPIP1* gene in potato provided drought and osmotic stress tolerance by maintaining overall plant water balance, photosynthesis, stomatal conductance, plant growth, carbon assimilation, storage and yield.	Wang et al. (2017)

Table 2 contd. ...

...Table 2 contd.

Sr. No.	Trait	Gene and genotype	Transgenic approach	Key findings	References
40.	Salt tolerance	*Arabidopsis* high-affinity potassium transporter gene (*AtHKT1*) cv. Shepody	*A. tumefaciens*	The study showed that the constitutive overexpression of *AtHKT1* reduced Na^+ accumulation in potato leaves and promoted the K^+/Na^+ homeostasis that minimizes osmotic imbalance, maintains photosynthesis and stomatal conductance, and increased plant productivity.	Wang et al. (2019)
41.	Starch phosphate content and quality	*Laforin* gene cv. Kardal	Overexpression	Expression of an (engineered) laforin in potato resulted in significantly higher phosphate content of starch. Modified starches exhibited altered granule morphology and size compared to the control.	Xu et al. (2017)
42.	Late blight and early blight resistance	Sesquiterpene cyclase (SC) *potato vetispiradiene synthase* (PVS) gene cv. Sayaka	RNAi	Potato antimicrobial sesquiterpenoid phytoalexins, lubimin and rishitin, showed resistance to late blight (*P. infestans*) and early blight (*A. solani*) pathogens. The study suggested that sesquiterpene cyclase-mediated compounds participate in pre-invasive resistance to *A. solani* and post-invasive resistance to pathogen *P. infestans*.	Yoshioka et al. (2019)
43.	Osmotic stress and lateral root growth	NAC transcription factor *StNAC262*, and miRNA *Stu-mi164* cv. Gannongshu 2 and Kexin 3	*A. tumefaciens*	Under osmotic (poly ethylene glycol) stress, results showed that transgenic potato plants overexpressing Stu-mi164 had reduced expression of *StNAC262*, decreased osmotic resistance and lower number of lateral roots. The study suggests the regulatory role of Stu-miRNAs in controlling plant response to osmotic stress.	Zhang et al. (2018)
44.	Starch quality	Sucrose non-fermenting-1-related protein kinase (*SnRK1*)-related genes cv. Desiree	Overexpression and RNAi *A. tumefaciens*	Identified five functional *SnRK1*-related genes in potato, including three novel genes, which encode one α-subunit isoform (*stKIN*), two β-subunit isoforms (*stKINβ1* and *stKINβ2*) and two γ-subunit isoforms (*stKINγ* and *stKINβγ*). The *stKIN* is the primary α subunit of *SnRK1* playing a key role in potato tuber development.	Zhang et al. (2018)
45.	Salt stress tolerance	Brassinosteroids biosynthesis gene *StDWF4* cv. Zihuabai	Overexpression and RNAi *A. tumefaciens*	The overexpression of *StDWF4* in potato enhanced salt tolerance with higher proline content, soluble protein, soluble sugar, superoxide dismutase activities, peroxidase and ascorbate peroxidise. However, its RNAi silenced expression depressed the salt tolerance.	Zhou et al. (2018)

46.	Colorado potato beetle resistance	Gene *cry3A* cv. Atlantic	Overexpression	Developed Colorado potato beetle (*Leptinotarsa decemlineata* Say) resistant transgenic potato plants by overexpressing *cry3A* gene. The mortality of the larvae was higher on the transgenic lines than in control lines.	Mi et al. (2015)
47.	Colorado potato beetle resistance	V-ATPase subunit A gene and Cry3Aa	RNAi (dsRNA)	The study documented high pesticide resistance, using dsRNA technology delivered through non-transgenic techniques. Information on the genetics of resistance and availability of dsRNA-resistant will provide insights on the design of resistance management in potato.	Mishra et al. (2021)
48.	Late blight resistance	*RB* gene (*S. bulbocastanum*) cvs. Kufri Jyoti, Kufri Bahar, and Katahdin transgenic event SP951	Overexpression and crossing	High degree of resistance (< 25% infection) was observed in Kufri Jyoti × SP951 derived seedlings than the seedlings of Kufri Bahar × SP951. This study suggests that the efficacy of R gene is not solely dependent on its presence in the variety but also on the genetic background of the recipient genotype.	Shandil et al. (2017)

to the tune of 15% have been estimated, which amounts to 6.7 million metric tons of potatoes. Though various management strategies that include host resistance, disease forecasting, use of chemicals, cultural practices, etc. are being employed to manage this disease, but *P. infestans* is such a notorious oomycete that evolves and adapts to the host background and new fungicide molecules rapidly within a few years of their release in the environment because of predominance of transposable elements in its genome.

In recent years, India and China emerged as the global leaders in potato production, together contributing about 50% of world production. Popularity of potato in these two Asian giants is largely due to remarkable productivity of this crop per unit area and time. If one thing can mar this happy situation, it is an epidemic of late blight. Causes of concerns about this disease in India are already evident. In addition, occurrence of both A1 and A2 mating type of *P. infestans*, resulting in sexual reproduction and survival through resilient oospores, have been reported that may give rise to immense variability in the pathogen population, thereby endangering durability of a cultivar. Moreover, this population is gradually becoming tolerant to higher doses of prophylactic fungicides. As a consequence of this hidden but serious population shift in *P. infestans*, Kufri Jyoti, the most popular Indian cultivar, has succumbed to this disease after a sustained performance for about 30 years. The other popular cultivar, Kufri Bahar, does not have any resistance to *P. infestans*. Together, these two cultivars along with Kufri Pukhraj occupy > 60% of potato area in India, creating an imminent danger. Race-specific major genes from the wild potato species, *Solanum demissum*, have been extensively used in resistance breeding programmes throughout the world, including India. However, efficacy of such major genes had been too short-lived to justify their deployment. Therefore, thrust in late blight breeding has now shifted to deployment of multi-gene, horizontal resistance. Identification of candidate genes responsible for horizontal resistance and their pyramiding is a formidable task that can't be achieved in the foreseeable future.

4.1.1 RB Gene Transgenics

The cloning of the *RB* gene opened up the possibility of using recombinant DNA technology to transfer the gene to commercially important and susceptible potato varieties to diversify and strengthen late blight resistance in cultivated potatoes. Late blight disease was managed by the use of *R* genes from the wild *Solanum* species into the cultivated potato. The *RB/Rpi-blb*1 gene was isolated from the wild *Solanum bulbocastanum* and cloned in pCLD04541 to be transferred into potato cv. Katahdin using *Agrobacterium tumefaciens* (Song et al., 2003; Vleeshouwers et al., 2008). This gene imparted broad spectrum of resistance against all known races of *P. infestans* to the RB-Katahdin lines, both in the greenhouse and in field experiments (Song et al., 2003; Kuhl et al., 2007; Halterman et al., 2008). A wild diploid potato species, *S. bulbocastanum* from Mexico and Guatemala, possessing very high degree of resistance to late blight controlled by classical resistance (*R*) genes, is an exception. *RB* gene conferring broad-spectrum resistance in *S. bulbocastanum* has been cloned and showed to provide stable late blight resistance over generations (Song et al., 2003). Similarly, several *RB*-homologous genes were identified in wild species, such as *S. chacoense*, *S. pinnatisectum*, *S. polyadenium*, *S. trifidum*, *S. cardiophyllum*, *S. lesteri*, *S. huancabambense*, *S. verrucosum*, *S. jamesii*, *S. polytrichon* and *S. stoloniferum*. This may serve as an important genomic resource for novel gene discovery for late blight resistance.

The event SP951 of the potato cultivar, Katahdin, contains the CC-NBS-LRR class of *R* gene (*RB*), isolated from the sexually incompatible wild species, *S. bulbocastanum*. This event was transferred to Indian Council of Agricultural Research from University of Wisconsin, USA through Agricultural Biotechnology Support Project II. ICAR-CPRI, Shimla carried out confined field trial (CFT) with permission from RCGM and validated late blight field resistance in this event under Indian conditions. Further, the *RB* gene was introgressed in F_1 hybrids by crossing the popular cultivar Kufri Jyoti (female) with SP951 (male) and identified promising transgenics lines with very high late blight resistance and other desirable traits. The hybrids were tested and two of the hybrids, namely KJ65 and KJ67, showed promising results consistently since the last seven to eight years. The site of integration and copy number of the inserted RB gene in the event SP951 has been determined through molecular analysis (Shandil et al., 2017) (Fig. 1).

Promising lines were selected on the basis of per cent infection on the last day (57th day after planting, DAP) of scoring as well as the area under the disease progress curve (AUDPC). AUDPC is a measure of disease development over the whole course of the epidemic. The lines showing 40% or less late blight infection at 57 DAP and AUDPC values below that of control varieties, Kufri Bahar and Kufri Jyoti, were selected for further evaluation. Therefore, *RB* transgene can be used effectively to reduce foliar late blight infection in cultivated potatoes. This enhanced resistance over the generations may be ascribed to the pooling of *R* gene activation or reshuffling of *R* genes in the background of Kufri Jyoti that imparts resistance against

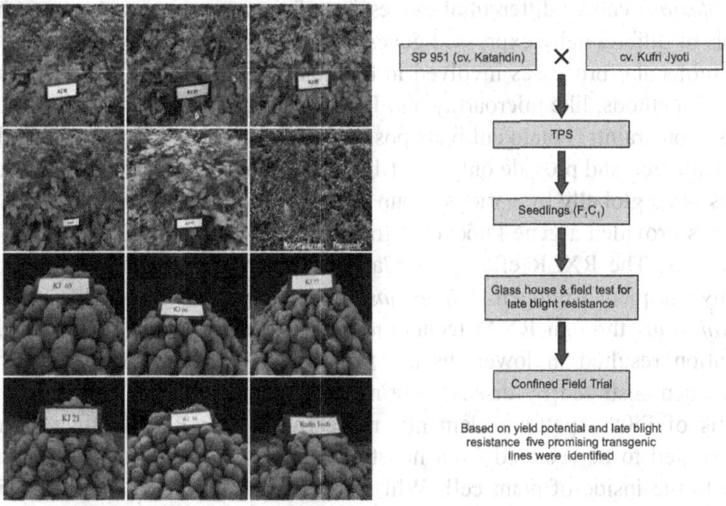

Fig. 1. Late blight resistant RB gene transgenics developed in India through crossing between cv. Katahdin Event SP951 and common cv. Kufri Jyoti, and selected for BRL1 trials (*Courtesy*: Photograph by Dr. Sundaresha S.).

Phytophthora infestans. This RB positive F_1 genotype showing different levels of late blight resistance will be useful for identifying promising downstream genes to build the future molecular breeding activities.

Expression of RB protein, like other R proteins in plants, is extremely low due to its intractable nature. Because of its intractable nature, RB protein could not be purified from either the transgenic Event SP951 or from the host species, *S. bulbocastanum*. As a consequence, no antibody could be produced for this protein. Similarly, no detectable level of recombinant RB protein could be expressed either in bacterial or yeast host. It posed a serious challenge to estimate RB protein in the transgenic events through Western blotting. Therefore, allergenicity of this protein was determined, using the bioinformatics tool, i.e., Allergen Online version 16. The tool did not identify any significant alignment with known allergens. PubMed searches also failed to show any published evidence indicating allergenic or toxic nature of the RB protein (Goodman, 2015). It is, therefore, argued that regulatory food safety study on pure RB protein cannot be undertaken due to the intractable nature of this common plant protein which is supposed to be non-toxic and non-allergenic. However, it would be possible to generate biosafety and food safety data of the Event SP951 or its F_1 progenies at the time of BRL-1 and BRL-2 trials. This RB lines with durable resistance levels would be useful for future resistance molecular breeding activities.

4.1.2 RNAi Transgenics

When potato and *P. infestans* come into contact, a complex and dynamic communication occurs between the two organisms. This activates the defense mechanism of potato and many specific genes may be induced. Several studies have demonstrated that the challenge with *P. infestans* leads to transcriptional activation of specific genes in potato. Furthermore, the activation of potato defense responses by *P. infestans* causes differential expression of numerous genes. A comprehensive analysis of differentially expressed genes could contribute to a better understanding of the molecular processes involved in the plant–pathogen interactions. Therefore, powerful methods, like microarray and RNAi approach, could be used to anticipate all these constraints. Potato cultivars possessing *R* genes are not resistant to all races of the pathogen and provide only short-lived resistance in the field as has been well demonstrated globally by various groups. Host-induced gene silencing of pathogen genes has provided a gene knockout strategy for a wide range of biotechnological applications. The RXLR effector *Avr3*a gene is largely responsible for virulence of oomycete plant pathogen *P. infestans*. We attempted to silence the *Avr3a* gene of *P. infestans* through RNAi technology (Thakur et al., 2015). The *P. infestans* inoculation resulted in lower disease progression and a reduction in pathogen load, as demonstrated by disease scoring and quantification of pathogen biomass in terms of Pi08 repetitive elements, respectively. The RXLR effector *Avr3*a is demonstrated to be secreted from haustoria and translocated from extrahaustorial matrix to the inside of plant cell (Whisson et al., 2007). This would explain why the higher accumulation of siRNA silencing molecules in the host cells led to the elevated endogenous silencing of target Avr3a gene. These observations suggest that the nature of silencing molecules involved in the HIGS of *P. infestans* seems to be siRNAs generated in the host.

Transgenic plants induced moderate silencing of *Avr3a* and the presence and/or expression of small interfering RNAs, as determined through northern hybridization, indicating siRNA targeted against *Avr3a* conferred moderate resistance to *P. infestans*. The single effector gene did not provide complete resistance against *P. infestans*. Although the *Avr3*a effector gene could confer moderate resistance, for complete resistance, the cumulative effect of effector genes in addition to *Avr3a* needs to be considered. Researchers have demonstrated that host-induced RNAi is an effective strategy for functional genomics in oomycetes. Whisson et al. (2012) suggested that RXLR effector genes are preferentially located in genomic regions heavily populated with transposons, and small RNAs involved in silencing can extend from the genome region spanning the effector and transposons. In future, further mining of the siRNA inhibitor and stacking of other effector genes on single or multiple constructs is required, and also to study the influence of nearby transposons on the evolution and expression of effectors that provide high levels and durable resistance against late blight.

Another technique which has been exploited in development of resistance against late blight is the RXLR effecter gene (*Avr3a*) of the pathogen that is responsible for pathogenicity. The silencing of pathogen effecter gene, responsible for pathogenicity through host-mediated technique, would prevent the pathogen establishment and spread. The *Avr3a* gene was targeted for silencing by using siRNA (small interfering RNA) and amiRNA (artificial micro RNA) method through binary construct developed for transformation of Indian potato cv. Kufri Pukhraj. Five lines of siRNA and three lines of amiRNA showing resistance against late blight up to 90% were identified and tested at the field level. Besides, RNA solution consisting of dsRNA of the target pathogenicity gene for silencing can be applied to plants through spraying to prevent the disease. This technology of dsRNA spray formulation to prevent the late blight disease is now emerging to control late blight disease.

4.2 Virus Resistance

Viral diseases are important for seed potato production system as they not only affect the yield loss but also cause the degeneration of seed quality. Around 30 different viruses are known to infect potato crop, but only a few are more important in contributing to degeneration of seed stock. The major important potato viruses are potato virus Y (PVY), potato leaf roll virus (PLRV), potato virus X (PVX) and potato apical leaf curl virus (PALCV). In addition to many known viruses infecting potato, newly emerging viral or virus-like diseases also threaten potato seed production. Due to vegetative mode of propagation of potato, many of the viruses accumulate in seed tubers, causing severe yield depression. PVY highly virulent mutant PVY^{NTN}, PLRV and the newly emerging tomato leaf curl New Delhi virus—ToLCNDV-potato, also called PALCV—is so serious that it cannot be managed by conventional breeding due to limited source of host resistance. Transgenic development by using pathogen-derived resistance is, therefore, being pursued for virus management.

4.2.1 Tomato Leaf Curl New Delhi Virus-potato (ToLCNDV-Potato)

Viruses are important pathogens which are ubiquitous and cause 80% losses in potato yield. Potato has been infected by more than 40 viruses and two viroids (Jeffries et al., 2005), of which so far only nine viruses and two viroids are of economic significance for potato. There are numerous variable factors, i.e., plant genetic diversity, biology, lifecycle of the host plant/ pathogen, vector species, biotype and environmental conditions, that affect the incidence and severity of viral diseases (Yi and Gray, 2020). For example, in India potato virus Y (PVY), potato leaf roll virus (PLRV), potato virus X (PVX), potato virus S (PVS), potato virus 68 A (PVA) and potato virus M (PVM) cause yield reduction up to 10–30%, depending upon the 69 intensity and vector pressure (Khurana, 2004; Kumar et al., 2017; Fox et al., 2017). The incidence of this virus has immensely increased year by year and it has captured the top position in Indian potato viruses during the last two decades. In the Indo-Gangetic Plains, the disease incidence is 40–100% which leads to heavy yield losses in potato (Lakra, 2002; Venkatasalam et al., 2011).

Potato apical leaf curl disease is an emerging Gemini viral disease in tropics and subtropics. Numerous data on PALCV incidence have been documented and are available in several published annual reports and research literature (Garg et al., 2001; Lakra, 2002; Usharani et al., 2003; Jeevalatha et al., 2013; Jeevalatha et al., 2018). PALCV-affected plants become severely stunted with apical leaf curl, crinkled leaves and conspicuous mosaic. Though potato genotypes with multiple disease resistance have been identified in India, using marker-assisted selection (Sharma et al., 2014), no specific source of resistance in potato has yet been reported for ToLCNDV-potato. Alternatively, advances in RNA biology open the way to create RNAi method to silence the viral particle, using the host as an RNAi machinery. RNAi occurs in most of the eukaryotes as a potent gene regulation phenomenon. In plants, RNAi is used as a generally applicable antiviral strategy. RNAi induced by double stranded RNA molecules, like short hairpins, short interfering RNAs and long distance dsRNA, have been developed as standard tools in gene function studies and as antiviral strategies. Tomar et al. (2018) initiated the isolation and cloning of the *AC1* gene of ToLCNDV from field-infected potato leaves, cloned and sequenced (JN393309). Transgenic plants encoding the hairpin loop *AC1* replicase gene was developed. Transgenic lines when inoculated with ToLCNDV-potato showed different levels of resistance. Transgene integration and copy number in selected transgenic lines were determined by qPCR and further confirmed by southern blot analysis. Though a reduction in viral titer was observed in transgenic lines, encoding either antisense or hairpin loop constructs of *AC1* gene, the latter transgenics showed most significant results through reduction in the level of symptom expression in glasshouse screening as well as real-time data of *in vivo* virus concentration. Potato apical leaf curl virus (PALCV) resistant transgenics, namely GTLC2-90 and GTLC2-127 of Kufri Badshah; KPLC2-13, KPLC2-37, KPLC2-53, KPLC2-54 of Kufri Pukhraj are multiplied under tissue culture. Out of these, two promising transgenic lines, KPLC2-53 and GTLC2-127, were screened through agro inoculation technique and the result showed that transgenic lines did not show ToLCNDV-potato symptoms as against non-transgenic plants. Through molecular characterization and successful clonal generation one event was identified from the Kufri Pukhraj and characterized for further biosafety research-level field trials.

These results suggest that gene-specific post-transcriptional gene silencing (PTGS) against ToLCNDV-potato was induced when AC1 gene was targeted. This result is supported by similar observations reported by Waterhouse et al. (1998) for rice and in *Arabidopsis* by Chuang and Meyerowitz (2000). Many other workers have also reported better efficiency of RNAi through transgene that express hpRNA than amplicon cassette alone (Smith et al., 2000; Wesley et al., 2001; Brummell et al., 2003; Fukusaki et al., 2004; Ifuku et al., 2003). There lies an important advantage of using hpRNA mediated silencing from the perspective of food safety, as no virus-derived mRNAs are accumulating in transgenic plants expressing hpRNA constructs, thereby reducing the risk of recombination or complementation events. Moreover, no viral protein is being produced in these plants, circumventing any fears concerning allergic response to novel proteins.

4.2.2 Potato Virus Y (PVY)

Coat protein (CP) gene of *potato virus Y* (PVY) was used for engineering pathogen-derived resistance against PVY disease in the Indian potato cultivar, Kufri Pukhraj. Identification of stability and integration of coat protein gene in Kufri Pukhraj transgenic lines, expressing PVY-CP through southern hybridization, was carried out. Copy number analysis of CP gene was performed according to standard protocol. The presence of *npt*II gene was detected by random prime labelling radioactive method, as mentioned by Amersham random prime labeling kit. Data collected was *npt*II gene copy number in the plant genome of each transgenic line. The southern blot analysis result shows the copy number of *npt*II gene in the selected promised transgenic (KPYS1, KPYS13, KPYS18, KPYS20, KPYS21, KPYS26 and KPYNT3). Therefore, by referring to *npt*II fragment, CP gene copy number was same among the transgenic lines. It confirmed as a co-integration of the gene in the selected transgenic lines, indicating the integration and stability of transgenes over the clonal propagation. The copy number analysis confirmed the integration of gene in the genome. These lines were subjected to bioassay by sap inoculation method. The lines were examined for their resistance against PVY by observing the symptom expression over control. Among the lines, KPYS-1, KPYS-13 and KPYS-20 showed resistance as the symptoms were not expressed in these selected lines. Finally, transgenic events KPYS-13 and KPYS-20 were selected, based on phenotype and tuber yield for further biosafety field trials.

4.3 Bacterial Wilt Resistance

Bacterial wilt caused by *Ralstonia solanacearum* is one of the most serious diseases of potato, with a wide host range of over 200 plant species. Potato yield loss due to bacterial wilt ranges between 33–90% in the world, while it is as high as 70% in India (Sagar et al., 2014). Bacterial wilt in India is caused by strains of phylotype I, II and IV of *Ralstonia solanacearum*. It is one of the most devastating diseases of potato, as observed in the Malwa region of Madhya Pradesh, in India. Bacterial wilt is another chronic disease problem that does not have any reliable source of resistance. Few available resistant germplasm (wild diploid sources) in potato have failed to provide durable resistance against Indian *R. solanacearum* races. The diploid potato (the two endosperm balance number, 2EBN) is difficult to introgress with the commercial cultivated tetraploid potato (4 EBN) and thus it restricts its use in resistance breeding programs. Thus eradication of this pathogen is a major challenge in potato production.

4.3.1 RNAi Transgenics

Recent studies have identified PAP2 as a crucial protein that controls pathogenesis in various plant species. In response to *R. solanacerum* infection, PAP2 acts as a negative regulator and makes it unavailable for ROS burst (Nakano et al., 2013, 2015). Therefore, PAP2 inhibits the accumulation of PA which further interferes with HR response in the host. In many cases, PAP2 works as a negative regulator by being temporarily inactive under pathogen infection to stimulate plant defense and adaptation. Therefore, PAP2 could be silenced, for regulating the defense action against *R. solanacearum* in potato.

An alternative approach for management of bacterial wilt caused by *Ralstonia solanacearum* in potato by silencing host susceptible gene was achieved. Since resistance host is not available, an alternative management approach identified and isolated disease susceptibility gene (responsible for degradation of reactive oxygen species and hypersensitive response, as a negative regulator of defense) and developed RNAi transgenic lines. Through this, a host susceptible gene phosphatidic acid phosphatase 2 (*PAP2*) was silenced, using RNAi technique and initial results have indicated imparting high level of resistance to the pathogen. RNAi lines were developed, showing an immune response to wilt symptoms and few lines showed delayed infection compared to control plants. The 80 Kufri Jyoti RNAi lines were developed and subjected to efficient, quick *in-vitro* hydroponic method of bio-assay to avoid the escape of the pathogen as well as the whole plant bio-assay, using the root inoculation method (Thakur et al., 2020). These lines were screened up to the third clonal generation and showed stable resistance response. The integration, expression and inheritance of three promising lines displayed enhanced resistance to wilt disease. A virulent colony assay and bacterial ooze test were used to authenticate the silencing of PAP2 in inhibiting the colonization and bacterial load.

There is still no clear mechanism to understand the conversion or competition of virulent to avirulent type in resistant genotypes. This might involve a complex host signaling cascade and pathogen effect or interaction inside the xylem vessel of the host system. It is important to identify and mine the PAP2 regulated PA intracellular signaling molecule to understand and assign its downstream defense pathway in *R. solanacearum* and potato host interaction. This system can be manipulated at a molecular level and used as an economic and eco-friendly disease-control method. Moreover, it is difficult for the pathogens to overcome PRR recognition as the PAMPs are conserved molecules (Lacombe et al., 2010; Lu et al., 2018). Thus, our investigation for imparting resistance in potato through the plant's immune system provides the novel genetic source in potato for bacterial wilt resistance, since none of the germplasm and cultivars have resistance loci for durable bacterial wilt resistance. This is the first evidence that provides the resistant source in potato for developing bacterial wilt resistance varieties. These sources would serve as future breeding material for the management of bacterial wilt disease in potatoes.

4.3.2 EBD Gene Transgenics

Therefore, an antimicrobial peptide gene, bovine enteric β-defensin (*EBD*) was used for conferring bacterial wilt resistance in potato. Transgenic lines of Kufri Badshah showed very high level of resistance to bacterial wilt in glasshouse screening. The gene has now been transferred to two commercial potato cultivars, Kufri Giriraj and Kufri Jyoti. Kufri Giriraj was selected because of its popularity in Shimla and Nilgiri hills where bacterial wilt is prevalent. Kufri Jyoti is a popular variety in eastern plains where bacterial wilt is endemic. Twenty-seven putative transgenic lines of Kufri Griraj and 12 lines of Kufri Jyoti have been developed.

4.4 Chip Quality with Reduced Cold-induced Sweetening

Cold storage of potato is an integral requirement of post-harvest handling of this semi-perishable crop. Unfortunately, potato accumulates high amounts of reducing sugars during cold-storage—a physiological phenomenon known as cold-induced sweetening. Processed products, like chips, French fries, etc. when prepared from sweetened potatoes develop brown coloration and are not preferred by the processing industry and consumers. Potatoes suitable for processing are not available throughout the year. This poses a serious impediment to fast emerging potato-processing industry in India. It is, therefore, important either to develop varieties that do not accumulate reducing sugars under cold storage or to improve the processing attributes of existing potato cultivars to meet the growing demand of processing units. The biggest hurdle towards development of cold-resistant potato cultivars through conventional breeding is lack of suitable germplasm for use as parents.

Genetic engineering of Indian popular potato cultivars for prevention of cold-induced sweetening offers an easier and cheaper alternative as potato is highly amenable to biotechnological tools. Cold-induced sweetening in potato is a complex physiological process that involves interplay of different cellular controls. Different strategies need to be adopted to get viable and stable reduction of cold-induced sweetening and also to prevent breaking down of any single strategy when the transgenic lines will be exposed to vagaries of nature in the field. Two different biotechnological approaches through metabolic engineering have been adopted to block the step leading to conversion of sucrose to reducing sugars, either by inhibiting the vacuolar invertase activity, the enzyme responsible for synthesis of reducing sugars, or silencing one or more genes encoding for enzymes responsible for sugar accumulation in potato, like vacuolar invertase (*INV*) and UDP Glucose Phosphatase (UGPase) at post-transcriptional level. A gene encoding tobacco invertase inhibitor (*Nt-Inhh*) has been used for *StINV* inhibition in potato by over-expressing the *Nt-Inhh* under the control of two different promoters, constitutive promoter, CaMV 35S and tuber-specific promoter, GBSS, while for the second strategy *StINV* and *UGPase* genes have been silenced, using the RNAi technique (both SiRNA and amiRNA). By inhibiting the expression of this gene, by using RNAi technology, it has proven that the transgenic plants prevent the cold-induced sweetening of potato tubers upon cold storage even up to 135 days. A total of seven transgenic lines of Kufri Chipsona-I with inhibited expression of INV were selected from more than 300 transgenic events and Confined Field Trials (CFT) at CPRS, Jalandhar (BT/BS/17/22/97-PID). Single line KChipInvRNAi-2214 was selected because it has significantly lower soluble sugars, superior chipping characters and on-par yield compared with the control. This transgenic event exhibited > 80% reduced hexose accumulation during cold storage. Moreover, during the frying process, the transgenic event KChipInvRNAi-2214 showed no browning up to 135 days of cold storage as compared to the wild type with significant reduction in acrylamide formation. These results suggest that inhibiting the expression of vacuolar invertase in potato tubers have greatly improved the processing quality.

4.5 Protein-rich Potato

Humans require a diverse and nutritionally well-balanced diet to maintain optimal health and depend largely on plants for their daily nutritional requirements. Moreover, a large proportion of the world's population is undernourished. Thus, nutritional improvement of crop plants is an urgent worldwide health issue as basic nutritional requirements for much of the world's population are still not met. Proteins, one of the principal constituents of a balanced diet, impart nutritional value to food due to their structural constituents, the amino acids. However, it is very rare in nature to find all of the essential amino acids in a single food crop. The United Nations declared 2008 as the 'International Year of the Potato', affirming the need to focus on the role that the potato can play in providing food security (United Nations General Assembly Resolution, 2005). Although in developing economies the majority of potato is used for direct consumption, a shift towards the use of potato in convenience foods, for example, in potato chips and fries, has dramatically increased in developed countries (FAOSTAT, 2005). Unfortunately, the nutritional quality of potato tubers is greatly compromised because they contain less protein and are deficient in lysine, tyrosine and the sulfur-containing amino acids (Jaynes et al., 1986). To guarantee a sufficient supply of quality protein in a diet consisting mainly of staple foods, such as potato, specific interventions in genetic engineering are an absolute necessity.

Potato tuber contains about 1–2% protein on the fresh weight basis, which is insufficient for a staple food. In terms of protein quality, potato protein has been given an intermediate score by WHO. This is because potato protein is deficient in certain essential amino acids, like methionine/cysteine, leucine, isoleucine and threonine. A collaborative project between ICAR-CPRI, Shimla and NCPGR, New Delhi to improve nutritional quality of potato protein by expressing the high quality seed storage protein (*AmA1*) of *Amaranthus hypochondriacus*, cloned and patented by both the institutes. *AmA*1 has great agricultural importance because it is a well-balanced protein in terms of amino acid composition, possessing even better values than recommended by the World Health Organization for a nutritionally rich protein. More importantly, because it is a non-allergenic protein that originated from an edible crop, the transgenic crops expressing *AmA*1 would have greater acceptability. Chakraborty et al. (2010) reported the development of transgenic potatoes with enhanced nutritive value by tuber-specific expression of a seed protein, *AmA*1 (*Amaranth Albumin* 1), in seven genotypic backgrounds suitable for cultivation in different agro-climatic regions. Analyses of the transgenic tubers revealed up to 60% increase in total protein content. In addition, the concentrations of several essential amino acids increased significantly in transgenic tubers, as they are otherwise limited in potato. Moreover, the transgenics also exhibited enhanced photosynthetic activity with a concomitant increase in total biomass. These results are striking because this genetic manipulation also resulted in a moderate increase in tuber yield. The comparative protein profiling suggests that the proteome rebalancing might cause increased protein content in transgenic tubers. Furthermore, the data on field performance and safety evaluation indicate that the transgenic potatoes are suitable for commercial cultivation. *In vitro* and *in vivo* studies on experimental animals demonstrate that the transgenic tubers are also safe for human consumption. Altogether, these results emphasize that the expression of *AmA*1 is a potential strategy for the nutritional improvement of food crops.

On the basis of gene expression analysis in putative transgenics, 40 lines were selected for preliminary evaluation of tuber yield at ICAR-CPRI RS in Modipuram. Based on tuber yield and expression of *AmA1* gene, 14 lines of seven potato cultivars were selected for further limited field trial at Modipuram and Jalandhar. Crude protein content of 14 transgenic lines along with their respective controls was estimated in dry powder made from peeled potato tubers by Kjeldahl's method at Modipuram. Out of the above nine promising transgenic lines selected on the basis of yield performance, increase in crude protein content was observed in Kufri Chipsona 1/18 (11.36%), Kufri Chipsona 1/21 (4.57%), Kufri Chipsona 2/15 (38.81%) and Kufri Badshah/5 (16.02%). Increase in protein content was also confirmed by NCPGR, New Delhi in case of Kufri Chipsona 1/18 (20.72%), Kufri Chipsona 2/15 (34.20%) and Kufri Badshah/5 (46.65%). Therefore, on the basis of improvement in protein content, three transgenic lines, viz. Kufri Chipsona 1/18, Kufri Chipsona 2/15 and Kufri Badshah/5 may be selected for further analysis.

Amino acid composition of 14 transgenic lines along with their respective controls was determined in freeze dried potato powder, using HPLC at NBPGR, New Delhi. In case of Kufri Chipsona 1/18, improvement only in lysine content (10%) was observed; other essential amino acid contents either decreased or remained unchanged. Improvement in valine (10%), methionone (5%), isoleuucine (22%), leucine (23%) and phenylalanine (17%) was observed in case of Kufri Chipsona 2/15 compared to non-transgenic control. Similarly, improvement in histidine (11%), arginine+threonine (12%), valine (12%), methionine (27%), isoleucine (14%), leucine (12%) and phenylalanine (29%) was observed in Kufri Badshah/5 compared to its non-transgenic control. Therefore, two transgenic lines, viz. Kufri Chipsona 2/115 and Kufri Badshah/5 may be selected for food safety analysis (Chakraborty et al., 2010). Very recently, with the intention of biofortification to improve nutrition quality of potato *StGalDH* for vitamin C, *StAN* for anthocyanin, are being over-expressed in potato to improve the respective nutrients. More efforts are in process on biofortification of potato with improved iron and zinc content.

4.6 Dwarf Plant Architecture

Potato plants, when grown under long days or high temperature conditions, accumulate gibberellic acid and grow tall, leading to reduced partitioning of dry matter in tubers. Potato varieties grown under short days during winter in the plains have high harvest index of about 80% whereas the same varieties, when grown under long days during summer in hills, grow very tall and have lower yields with an harvest index of about 50%. To reduce internal gibberellic acid content and consequently excessive vegetative growth, we have silenced *GA20-oxidase1* gene (involved in the synthesis of gibberellic acid) through Post Transcriptional Gene Silencing (PTGS) to obtain plants with reduced GA content. Glass house trials were conducted with 75 IRGA, 35 sense and 34 antisense plants. Based on the yield performance for two years as well as plant height, 10 IRGA lines, five sense lines and five antisense lines were selected for further trials.

4.7 Other Traits

Many other characters have been targeted for improving the yield in potato through transgenics approach. For example, potato tuber moth is an insect pest of potato prevalent in certain pockets where potato is stored under country storage without refrigeration or where no cold storage facility is available. No conventional source of resistance has been reported so far against this pest. Genetic engineering, therefore, is an alternative for developing resistance cultivars. The synthetic *Cry1A*b gene of *Bacillus thuringiensis* was deployed to confer resistance against potato tuber moth. Transgenic tubers of five lines (KB-1, KB-22, KS-6, KJW-3 and KL-2) showed good level of resistance on laboratory evaluation. However, PTM resistance in transgenic tubers was lost with increase in storage time. Since *Cry9Aa2* was the most effective Bt toxin for PTM, attempts were made to check the efficacy of the native gene by chloroplast transformation and was found effective in potato.

5. Conclusion

The International Food Policy Research Institute launched 2020 Vision—a call for a new 'Green Revolution'. A total of 1.02 billion people, residing mostly in the developing world, suffer from availability of needed food, including under-nutrition. Therefore, to meet the demand of poverty population, advanced biotechnology tools are necessary to boost food-based crops improvement research programs, including rice and wheat. The biotechnology-based strategies to modify genome, especially seed crops, are the major challenges, where homozygosity is the concern, whereas in case of vegetative propagated potatoes, maintenance of heritability of gene via tubers is easy and does not require backcrossing to maintain homozygosity. The benefits of biotech potato, such as limited gene flow to conventionally grown crops and weedy relatives according to several study reports suggest sexual incompatibility between cultivated potato and wild relatives. This opportunity makes easy to justify environmental risk-assessment studies.

The wide advantages of transgenic crops for a society to solve food security or nutrition security issues have been well established. Many added GM crops and foods benefits, such as higher nutritional value, herbicide tolerance, virus resistance, tolerance to various abiotic stresses, increase the shelf life. There is a need the world over to carry on its GM crop research program to sustain its food and nutrition security targets. Ideally, labs should continue research on GM crop and a regulatory body should develop strategies of deregulation along with building basic infrastructure facilities and preparing stringent biosafety and marketing guidelines. Although Indian portals, like GEAC, IGMORIS (Indian GMO Research Information System) and Biosafety Clearing House are doing their bit for assessing biosafety and their regulation of GM crops, there is an urgent need to build a single window system and online portal for assessment, regulations and approval of GM crops.

Acknowledgement

I am thankful to the Director, ICAR-Central Potato Research Institute, Shimla, and scientists/technicians/research fellows and other colleagues of the institute for their support under the institute research projects on biotechnology, germplasm, breeding, and seed research on aeroponics. I am also grateful to the funding agencies for support under the externally funded projects (CABin, ICAR-IASRI, New Delhi; ICAR-LBS Young Scientist Award Project, and DBT, Government of India).

References

Arif, M., Azhar, U., Arshad, M., Zafar, Y., Mansoor, S. and Asad, S. (2012). Engineering broad-spectrum resistance against RNA viruses in potato. *Transgenic Res.*, 21(2): 303–311.

Bakhsh, A. (2020). Development of efficient, reproducible and stable agrobacterium-mediated genetic transformation of five potato cultivars. *Food Tech. Biotechnol.*, 58(1): 57–63.

Boher, P., Soler, M., Fernández-Piñán, S., Torrent, X., Müller, S.Y. et al. (2021). Silencing of StRIK in potato suggests a role in periderm related to RNA processing and stress. *BMC Plant Biol.*, 21(1): 409.

Boschi, F., Schvartzman, C., Murchio, S., Ferreira, V., Siri, M.I. et al. (2017). Enhanced bacterial wilt resistance in potato through expression of *Arabidopsis* EFR and introgression of quantitative resistance from *Solanum commersonii*. *Front. Plant Sci.*, 8: 1642.

Bouaziz, D., Pirrello, J., Charfeddine, M., Hammami, A., Jbir, R. et al. (2013). Overexpression of StDREB1 transcription factor increases tolerance to salt in transgenic potato plants. *Mol. Biotechnol.*, 54(3): 803–817.

Brummell, D.A., Balint-Kurti, P.J., Harpster, M.H., Palys, J.M., Oeller, P.W. and Gutterson, N. (2003). Inverted repeat of a heterologous-untranslated region for high efficiency, high throughput gene silencing. *Plant J.*, 33: 793–800.

Brummell, D.A., Watson, L.M., Zhou, J., McKenzie, M.J., Hallett, I.C. et al. (2015). Overexpression of Starch Branching Enzyme II increases short-chain branching of amylopectin and alters the physicochemical properties of starch from potato tuber. *BMC Biotechnol.*, 15: 28.

Byarugaba, A.A., Baguma, G., Jjemba, D.M., Faith, A.K., Wasukira, A. et al. (2021). Comparative phenotypic and agronomic assessment of transgenic potato with 3R-gene stack with complete resistance to late blight disease. *Biology*, 10(10): 952.

Chakraborty, S., Chakraborty, N., Agrawal, L., Ghosh, S., Narula, K. et al. (2010). Next-generation protein-rich potato expressing the seed protein gene *AmA1* is a result of proteome rebalancing in transgenic tuber. *Proc. Natl. Acad. Sci., U.S.A.*, 107: 17533–17538.

Charfeddine, M., Charfeddine, S., Ghazala, I., Bouaziz, D. and Bouzid, R.G. (2019). Investigation of the response to salinity of transgenic potato plants overexpressing the transcription factor StERF94. *J. Biosci.*, 44(6): 141.

Chuang, C.F. and Meyerowitz, E.M. (2000). Specific and heritable genetic interference by double stranded RNA in Arabidopsis thaliana. *Proc. Natl. Acad. Sci., U.S.A.*, 97: 4985–4990.

Clasen, B.M., Stoddard, T.J., Luo, S., Demorest, Z.L., Li, J. et al. (2016). Improving cold storage and processing traits in potato through targeted gene knockout. *Plant Biotechnol. J.*, 14(1): 169–176.

Dinh, P.T., Brown, C.R. and Elling, A.A. (2014). RNA interference of effector gene Mc16D10L confers resistance against *Meloidogyne chitwoodi* in *Arabidopsis* and potato. *Phytopathology*, 104(10): 1098–1106.

Dinh, P.T., Zhang, L., Mojtahedi, H., Brown, C.R. and Elling, A.A. (2015). Broad *Meloidogyne* resistance in potato based on RNA interference of effector gene 16D10. *J. Nematol.*, 47(1): 71–78.

Dobnik, D., Lazar, A., Stare, T., Gruden, K., Vleeshouwers, V.G. and Žel, J. (2016). *Solanum venturii*, a suitable model system for virus-induced gene silencing studies in potato reveals StMKK6 as an important player in plant immunity. *Plant Method.*, 12: 29.

Dou, H., Xv, K., Meng, Q., Li, G. and Yang, X. (2014). Potato plants ectopically expressing *Arabidopsis thaliana* CBF3 exhibit enhanced tolerance to high-temperature stress. *Plant, Cell Environ.*, 38(1): 61–72.

Duan, H., Richael, C. and Rommens, C.M. (2012). Overexpression of the wild potato eIF4E-1 variant Eval elicits Potato virus Y resistance in plants silenced for native eIF4E-1. *Transgenic Res.*, 21(5): 929–938.

FAOSTAT. (2005). *Food and Agricultural Organization of the United Nations Statistical Database.* Available at http://www.faostat.fao.org; accessed March 8, 2010.

Ferreira, S.J., Senning, M., Fischer-Stettler, M., Streb, S., Ast, M. et al. (2017). Simultaneous silencing of isoamylases ISA1, ISA2 and ISA3 by multi-target RNAi in potato tubers leads to decreased starch content and an early sprouting phenotype. *PLoS ONE*, 12(7): e0181444.

Fox, A., Collins, L.E., Macarthur, R., Blackburn, L.F. and Northing, P. (2017). New aphid vectors and efficiency of transmission of *Potato virus A* and strains of *Potato virus Y* in the UK. *Plant Pathol.*, 66: 325–335.

Fukusaki, E.I., Kawasaki, K., Kajiyama, S., An, C.I. and Suzuki, K. (2004). Flower color modulations of *Torenia hybrida* by down regulation of chalcone synthase genes with RNA interference. *J. Biotech.*, 111: 229–240.

Garg, I.D., Khurana, S.M.P., Kumar, S. and Lakra, B.S. (2001). Association of Gemini virus with potato apical leaf curl in India, and its immune-electron microscopic detection. *Potato J.*, 28: 227–232.

Ghislain, M., Byarugaba, A.A., Magembe, E., Njoroge, A., Rivera, C. et al. (2019). Stacking three late blight resistance genes from wild species directly into African highland potato varieties confers complete field resistance to local blight races. *Plant Biotechnol. J.*, 17(6): 1119–1129.

Gonsalves, D. (1998). Control of Papaya ringspot virus in papaya: A case study. *Annu. Rev. Phytopathol.*, 36: 415–437.

González-Romero, M.E., Rivera, C., Cancino, K., Geu-Flores, F., Cosio, E.G. et al. (2021). Bioengineering potato plants to produce benzylglucosinolate for improved broad-spectrum pest and disease resistance. *Transgenic Res.*, 30(5): 649–660.

Goodman, R.E. (2015). Bioinformatic analysis and literature search for potential allergenicity and toxicity test of late blight resistance potato event SP951, with *RB* and *Npt*II proteins, Food Allergy Research and Resource program, Department of Food Science and Technology, University of Nebraska, *Study No. RFG-LBR-2015*, pp. 1–26.

Halterman, D.A., Kramer, L.C., Wielgus, S. and Jiang, J.M. (2008). Performance of transgenic potato containing the late blight resistance gene *RB*. *Plant Dis.*, 92: 339–343.

Hameed, A., Tahir, M.N., Asad, S., Bilal, R., Van Eck, J. et al. (2017). RNAi-mediated simultaneous resistance against three RNA viruses in potato. *Mol. Biotechnol.*, 59(2-3): 73–83.

Horn, P., Santala, J., Nielsen, S.L., Hühns, M., Broer, I. and Valkonen, J.P. (2014). Composite potato plants with transgenic roots on non-transgenic shoots: A model system for studying gene silencing in roots. *Plant Cell Rep.*, 33(12): 1977–1992.

Hussain, T., Aksoy, E., Çalışkan, M.E. and Bakhsh, A. (2019). Transgenic potato lines expressing hairpin RNAi construct of molting-associated EcR gene exhibit enhanced resistance against Colorado potato beetle (*Leptinotarsa decemlineata*, Say). *Transgenic Res.*, 28(1): 151–164.

Ifuku, K., Yamamoto, Y. and Sato F. (2003). Specific RNA interference in psbP genes encoded by a multigene family in *Nicotiana tabacum* with a short 3′-untranslated sequence. *Biosci. Biotechnol. Biochem.*, 67: 107–113.

Jahan, S.N., Åsman, A.K., Corcoran, P., Fogelqvist, J., Vetukuri, R.R. and Dixelius, C. (2015). Plant-mediated gene silencing restricts growth of the potato late blight pathogen *Phytophthora infestans*. *J. Exp. Bot.*, 66(9): 2785–2794.

Jahromi, M.G., Rahnama, H., Mousavi, A. and Safarnejad, M.R. (2022). Comparative evaluation of resistance to potato virus Y (PVY) in three different RNAi-based transgenic potato plants. *Transgenic Res.*, 10.1007/s11248-022-00302-0.

James, C. (2015). Global status of commercialized biotech/GM crops: 2014. *ISAAA Brief*, p. 49.

Jaynes, J.M., Yang, M.S., Espinoza, N. and Dodds, J.H. (1986). Plant protein improvement by genetic engineering: use of synthetic genes. *Trends Biotechnol.*, 4: 314–320.

Jeevalatha, A., Kaundal, P., Venkatasalam, E.P., Chakrabarti, S.K. and Singh, B.P. (2013). Uniplex and duplex PCR detection of geminivirus associated with potato apical leaf curl disease in India. *J. Virol. Methods*, 193: 62–67.

Jeevalatha, A., Kaundal, P., Kumar, R., Raigond, B., Kumar, R. et al. (2018). Optimized loop-mediated isothermal amplification assay for tomato leaf curl New Delhi virus-potato detection in potato leaves and tubers. *Eur. J. Plant Pathol.*, 150: 565–573.

Jeffries, C., Barker, H. and Khurana, S.M.P. (2005). Potato viruses and viroids. *In*: Gopal, J. and Khurana, S.M.P. (eds,). *Handbook of Potato Production, Improvement and Post-harvest Management*. The Haworth's Food Products Press, New York, USA.

Jo, K.R., Kim, C.J., Kim, S.J., Kim, T.Y., Bergervoet, M. et al. (2014). Development of late blight resistant potatoes by cisgene stacking. *BMC Biotechnol.*, 14: 50.

Khurana, S.M.P. (2004). Potato viruses and their management. pp. 389–440. *In*: Naqvi, S.A.M.H. (ed.). *Diseases of Fruits and Vegetables: Diagnosis and Management*. Kluwer Academic, Dordrecht, Boston and London.

Kikuchi, A., Huynh, H.D., Endo, T. and Watanabe, K. (2015). Review of recent transgenic studies on abiotic stress tolerance and future molecular breeding in potato. *Breed. Sci.*, 65(1): 85–102.

Kuhl, J.C., Zarka, K., Coombs, J., Kirk, W.W. and Douches, D.S. (2007). Late blight resistance of *RB* transgenic potato lines. *J. Am Soc. Hortic. Sci.*, 132: 783–789.

Kumar, R., Jeevalatha, A., Raigond, B., Kumar, R., Sharma, S. and Nagesh, M. (2017). A multiplex RT PCR assay for simultaneous detection of five viruses in potato. *J. Plant Pathol.*, 99: 37–45.

Lacombe, S., Rougon-Cardoso, A., Sherwood, E., Peeters, N., Dahlbeck, D. et al. (2010). Interfamily transfer of a plant pattern-recognition receptor confers broad-spectrum bacterial resistance. *Nat. Biotech.*, 28: 365–369.

Lakra, B.S. (2002). Leaf curl: A threat to potato crop in Haryana. *J. Mycol. Plant Pathol.*, 32: 367.

Lehretz, G.G., Sonnewald, S., Lugassi, N., Granot, D. and Sonnewald, U. (2021). Future-proofing potato for drought and heat tolerance by overexpression of hexokinase and SP6A. *Front. Plant Sci.*, 11: 614534.

Li, F., Bian, C.S., Xu, J.F., Pang, W.F., Liu, J. et al. (2015). Cloning and functional characterization of SAD genes in potato. *PLoS ONE*, 10(3): e0122036.

Lu, H., Lema, A.S., Planas-Marques, M., Alonso-Diaz, A., Valls, M. and Col, N.S. (2018). Type III secretion-dependent and -independent phenotypes caused by *Ralstonia solanacearum* in Arabidopsis Roots. *Mol. Plant Microbe Interact.*, 31: 175–184.

Mao, J. and Zeng, F. (2014). Plant-mediated RNAi of a gap gene-enhanced tobacco tolerance against the *Myzus persicae*. *Transgenic Res.*, 23(1): 145–152.

Martínez-Prada, M.D.M., Curtin, S.J. and Gutiérrez-González, J.J. (2021). Potato improvement through genetic engineering. *GM Crops Food*, 12(1): 479–496.

McCue, K.F., Gardner, E., Chan, R., Thilmony, R. and Thomson, J. (2019). Transgene stacking in potato using the GAANTRY system. *BMC Res. Notes*, 12(1): 457.

Mi, X., Ji, X., Yang, J., Liang, L., Si, H. et al. (2015). Transgenic potato plants expressing cry3A gene confer resistance to Colorado potato beetle. *C.R. Biol.*, 338(7): 443–450.

Miller, J. and Bradford, K. (2010). The regulatory bottleneck for biotech specialty crops. *Nat. Biotechnol.*, 28: 1012–1014.

Mishra, S., Dee, J., Moar, W., Dufner-Beattie, J., Baum, J. et al. (2021). Selection for high levels of resistance to double-stranded RNA (dsRNA) in Colorado potato beetle (*Leptinotarsa decemlineata* Say) using non-transgenic foliar delivery. *Sci. Rep.*, 11(1): 6523.

Mohammed, A., Diab, M.R., Abdelsattar, M. and Khalil, S. (2017). Characterization and RNAi-mediated knockdown of Chitin Synthase A in the potato tuber moth, *Phthorimaea operculella. Sci. Rep.*, 7(1): 9502.

Nahirñak, V., Almasia, N.I., González, M.N., Massa, G.A., Décima Oneto, C.A. et al. (2022). State of the art of genetic engineering in potato: from the first report to its future potential. *Front. Plant Sci.*, 12: 768233.

Nakano, M., Nishihara, M., Oshioka, H.Y., Takahashi, H., Sawasaki, T. et al. (2013). Suppression of DS1 Phosphatidic Acid Phosphatase confirms resistance to *Ralstonia solanacearum* in *Nicotiana benthamiana. PLoS ONE*, 8: e75124.

Nakano, M., Yoshioka, H., Ohnishic, K., Hikichia, Y. and Kiba, A. (2015). Cell death-inducing stresses are required for defense activation in DS1-phosphatidic acid phosphatase-silenced *Nicotiana benthamiana. J. Plant Physiol.*, 184: 15–19.

Orbegozo, J., Solorzano, D., Cuellar, W.J., Bartolini, I., Roman, M.L. et al. (2016). Marker-free PLRV resistant potato mediated by Cre-loxP excision and RNAi. *Transgenic Res.*, 25(6): 813–828.

Ottaviani, M.-P., Cate, C.H.H.T. and van Vloten-Doting, L. (1992). Expression of introduced genes after tuber propagation of transgenic potato plants. *Plant Breed.*, 109: 89–96.

Pasare, S.A., Ducreux, L., Morris, W.L., Campbell, R., Sharma, S.K. et al. (2013). The role of the potato (*Solanum tuberosum*) CCD8 gene in stolon and tuber development. *New Phytol.*, 198(4): 1108–1120.

Petrov, N., Stoyanova, M., Andonova, R. and Teneva, A. (2015). Induction of resistance to potato virus Y strain NTN in potato plants through RNAi. *Biotechnol. Equip.*, 29(1): 21–26.

Qi, X., Tang, X., Liu, W., Fu, X., Luo, H. et al. (2020). A potato RING-finger protein gene StRFP2 is involved in drought tolerance. *Plant Physiol. Biochem.*, 146: 438–446.

Sagar, V., Gurjar, M.S., Arjunan, J., Bakade, R.R., Chakrabarti, S.K. et al. (2014). Phylotype analysis of Ralstonia solanacearum strains causing potato bacterial wilt in Karnataka in India. *African J. Microbiol. Res.*, 8: 1277–1281.

Sanju, S., Sundaresha, S., Thakur, A., Shukla, P.K., Srivastava, N. et al. (2015). Host-induced gene silencing of Phytophthora infestans effector gene *Avr3a* in potato. *Funct. Integr. Genom.*, 15: 697–607.

Shandil, R.K., Chakrabarti, S.K., Singh, B.P., Sharma, S., Sundaresha, S. et al. (2017). Genotypic background of the recipient plant is crucial for conferring *RB* gene mediated late blight resistance in potato. *BMC Genetics*, 18(1): 22.

Sharma, R., Bhardwaj, V., Dalamu, Kaushik, S.K., Singh, B.P. et al. (2014). Identification of elite potato genotypes possessing multiple disease resistance genes through molecular approaches. *Sci. Hortic.*, 179: 204–211.

Shi, J., Wang, J., Wang, N., Zhou, H., Xu, Q. and Yan, G. (2019). Overexpression of StGA2ox1 gene increases the tolerance to abiotic stress in transgenic potato (*Solanum tuberosum* L.) plants. *Appl. Biochem. Biotechnol.*, 187(4): 1204–1219.

Smith, N., Singh, S., Wang, M.B., Stoutjesdijk, P., Green, A. and Waterhouse, P.M. (2000). Total silencing by intron-spliced hairpin RNAs. *Nature*, 407: 319–320.

Song, J.Q., Bradeen, J.M., Naess, S.K., Raasch, J.A., Wielgus, S.M. et al. (2003). Gene *RB* cloned from *Solanum bulbocastanum* confers broad spectrum resistance to potato late blight. *Proc. Natl. Acad. Sci.*, 100: 9128–9133.

Sun, K., Wolters, A.M., Loonen, A.E., Huibers, R.P., van der Vlugt, R. et al. (2016). Down-regulation of Arabidopsis DND1 orthologs in potato and tomato leads to broad-spectrum resistance to late blight and powdery mildew. *Transgenic Res.*, 25(2): 123–138.

Sun, K., Wolters, A.M., Vossen, J.H., Rouwet, M.E., Loonen, A.E. et al. (2016). Silencing of six susceptibility genes results in potato late blight resistance. *Transgenic Res.*, 25(5): 731–742.

Tang, X., Ghimire, S., Liu, W., Fu, X., Zhang, H. et al. (2020). Potato E3 ubiquitin ligase PUB27 negatively regulates drought tolerance by mediating stomatal movement. *Plant Physiol. Biochem.*, 154: 557–563.

Thakur, A., Sanju, S., Sundaresha, S. Sharma, S. and Singh, B.P. (2015). Artificial microRNA mediated gene silencing of *Phytophthora infestans AVr3a* gene for imparting late blight resistance in potato. *Plant Pathol. J.*, 14: 1–12.

Thakur, K., Sagar, V., Sundaresha, S., Salaria, N., Goutam, U. and Sharma, S. (2020). An efficient and quick *in-vitro* method of Evaluation of potato genotypes against *Ralstonia solanacearum*. *Plant Cell Biotechnol. Mol. Biol.*, 21: 46–52.

Tomar, G., Chakrabarti, S.K., Sharma, N.N., Jeevalatha, A., Sundaresha, S. et al. (2018). RNAi-based transgene conferred extreme resistance to the geminivirus causing apical leaf curl disease in potato. *Plant Biotechnol. Rep.*, 12: 195–205.

United Nations General Assembly Resolution. (2005). Available at: http://www.un.org/depts/dhl/resguide/r60.htm (Resolution No. A/RES/60/191); accessed March 8, 2010.

Upadhyaya, D.C., Bagri, D.S., Upadhyaya, C.P., Kumar, A., Thiruvengadam, M. and Jain, S.K. (2021). Genetic engineering of potato (*Solanum tuberosum* L.) for enhanced α-tocopherols and abiotic stress tolerance. *Physiol. Plant.*, 173(1): 116–128.

Usharani, K.S., Surendranath, B., Khurana, S.M.P., Garg, I.D. and Malathi, V.G. (2003). Potato leaf curl—A new disease of potato in northern India caused by a strain of *Tomato leaf curl New Delhi virus*. *New Dis. Rep.*, 8: 2.

Venkatasalam, E.P., Singh, S., Sivalingam, P.N., Malathi, V.G. and Garg, I.D. (2011). Polymerase chain reaction and nucleic acid spot hybridization detection of begomovirus(es) associated with apical leaf curl disease of potato. *Arch. Phytopathol. Plant Prot.*, 44: 987–992.

Vleeshouwers, V.G.A.A., Rietman, H., Krenek, P., Champouret, N., Young, C. et al. (2008). Effector genomics accelerates discovery and functional profiling of potato disease resistance and *Phytophthora infestans* avirulence genes. *PLoS ONE*, 3: 1–10.

Wang, L., Liu, Y., Feng, S., Yang, J., Li, D. and Zhang, J. (2017). Roles of plasmalemma aquaporin gene *StPIP1* in enhancing drought tolerance in potato. *Front. Plant Sci.*, 8: 616.

Wang, L., Liu, Y., Li, D., Feng, S., Yang, J. et al. (2019). Improving salt tolerance in potato through overexpression of *AtHKT1* gene. *BMC Plant Biol.*, 19(1): 357.

Waterhouse, P.M., Graham, M.W. and Wang, M.B. (1998). Virus resistance and gene silencing in plants can be induced by simultaneous expression of sense and antisense RNA. *Proc. Natl. Acad. Sci., U.S.A.*, 96: 13959–13964.

Wesley, V., Helliwell, C.A., Smith, N.A., Wang, M.B., Rouse, D.T. et al. (2001). Construct design for efficient, effective and high throughput gene silencing in plants. *Plant J.*, 27: 581–591.

Whisson S.C., Boevink, P.C., Moleleki, L., Avrova, A.O., Morales, J.G. et al. (2007). A translocation signal for delivery of oomycete effector proteins into host plant cells. *Nature*, 450: 115–118.

Whisson, S., Vetukuri, R., Avrova, A. and Dixelius, C. (2012). Can silencing of transposons contribute to variation in effector gene expression in *Phytophthora infestans*? *Mobile Genet. Element.*, 2(2): 110–114.

Xu, X., Huang, X.F., Visser, R.G. and Trindade, L.M. (2017). Engineering potato starch with a higher phosphate content. *PLoS ONE*, 12(1): e0169610.

Yi, X.U. and Gray, S.W. (2020). Aphids and their transmitted potato viruses: A continuous challenges in potato crops. *J. Integr. Agric.*, 19: 367–375.

Yoshioka, M., Adachi, A., Sato, Y., Doke, N., Kondo, T. and Yoshioka, H. (2019). RNAi of the sesquiterpene cyclase gene for phytoalexin production impairs pre- and post-invasive resistance to potato blight pathogens. *Mol. Plant Pathol.*, 20(7): 907–922.

Zhang, L., Yao, L., Zhang, N., Yang, J., Zhu, X. et al. (2018). Lateral root development in potato is mediated by Stu-mi164 regulation of NAC transcription factor. *Front. Plant Sci.*, 9: 383.

Zhang, Y. and Huang, B. (2018). Identification of multiple genes encoding SnRK1 subunits in potato tuber. *PLoS ONE*, 13(7): e0200321.

Zhou, X., Zhang, N., Yang, J., Tang, X., Wen, Y. and Si, H. (2018). Functional analysis of *StDWF4* gene in response to salt stress in potato. *Plant Physiol. Biochem.*, 125: 63–73.

Index

A

B

C

R

S

T

About the Author

Dr. Jagesh Kumar Tiwari, Ph.D., Senior Scientist (Horticulture-Vegetable Science), ICAR-Central Potato Research Institute, Shimla, Himachal Pradesh, India has over 14 years of research experience (since 2008) on potato biotechnology, particularly potato genetic resource management, wild species characterization, tissue culture, pre-breeding through somatic hybridization, cell biology, breeding, molecular markers, gene/QTL mapping, genome sequencing, functional genomics for late blight resistance, tuberization and nitrogen use efficiency and genome resequencing dihaploid (C-13) potato and aphid (*A. solani*). Presently, he is posted at ICAR-Indian Institute of Vegetable Research, Varanasi (Uttar Pradesh, India), and now working on breeding and genomics of vegetable crops. He has done post-doctorate at the University of Adelaide, Australia (2015), and NAIP International Training on MAS at International Potato Centre (CIP), Lima, Peru (2010). He is the recipient of awards like IAHS (formerly HSI)-DP Ghosh Memorial Young Scientist Award (2021), ICAR-Lal Bahadur Shastri Outstanding Young Scientist Award (2020), ICAR Hari Om Ashram Trust Award (2019), NAAS Associateship Award (2018), ICAR-CPRI Best Worker Award (Scientific category) (2018), Dr. S. Ramanujam Award (2016), Endeavor Post-doctoral Research Fellowship (Australia) Award (2015), IPA Chandra Prabha Singh Young Scientist Award (2014), best research paper awards and a few more. He is the Fellow of Indian Society of Vegetable Science (ISVS), Associate of National Academy of Agricultural Sciences (NAAS) and Life Member of Indian Potato Association (IPA), National Academy of Biological Sciences (NABS), Indian Academy of Horticultural Sciences (IAHS, formerly HSI), and Indian Society of Vegetable Science (ISVS). He has published over 90 research papers/reviews articles in national/international peer reviewed journals, like *Scientific Reports, Frontiers in Plant Science, Plant Physiology and Biochemistry, PLoS One, Journal of Integrative Agriculture, Plant Growth Regulation, Functional Plant Biology, Plant Cell Tissue and Organ Culture, Genome, Electronic Journal of Biotechnology, Scientia Horticulturae, 3 Biotech, Acta Physiologia Plantarum, Plant Molecular Biology Reporter, Molecular Biology Reports, Physiology and Molecular Biology of Plants, Plant Breeding, Journal of Asia-Pacific Entomology* and *Potato Research* so on. Earlier he has authored/edited four books by NIPA, New Delhi (edited); Malhotra Publishing House, New Delhi (authored); ICAR-CPRI, Shimla (authored) and *The Potato Genome* by Springer Nature (edited) and published over 15 book chapters.

About the Author

[illegible]

Printed in the United States
by Baker & Taylor Publisher Services